Physical Constants

ATOMIC MASS UNIT	amu	1.66×10^{-27} kg $= 931.5$ Mev
AVOGADRO'S NUMBER	N_0	6.0225×10^{26} kmoles^{-1}
BOLTZMANN'S CONSTANT	k	1.380×10^{-23} joule/^{0}K
ELECTRON CHARGE	e	1.602×10^{-19} coul
ELECTRON REST MASS	m_e	9.109×10^{-31} kg $= 0.00055$ amu
ELECTRON VOLT	ev	1.602×10^{-19} joule
NEUTRON REST MASS	m_n	1.675×10^{-27} kg $= 1.00867$ amu
PERMEABILITY OF FREE SPACE	μ_0	1.257×10^{-6} weber/amp-m $\left(\frac{\mu_0}{4\pi} = 10^{-7}\text{ weber/amp-m}\right)$
PERMITTIVITY OF FREE SPACE	ε_0	8.854×10^{-12} coul^2/newton-m^2 $\left(\frac{1}{4\pi\varepsilon_0} = 9 \times 10^9\text{ newton-m}^2\text{/coul}^2\right)$
PLANCK'S CONSTANT	h	6.626×10^{-34} joule-sec
	$\hbar = \frac{h}{2\pi}$	1.054×10^{-34} joule-sec
PROTON REST MASS	m_p	1.672×10^{-27} kg $= 1.00728$ amu
RYDBERG CONSTANT	R	1.097×10^{7} m^{-1}
SPEED OF LIGHT IN FREE SPACE	c	2.988×10^{6} m/sec

$1\text{eV} = 1.60 \times 10^{-19}$ joule

CONCEPTS OF MODERN PHYSICS

McGraw-Hill Series in Fundamentals of Physics

Introductory Program

E. U. Condon, Editor

FUNDAMENTALS OF MECHANICS AND HEAT

by Hugh D. Young

FUNDAMENTALS OF ELECTRICITY AND MAGNETISM

by Arthur F. Kip

FUNDAMENTALS OF OPTICS AND MODERN PHYSICS

by Hugh D. Young, in preparation

FUNDAMENTALS OF OPTICS

in preparation

CONCEPTS OF MODERN PHYSICS

by Arthur Beiser

Concepts of Modern Physics

REVISED EDITION

Arthur Beiser

FORMERLY ASSOCIATE PROFESSOR OF PHYSICS
NEW YORK UNIVERSITY

McGRAW-HILL BOOK COMPANY

NEW YORK ST. LOUIS SAN FRANCISCO
TORONTO LONDON SYDNEY

CONCEPTS OF MODERN PHYSICS

 Library of Congress Catalog Card Number: 66-23389

ISBN 07-004348-5

4567890 HDBP 75432

Text illustrations by Felix Cooper

PREFACE

This book is intended for use with one-semester courses in modern physics that have elementary classical physics and calculus as prerequisites. The first topics considered are relativity and quantum theory, which provide a framework for understanding the physics of atoms and nuclei. The theory of the atom is then developed with emphasis on elementary quantum-mechanical notions, and is followed by a discussion of the properties of aggregates of atoms. Finally, atomic nuclei and the related subject of elementary particles are discussed. The balance in the book leans more toward ideas than toward experimental methods and practical applications, because I believe that the beginning student is better served in his introduction to modern physics by a conceptual framework than by a mass of individual details. However, all physical theories live or die by the sword of experiment, and it seemed appropriate to present certain extended derivations, such as those of the Rutherford scattering formula, the quantum-mechanical treatment of the hydrogen atom, the various distribution laws of statistical mechanics, and the theory of alpha decay, in order to demonstrate exactly how an abstract concept can be related to experimental measurements. Further to this end, special descriptive passages have been inserted to accompany the regular text where appropriate.

PREFACE TO REVISED EDITION

In preparing this edition of *Concepts of Modern Physics,* the treatment of several topics has been expanded, parts of the original text have been rewritten to improve their clarity and to incorporate recent developments, and more problems have been added. Although the book is now too long to be comfortably covered in a single term, there is scope for an instructor to fashion the type of course he prefers, whether a general survey or a deeper inquiry into selected aspects of the subject. The major additions concern the harmonic oscillator in wave mechanics, the vector model of the atom, X-ray spectra, radiative transitions from the point of view of the Einstein A and B coefficients (including a glance at masers and lasers), the theory of the deuteron, and, inevitably, elementary particles.

A number of users of the original text were kind enough to send me their comments, which have been of great help in preparing this edition. I am especially grateful in this respect to E. W. Burke, Jr., B. Chasan, C. T. Chu, M. E. Cox, M. J. Dresser, G. Gutsche, A. Patitsas, H. Pendleton, and W. F. Williams.

ARTHUR BEISER

INTRODUCTION

By all odds, this last third of the twentieth century will be the most exciting period in human history for a student starting the study of physics. This is true whether he intends to become a research physicist or to pursue a research career in chemistry or in biology, whether he intends a medical career or plans to work in any one of the many specialized branches of modern science-based technology. It is also true whether he intends to follow the law, business, or public service, or even if he "merely" wishes to become a liberally educated person.

Thirty-five years ago, when America was caught in the throes of a great depression, the entire Federal budget was less than one billion dollars a year, and no Federal funds whatever were available for the support of graduate study in science or of university research projects in physics. A few industries were aware of the value of basic research but their efforts were confined to small staffs working in poorly equipped laboratories. In those days one never saw in the journals or newspapers even one tiny personnel advertisement which said, "Physicist Wanted."

Physicists in 1931 had had but fifteen years in which to exploit the ideas of Niels Bohr concerning the electronic structure of atoms, and a mere five years in which to grapple with the revolutionary principles of quantum mechanics. They were just beginning to take up the experimental study of the structure of the nucleus, aided by the development of the cyclotron by Ernest Lawrence and of electrostatic generators by Robert van de Graaf. In those far-off days, high-energy physics meant the use of particle energies of the order of one million electron volts. The success (1928) of the barrier-leakage model of α-particle radioactivity had indicated the applicability of quantum mechanics to the nucleus, but nuclear physics did not spring fully into being until 1932 with the discoveries in that year of the neutron, the deuteron, and the positron. Although X-ray diffraction was being used to elucidate crystal structures, there was no solid-state physics and no solid-state technology in the sense in which these terms are used today.

In contrast with these ancient times, the Federal government today spends upward of fifteen billion dollars a year on scientific research and development, nearly all of it directly involving the principles and concepts of modern atomic physics. Industry alone is spending a comparable total amount, of which one billion dollars a year can be fairly classified as basic research.

All over America very large and amazingly well equipped new physics research laboratories have been and are being built. Now journals and newspapers abound with large and enticingly worded advertisements imploring physicists to join the staffs of these new laboratories.

For a very knowledgeable account of this situation, every beginning student and every professor of physics should read carefully the report *Physics: Survey and Outlook** of the Physics Survey Committee under the chairmanship of Professor George Pake, provost of Washington University, St. Louis. This gives a vivid picture of the problems that lie ahead for physics research and of the plans being made for a massive attack on them with the support of American government and industry. It also shows why the study and teaching of physics are now such exciting pursuits and why so much effort is going into the modernization of teaching materials with which to help the next generation of students become aware of the opportunities that lie before it.

One result of this effort to have textbooks reflect the new importance of physics is the series of which this book is a member. Based on the premise that it is important that all students understand the methods and principles of physics and that the old one-year general physics course is inadequate, this series provides text materials for a modern presentation of fundamental physics in three or four semesters. This particular volume, *Concepts of Modern Physics* by Professor Arthur Beiser, is intended for the last term of a four-semester course in fundamental physics. It provides a sound and appropriately rigorous introduction to the progress that has been made in learning the nature of atoms and atomic nuclei during the present century.

Although it is written to be used independently, this textbook was particularly planned to be used in conjunction with Professor Arthur F. Kip's *Fundamentals of Electricity and Magnetism* and Professor Hugh D. Young's *Fundamentals of Mechanics and Heat,* companion volumes now in wide use. To bridge the gap between these two volumes and *Concepts of Modern Physics,* a fourth volume on optics is envisioned. For three-semester courses, Professor Young's *Fundamentals of Optics and Modern Physics,* now in production, will complete the sequence.

E. U. CONDON
Professor of Physics and
Fellow of the Joint Institute
for Laboratory Astrophysics
University of Colorado

* Published by the National Academy of Sciences, National Research Council, Washington, D.C., 1966. 2 vols.

CONTENTS

1

BASIC CONCEPTS

THE SPECIAL THEORY OF RELATIVITY 1

Our study of modern physics begins with a consideration of the special theory of relativity. This is a logical starting point, since all of physics is ultimately concerned with measurement and relativity involves an analysis of how measurements depend upon the observer as well as upon what is observed. From relativity emerges a new mechanics in which there are intimate relationships between space and time, mass and energy. Without these relationships it would be impossible to understand the microscopic world within the atom whose elucidation is the central problem of modern physics.

1.1 The Michelson-Morley Experiment

The wave theory of light was devised and perfected several decades before the electromagnetic nature of the waves became known. It was reasonable for the pioneers in optics to regard light waves as undulations in an all-pervading elastic medium called the *ether*, and their successful explanation of diffraction and interference phenomena in terms of ether waves made the notion of the ether so familiar that its existence was accepted without question. Maxwell's development of the electromagnetic theory of light in 1864 and Hertz's experimental confirmation of it in 1887 deprived the ether of most of its properties, but nobody at the time seemed willing to discard the fundamental idea represented by the ether: that light propagates relative to some sort of universal frame of reference. Let us consider an example of what this idea implies with the help of a simple analogy.

Figure 1-1 is a sketch of a river of width D which flows with the speed v. Two boats start out from one bank of the river with the same speed **V**. Boat A crosses the river to a point on the other bank directly opposite the starting point and then returns, while boat B heads downstream for the distance D and then returns to the starting point. Let us calculate the time required for each round trip.

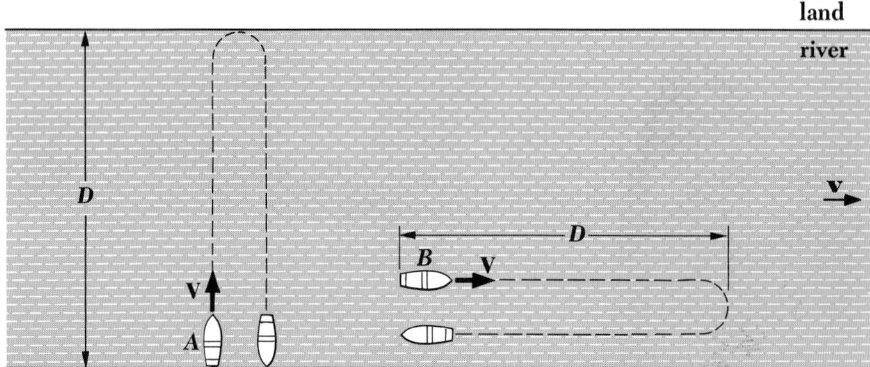

FIGURE 1-1 Boat A goes directly across the river and returns to its starting point, while boat B heads downstream for an identical distance and then returns.

We begin by considering boat A. If A heads perpendicularly across the river, the current will carry it downstream from its goal on the opposite bank (Fig. 1-2). It must therefore head somewhat upstream in order to compensate for the current. In order to accomplish this, its upstream component of velocity should be exactly $-\mathbf{v}$ in order to cancel out the river current $\mathbf{v}$,

FIGURE 1-2 Boat A must head upstream in order to compensate for the river current.

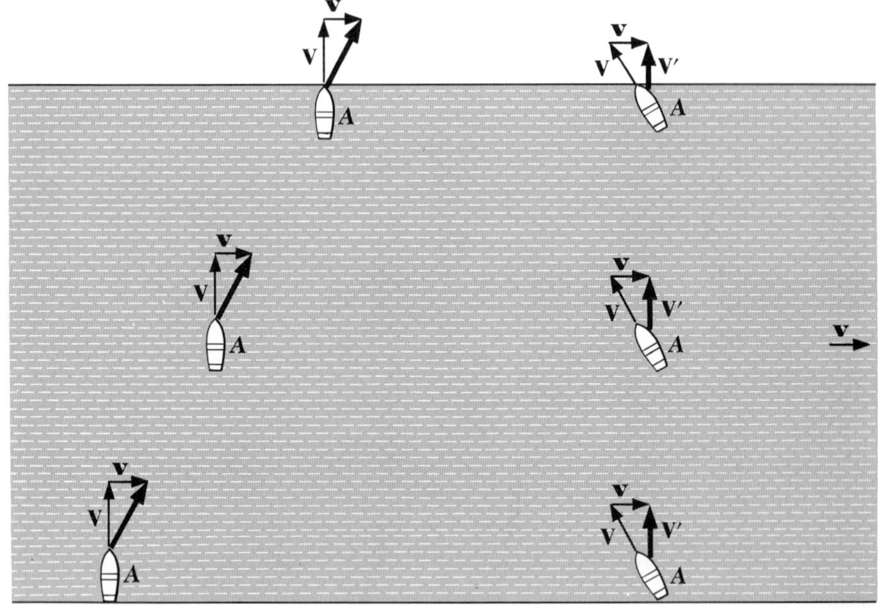

leaving the component $\mathbf{V}'$ as its net speed across the river. From Fig. 1-2 we see that these speeds are related by the formula

$$V^2 = V'^2 + v^2$$

so that the actual speed with which boat A crosses the river is

$$\begin{aligned} V' &= \sqrt{V^2 - v^2} \\ &= V\sqrt{1 - v^2/V^2} \end{aligned}$$

Hence the time for the initial crossing is the distance D divided by the speed V'. Since the reverse crossing involves exactly the same amount of time, the total round-trip time t_A is twice D/V', or

1.1 $$t_A = \frac{2D/V}{\sqrt{1 - v^2/V^2}}$$

The case of boat B is somewhat different. As it heads downstream, its speed relative to the shore is its own speed V *plus* the speed v of the river (Fig. 1-3), and it travels the distance D downstream in the time

$$\frac{D}{V + v}$$

On its return trip, however, B's speed relative to the shore is its own speed V *minus* the speed v of the river. It therefore requires the longer time

$$\frac{D}{V - v}$$

to travel upstream the distance D to its starting point. The total round-trip time t_B is the sum of these times, namely,

$$t_B = \frac{D}{V + v} + \frac{D}{V - v}$$

FIGURE 1-3 The speed of boat B downstream relative to the shore is increased by the speed of the river current while its speed upstream is reduced by the same amount.

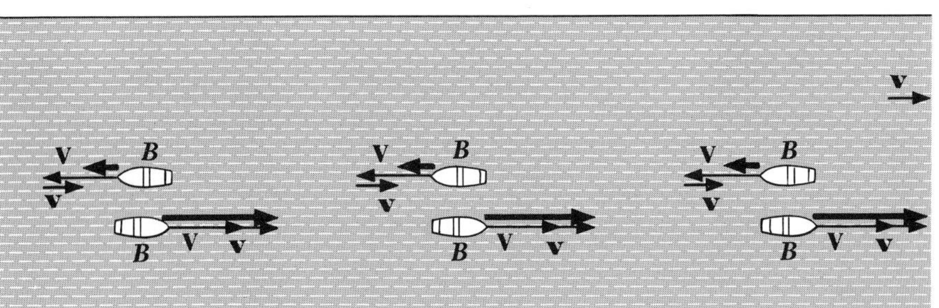

Using the common denominator $(V + v)(V - v)$ for both terms,

$$t_B = \frac{D(V - v) + D(V + v)}{(V + v)(V - v)}$$

$$= \frac{2DV}{V^2 - v^2}$$

1.2 $$= \frac{2D/V}{1 - v^2/V^2}$$

which is greater than t_A, the corresponding round-trip time for the other boat.

The ratio between the times t_A and t_B is

1.3 $$\frac{t_A}{t_B} = \sqrt{1 - v^2/V^2}$$

If we know the common speed V of the two boats and measure the ratio t_A/t_B, we can determine the speed v of the river.

The reasoning used in this problem may be transferred to the analogous problem of the passage of light waves through the ether. If there is an ether pervading space, we move through it with at least the 3×10^4 m/sec (18.5 mi/sec) speed of the earth's orbital motion about the sun; if the sun is also in motion, our speed through the ether is even greater (Fig. 1-4). From the point of view of an observer on the earth, the ether is moving past the earth.

FIGURE 1-4 Motions of the earth through a hypothetical ether.

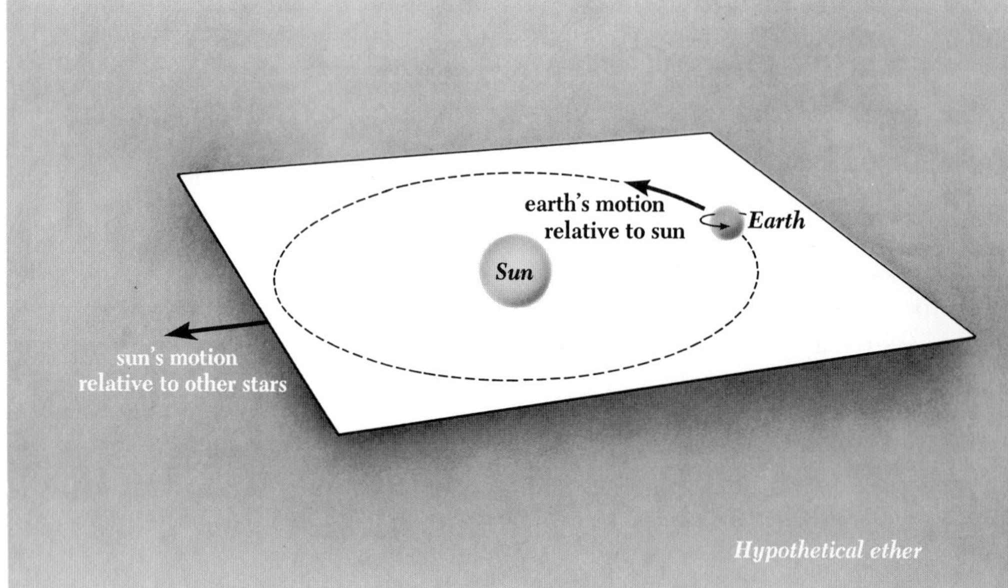

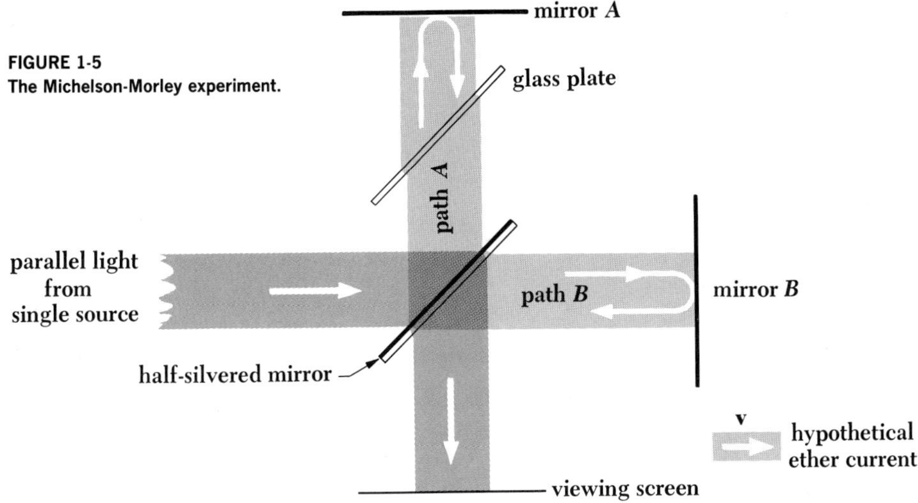

FIGURE 1-5
The Michelson-Morley experiment.

To detect this motion, we can use the pair of light beams formed by a half-silvered mirror instead of a pair of boats (Fig. 1-5). One of these light beams is directed to a mirror along a path perpendicular to the ether current, while the other goes to a mirror along a path parallel to the ether current. The optical arrangement is such that both beams return to the same viewing screen. The purpose of the clear glass plate is to ensure that both beams pass through the same thicknesses of air and glass.

If the path lengths of the two beams are *exactly* the same, they will arrive at the screen in phase and will interfere constructively to yield a bright field of view. The presence of an ether current in the direction shown, however, would cause the beams to have different transit times in going from the half-silvered mirror to the screen, so that they would no longer arrive at the screen in phase but would interfere destructively. In essence this is the famous experiment performed in 1887 by the American physicists Michelson and Morley.

In the actual experiment the two mirrors are not perfectly perpendicular, with the result that the viewing screen appears crossed with a series of bright and dark interference fringes due to differences in path length between adjacent light waves (Fig. 1-6). If either of the optical paths in the apparatus is varied in length, the fringes appear to move across the screen as reinforcement and cancellation of the waves succeed one another at each point. The stationary apparatus, then, can tell us nothing about any time difference between the two paths. When the apparatus is rotated by 90°, however, the two paths change their orientations relative to the hypothetical ether stream, so that the beam formerly requiring the time t_A for the round trip now

FIGURE 1-6 Fringe pattern observed in Michelson-Morley experiment.

requires t_B and vice versa. If these times are different, the fringes will move across the screen during the rotation.

Let us calculate the fringe shift expected on the basis of the ether theory. From Eqs. 1.1 and 1.2 the time difference between the two paths owing to the ether drift is

$$\Delta t = t_B - t_A$$
$$= \frac{2D/V}{1 - v^2/V^2} - \frac{2D/V}{\sqrt{1 - v^2/V^2}}$$

Here v is the ether speed, which we shall take as the earth's orbital speed of 3×10^4 m/sec, and V is the speed of light c, where $c = 3 \times 10^8$ m/sec. Hence

$$\frac{v^2}{V^2} = \frac{v^2}{c^2}$$
$$= 10^{-8}$$

which is much smaller than 1. According to the binomial theorem,

$$(1 \pm x)^n = 1 \pm nx + \frac{n(n-1)x^2}{2!} \pm \frac{n(n-1)(n-2)x^3}{3!} + \ldots$$

which is valid for $x^2 < 1$. When x is extremely small compared with 1,

$$(1 \pm x)^n \approx 1 \pm nx$$

We may therefore express Δt to a good approximation as

$$\Delta t = \frac{2D}{c}\left[\left(1 + \frac{v^2}{c^2}\right) - \left(1 + \frac{1}{2}\frac{v^2}{c^2}\right)\right]$$
$$= \left(\frac{D}{c}\right)\left(\frac{v^2}{c^2}\right)$$

Here D is the distance between the half-silvered mirror and each of the other mirrors. The path difference d corresponding to a time difference Δt is

$$d = c\,\Delta t$$

If d corresponds to the shifting of n fringes,

$$d = n\lambda$$

where λ is the wavelength of the light used. Equating these two formulas for d, we find that

$$\begin{aligned} n &= \frac{c\,\Delta t}{\lambda} \\ &= \frac{Dv^2}{\lambda c^2} \end{aligned}$$

In the actual experiment Michelson and Morley were able to make D about 10 m in effective length through the use of multiple reflections, and the wavelength of the light they used was about 5,000 A (1 A = 10^{-10} m). The expected fringe shift in each path when the apparatus is rotated by 90° is therefore

$$\begin{aligned} n &= \frac{Dv^2}{\lambda c^2} \\ &= \frac{10\text{ m} \times (3 \times 10^4\text{ m/sec})^2}{5 \times 10^{-7}\text{ m} \times (3 \times 10^8\text{ m/sec})^2} \\ &= 0.2\text{ fringe} \end{aligned}$$

Since both paths experience this fringe shift, the total shift should amount to $2n$ or 0.4 fringe. A shift of this magnitude is readily observable, and therefore Michelson and Morley looked forward to establishing directly the existence of the ether.

To everybody's surprise, *no fringe shift whatever* was found. When the experiment was performed at different seasons of the year and in different locations, and when experiments of other kinds were tried for the same purpose, the conclusions were always identical: no motion through the ether was detected.

The negative result of the Michelson-Morley experiment had two consequences. First, it rendered untenable the hypothesis of the ether by demonstrating that the ether has no measurable properties—an ignominious end for what had once been a respected idea. Second, it suggested a new physical principle: the speed of light in free space is the same everywhere, regardless of any motion of source or observer.

1.2 The Special Theory of Relativity

We mentioned earlier the role of the ether as a universal frame of reference with respect to which light waves were supposed to propagate. Whenever we speak of "motion," of course, we really mean "motion relative to a frame of reference." The frame of reference may be a road, the earth's surface, the sun, the center of our galaxy; but in every case we must specify it. Stones dropped in Bermuda and in Perth, Australia, both fall "down," and yet the two move in exactly opposite directions relative to the earth's center. Which is the correct location of the frame of reference in this situation, the earth's surface or its center? The answer is that *all* frames of reference are equally correct, although one may be more convenient to use in a specific case. *If* there were an ether pervading all space, we could refer all motion to it, and the inhabitants of Bermuda and Perth would escape from their quandary. The absence of an ether, then, implies that there is no universal frame of reference, so that all motion exists solely relative to the person or instrument observing it. If we are in a free balloon above a uniform cloud bank and see another free balloon change its position relative to us, we have no way of knowing which balloon is "really" moving (Fig. 1-7). Should we be isolated in the universe, there would be no way in which we could determine whether we are in motion or not, because without a frame of reference the concept of motion has no meaning.

The theory of relativity resulted from an analysis of the physical consequences implied by the absence of a universal frame of reference. The special theory of relativity, developed by Albert Einstein in 1905, treats problems involving the motion of frames of reference at constant velocity (that is, both constant speed and constant direction) with respect to one another; the general theory of relativity, proposed by Einstein a decade later, treats problems involving frames of reference accelerated with respect to one another. The special theory has had a profound influence on all of physics, and we shall restrict ourselves to it.

The special theory of relativity is based upon two postulates. The first states that **the laws of physics may be expressed in equations having the same form in all frames of reference moving at constant velocity with respect to one another.** This postulate expresses the absence of a universal frame of reference. If the laws of physics had different forms for different observers in relative motion, it could be determined from these differences which objects are "stationary" in space and which are "moving." But because there is no universal frame of reference, this distinction does not exist in nature; hence the above postulate.

The second postulate of special relativity states that **the speed of light in free space has the same value for all observers, regardless of their state of**

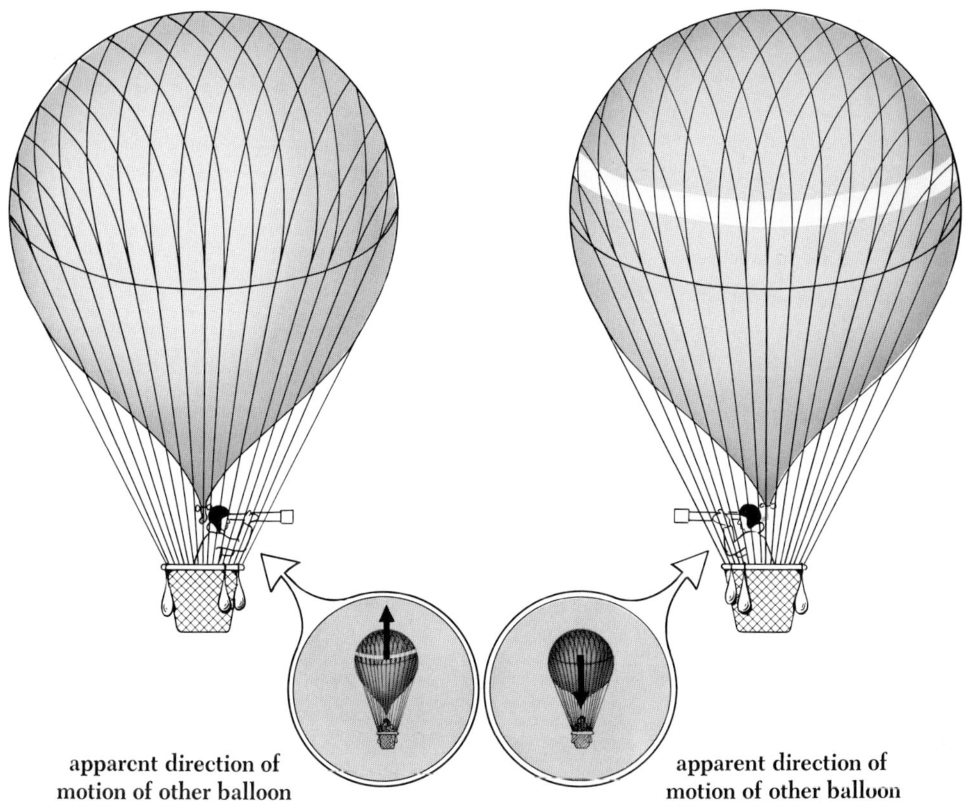

FIGURE 1-7 All motion is relative to the observer.

motion. This postulate follows directly from the result of the Michelson-Morley experiment (and others).

At first sight these postulates hardly seem radical. Actually they subvert almost all of the intuitive concepts of time and space we form on the basis of our daily experience. A simple example will illustrate this statement. In Fig. 1-8 we have the two boats A and B once more, with boat A stationary in the water while boat B drifts at the constant velocity $\mathbf{v}$. There is a dense fog present, and so on neither boat does the observer have any idea which is the moving one. At the instant that B is abreast of A, a flare is fired. The light from the flare travels uniformly in all directions, according to the second postulate of special relativity. An observer on either boat must see a sphere of light expanding with *himself* at its center, according to the first postulate of special relativity, even though one of them is changing his position with respect to the point where the flare went off. The observers cannot detect which of them is undergoing such a change in position since the fog eliminates any frame of reference other than each boat itself, and so, since the

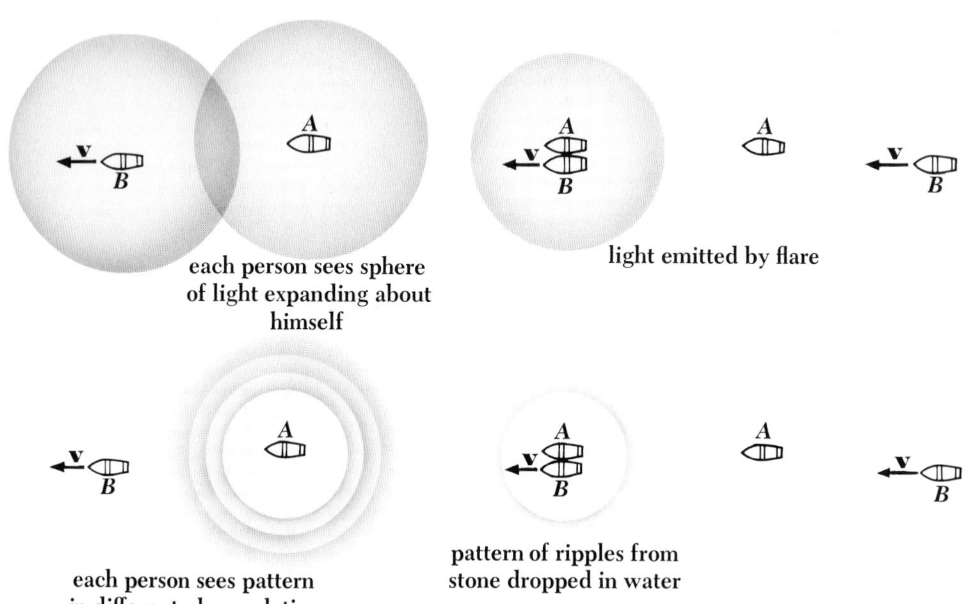

FIGURE 1-8 Relativistic phenomena differ from everyday experience.

speed of light is the same for both of them, they must both see the identical phenomenon.

Why is the situation of Fig. 1-8 unusual? Let us consider a more familiar analog. The boats are at sea on a clear day and somebody on one of them drops a stone into the water when they are abreast of each other. A circular pattern of ripples spreads out, as at the bottom of Fig. 1-8, *which appears different* to observers on each boat. Merely by observing whether or not he is at the center of the pattern of ripples, each observer can tell whether he is moving relative to the water or not. Water is in itself a frame of reference, and an observer on a boat moving through it measures ripple speeds with respect to himself that are different in different directions, in contrast to the uniform ripple speed measured by an observer on a stationary boat. It is important to recognize that motion and waves *in water* are entirely different from motion and waves *in space;* water is in itself a frame of reference while space is not, and wave speeds in water vary with the observer's motion while wave speeds of light in space do not.

The only way of interpreting the fact that observers in the two boats in our example perceive identical expanding spheres of light is to regard the coordinate system of each observer, from the point of view of the other, as being affected by their relative motion. When this idea is developed, using only accepted laws of physics and Einstein's postulates, we shall see that many peculiar effects are predicted. One of the triumphs of modern physics is the experimental confirmation of these effects.

1.3 The Galilean Transformation

Let us suppose that we are in a frame of reference S and find that the coordinates of some event that occurs at the time t are x, y, z. An observer located in a different frame of reference S′ which is moving with respect to S at the constant velocity v will find that the same event occurs at the time t' and has the coordinates x', y', z'. (In order to simplify our work, we shall assume that $\mathbf{v}$ is in the $+x$ direction, as in Fig. 1-9.) How are the measurements x, y, z, t related to x', y', z', t'?

At first glance the answer seems obvious enough. If time in both systems is measured from the instant when the origins of S and S′ coincided, measurements in the x direction made in S will exceed those made in S′ by the amount vt, which represents the distance that S′ has moved in the x direction. That is

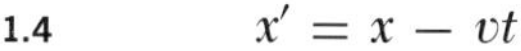

1.4 $$x' = x - vt$$

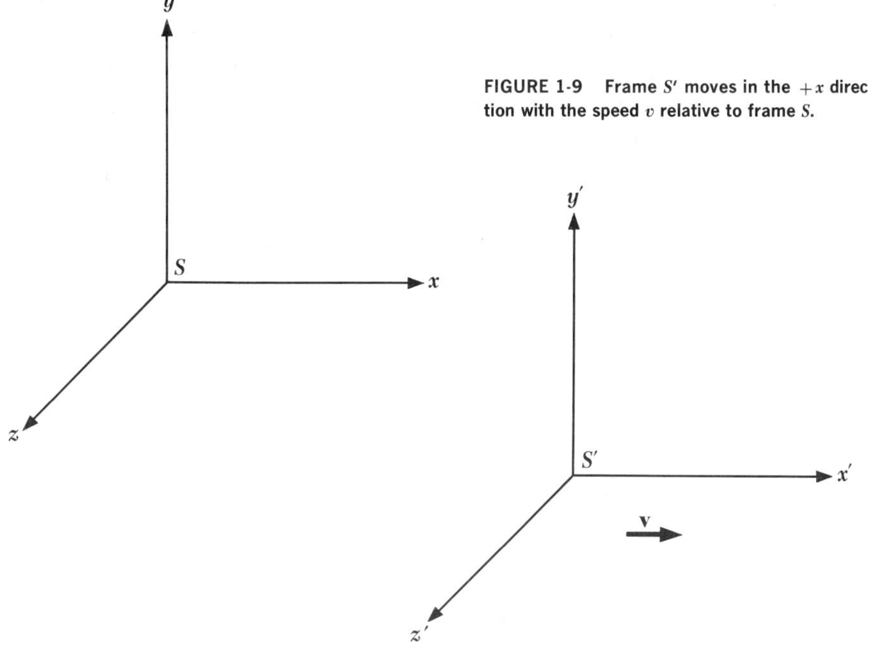

FIGURE 1-9 Frame S' moves in the $+x$ direction with the speed v relative to frame S.

There is no relative motion in the y and z directions, and so

$$y' = y \tag{1.5}$$

$$z' = z \tag{1.6}$$

In the absence of any indication to the contrary in our everyday experience, we further assume that

$$t' = t \tag{1.7}$$

The set of Eqs. 1.4 to 1.7 is known as the *Galilean transformation.*

To convert velocity components measured in the S frame to their equivalents in the S′ frame according to the Galilean transformation, we simply differentiate Eqs. 1.4 to 1.6 with respect to time:

$$v_x' = \frac{dx'}{dt'} = v_x - v \tag{1.8}$$

$$v_y' = \frac{dy'}{dt'} = v_y \tag{1.9}$$

$$v_z' = \frac{dz'}{dt'} = v_z \tag{1.10}$$

While the Galilean transformation and the velocity transformation it leads to are both in accord with our intuitive expectations, they violate both of the postulates of special relativity. The first postulate calls for identical equations of physics in both the S and S′ frames of reference, but the fundamental equations of electricity and magnetism assume very different forms when Eqs. 1.4 to 1.7 are used to convert quantities measured in one frame into their equivalents in the other. The second postulate calls for the same value of the speed of light c whether determined in S or S′. If we measure the speed of light in the x direction in the S system to be c, however, in the S′ system it will be

$$c' = c - v$$

according to Eq. 1.8. Clearly a different transformation is required if the postulates of special relativity are to be satisfied.

1.4 The Lorentz Transformation

We shall now develop a set of transformation equations directly from the postulates of special relativity. A reasonable guess as to the kind of relationship between x and x' is

$$x' = k(x - vt) \tag{1.11}$$

where k is a factor of proportionality that does not depend upon either x or t but may be a function of v. The choice of Eq. 1.11 follows from several considerations: it is linear in x and x', so that a single event in frame S corresponds to a single event in frame S′, as it must; it is simple, and a simple solution to a problem should always be explored first; and it has the possibility of reducing to Eq. 1.4, which we know to be correct in ordinary mechanics. Because the equations of physics must have the same form in both S and S′, we need only change the sign of v (in order to take into account the difference in the direction of relative motion) to write the corresponding equation for x in terms of x' and t':

1.12 $$x = k(x' + vt')$$

The factor k must be the same in both frames of reference since there is no difference between S and S′ other than in the sign of v.

As in the case of the Galilean transformation, there is nothing to indicate that there might be differences between the corresponding coordinates y, y', and z, z' which are normal to the direction of v. Hence we again take

1.13 $$y' = y$$

1.14 $$z' = z$$

The time coordinates t and t', however, are *not* equal. We can see this by substituting the value of x' given by Eq. 1.11 into Eq. 1.12. We obtain

$$x = k^2(x - vt) + kvt'$$

from which we find that

1.15 $$t' = kt + \left(\frac{1 - k^2}{kv}\right)x$$

Equations 1.11, 1.13, 1.14, and 1.15 constitute a coordinate transformation that satisfies the first postulate of special relativity.

The second postulate of relativity enables us to evaluate k. At the instant $t = 0$, the origins of the two frames of reference S and S′ are in the same place, according to our initial conditions, and $t' = 0$ then also. Suppose that a flare is set off at the common origin of S and S′ at $t = t' = 0$, and the observers in each system proceed to measure the speed with which the light from it spreads out. Both observers must find the same speed c, which means that in the S frame

1.16 $$x = ct$$

while in the S′ frame

1.17 $$x' = ct'$$

Substituting for x' and t' in Eq. 1.17 with the help of Eqs. 1.11 and 1.15,

$$k(x - vt) = ckt + \left(\frac{1 - k^2}{kv}\right)cx$$

and solving for x,

$$x = \frac{ckt + vkt}{k - \left(\dfrac{1 - k^2}{kv}\right)c}$$

$$= ct\left[\frac{k + \dfrac{v}{c}k}{k - \left(\dfrac{1 - k^2}{kv}\right)c}\right]$$

$$= ct\left[\frac{1 + \dfrac{v}{c}}{1 - \left(\dfrac{1}{k^2} - 1\right)\dfrac{c}{v}}\right]$$

This expression for x will be the same as that given by Eq. 1.16, namely, $x = ct$, provided that the quantity in the brackets equals 1. Therefore

$$\frac{1 + \dfrac{v}{c}}{1 - \left(\dfrac{1}{k^2} - 1\right)\dfrac{c}{v}} = 1$$

and

1.18 $$k = \frac{1}{\sqrt{1 - v^2/c^2}}$$

Inserting the above value of k in Eqs. 1.11 and 1.15, we have for the complete transformation of measurements of an event made in S to the corresponding measurements made in S′ the equations

1.19 $$x' = \frac{x - vt}{\sqrt{1 - v^2/c^2}}$$ **Lorentz transformation**

1.20 $$y' = y$$ **Lorentz transformation**

1.21 $$z' = z$$ **Lorentz transformation**

1.22 $$t' = \frac{t - \dfrac{vx}{c^2}}{\sqrt{1 - v^2/c^2}}$$ **Lorentz transformation**

Equations 1.19 to 1.22 comprise the *Lorentz transformation*. They were first obtained by the Dutch physicist H. A. Lorentz, who showed that the basic formulas of electromagnetism are the same in all frames of reference in uniform relative motion only when these transformation equations are used. It was not until a number of years later that Einstein discovered their full significance.

In order to transform measurements from S′ to S, the only changes in the Lorentz transformation equations that need be made are to exchange primed for unprimed quantities and vice versa, and to replace v by $-v$. Thus the inverse Lorentz transformation is

1.23 $$x = \frac{x' + vt'}{\sqrt{1 - v^2/c^2}}$$ Inverse Lorentz transformation

1.24 $$y = y'$$ Inverse Lorentz transformation

1.25 $$z = z'$$ Inverse Lorentz transformation

1.26 $$t = \frac{t' + \dfrac{vx'}{c^2}}{\sqrt{1 - v^2/c^2}}$$ Inverse Lorentz transformation

Two obvious aspects of the Lorentz transformation are worth noting. The first is that measurements of time as well as of position depend upon the frame of reference of the observer, so that two events which occur simultaneously in one frame at different places need not be simultaneous in another. The second is that the Lorentz equations reduce to the ordinary Galilean equations 1.4 to 1.7 when the relative velocity v of S and S′ is small compared with the velocity of light c. Therefore we may anticipate that the peculiar consequences of special relativity to be explored in the remainder of this chapter will not be apparent except under circumstances where enormous velocities are encountered.

1.5 The Lorentz-FitzGerald Contraction

A rod is lying along the x axis of a frame of reference S. We determine the coordinates of its ends to be x_1 and x_2, and so we conclude that the length L_0 of the rod is

1.27 $$L_0 = x_2 - x_1$$

L_0 is the rod's length as measured in a frame of reference in which it is at rest. Suppose that the same quantity is determined from a frame of reference S′ which is moving parallel to the rod with the velocity v: will the length L measured in S′ be the same as the length L_0 that was measured in S? In more familiar terms, if a rod is lying beside a highway, does it seem to have the same length to someone speeding past in a car as it does to someone standing on the ground?

In order to find L, we use the Lorentz transformation to go from the coordinates x_1 and x_2 in the stationary (relative to the rod) frame S to the corresponding coordinates x_1' and x_2' in the moving (relative to the rod) frame S′. From Eq. 1.23 we have

$$x_1 = \frac{x_1' + vt'}{\sqrt{1 - v^2/c^2}}$$

$$x_2 = \frac{x_2' + vt'}{\sqrt{1 - v^2/c^2}}$$

and so

$$L_0 = x_2 - x_1$$

$$= \frac{x_2' - x_1'}{\sqrt{1 - v^2/c^2}}$$

By definition

1.28 $$L = x_2' - x_1'$$

which means that

$$L_0 = \frac{L}{\sqrt{1 - v^2/c^2}}$$

or

1.29 $$L = L_0\sqrt{1 - v^2/c^2}$$ **Lorentz-FitzGerald contraction**

The length of an object in motion with respect to an observer appears to the observer to be shorter than when it is at rest with respect to him, a phenomenon known as the *Lorentz-FitzGerald contraction.*

Because the relative velocity of the two frames S and S′ appears only as v^2 in Eq. 1.29, it does not matter which frame we call S and which S′. If we find that the length of a rocket is L_0 when it is on its launching pad, we will find from the ground that its length L when moving with the speed v is

$$L = L_0\sqrt{1 - v^2/c^2}$$

while to a man in the rocket, objects on the earth behind him appear shorter than they did when he was on the ground by the same factor $\sqrt{1 - v^2/c^2}$. The length of an object is a maximum when measured in a reference frame in which it is stationary, and its length is less when measured in a reference frame in which it is moving.

The ratio between L and L_0 in Eq. 1.27 is the same as that in Eq. 1.3 when it is applied to the times of travel of the two light beams, so that we might be tempted to consider the Michelson-Morley result evidence for the contraction of the length of their apparatus in the direction of the earth's motion. This interpretation was tested by Kennedy and Thorndike in a similar experiment using an interferometer with arms of unequal length. They also found no fringe shift, which means that these experiments must be considered evidence for the absence of an ether and not for actual contractions of the apparatus. The relativistic length contraction is negligible for ordinary speeds, but it is an important effect at speeds close to the speed of light. A speed of 1,000 mi/sec seems enormous to us, and yet it results in a shortening in the direction of motion to only

$$\begin{aligned}\frac{L}{L_0} &= \sqrt{1 - \frac{v^2}{c^2}} \\ &= \sqrt{1 - \frac{(1{,}000 \text{ mi/sec})^2}{(186{,}000 \text{ mi/sec})^2}} \\ &= 0.999985 \\ &= 99.9985 \text{ per cent}\end{aligned}$$

of the length at rest. On the other hand, a body traveling at 0.9 the speed of light is shortened to

$$\begin{aligned}\frac{L}{L_0} &= \sqrt{1 - \frac{(0.9c)^2}{c^2}} \\ &= 0.436 \\ &= 43.6 \text{ per cent}\end{aligned}$$

of the length at rest, a significant change.

The Lorentz-FitzGerald contraction occurs only in the direction of the relative motion: if v is parallel to x, the y and z dimensions of a moving object are the same in both S and S′.

An actual photograph of an object in very rapid relative motion might reveal a somewhat different distortion, depending upon the direction from which the object is viewed and the ratio v/c. The reason for this effect is that light reaching the camera (or eye for that matter) from the more distant parts of the object was emitted earlier than that coming from the nearer parts; the camera "sees" a picture that is actually a composite, since the object was at different locations when the various elements of the single image that reaches the film left it. This effect alters the Lorentz-FitzGerald contraction by extending the apparent length of a moving object in the direction of motion. A three-dimensional body, such as a cube, may be seen as rotated in orientation as well as changed in shape, again depending upon the position of the observer and the value of v/c. This visual effect must be distinguished from the Lorentz-FitzGerald contraction itself, which is a physical phenomenon.

If there were no Lorentz-FitzGerald contraction, the appearance of a moving body would be different from what it is at rest, but in another way.

It is interesting to note that the above approach to the visual appearance of rapidly moving objects was not made until 1959, 54 years after the publication of the special theory of relativity.

1.6 Time Dilation

Time intervals, too, are affected by relative motion. Clocks moving with respect to an observer appear to tick less rapidly than they do when at rest with respect to him. If we, in the S frame, observe the length of time t some event requires in a frame of reference S′ in motion relative to us, our clock will indicate a longer time interval than the t_0 determined by a clock in the moving frame. This effect is called *time dilation.*

To see how time dilation comes about, let us imagine a clock at the point x' in the moving frame S′. When an observer in S′ finds that the time is t_1', an observer in S will find it to be t_1, where, from Eq. 1.26,

$$t_1 = \frac{t_1' + \dfrac{vx'}{c^2}}{\sqrt{1 - v^2/c^2}}$$

After a time interval of t_0 (to him), the observer in the moving system finds that the time is now t_2' according to his clock. That is,

1.30 $$t_0 = t_2' - t_1'$$

The observer in S, however, measures the end of the same time interval to be

$$t_2 = \frac{t_2' + \dfrac{vx'}{c^2}}{\sqrt{1 - v^2/c^2}}$$

so to him the duration of the interval t is

1.31 $$\begin{aligned} t &= t_2 - t_1 \\ &= \frac{t_2' - t_1'}{\sqrt{1 - v^2/c^2}} \end{aligned}$$

or

1.32 $$t = \frac{t_0}{\sqrt{1 - v^2/c^2}}$$ **Time dilation**

A stationary clock measures a longer time interval between events occurring in a moving frame of reference than does a clock in the moving frame.

A striking illustration of both the time dilation of Eq. 1.32 and the length contraction of Eq. 1.29 occurs in the decay of unstable particles called *μ mesons,* whose properties we shall discuss in greater detail later. For the moment our interest lies in the fact that a μ meson decays into an electron an average of 2×10^{-6} sec after it comes into being. Now μ mesons are created high in the atmosphere by fast cosmic-ray particles arriving at the earth from space, and reach sea level in profusion. Such mesons have a typical speed of 2.994×10^8 m/sec, which is 0.998 of the velocity of light c. But in $t_0 = 2 \times 10^{-6}$ sec, the meson's mean lifetime, they can travel a distance of only

$$\begin{aligned} y &= vt_0 \\ &= 2.994 \times 10^8 \text{ m/sec} \times 2 \times 10^{-6} \text{ sec} \\ &= 600 \text{ m} \end{aligned}$$

while they are actually created at altitudes more than 10 times greater than this.

We can resolve the meson paradox by using the results of the special theory of relativity. Let us examine the problem from the frame of reference of the meson, in which its lifetime is 2×10^{-6} sec. While the meson lifetime is unaffected by the motion, its distance to the ground appears shortened by the factor

$$\frac{y}{y_0} = \sqrt{1 - v^2/c^2}$$

That is, while we, on the ground, measure the altitude at which the meson is produced as y_0, the meson "sees" it as y. If we let y be 600 m, the maximum distance the meson can go *in its own frame of reference* at the speed $0.998c$ before decaying, we find that the corresponding distance y_0 in *our reference frame* is

$$\begin{aligned} y_0 &= \frac{y}{\sqrt{1 - v^2/c^2}} \\ &= \frac{600}{\sqrt{1 - \dfrac{(0.998c)^2}{c^2}}} \text{ m} \\ &= \frac{600}{\sqrt{1 - 0.996}} \text{ m} \\ &= \frac{600}{0.063} \text{ m} \\ &= 9{,}500 \text{ m} \end{aligned}$$

Hence, despite their brief lifespans, it is possible for the mesons to reach the ground from the considerable altitudes at which they are actually formed.

Now let us examine the problem from the frame of reference of an observer on the ground. From the ground the altitude at which the meson is produced is y_0, but its lifetime in *our reference frame* has been extended, owing to the relative motion, to the value

$$
\begin{aligned}
t &= \frac{t_0}{\sqrt{1 - v^2/c^2}} \\
&= \frac{2 \times 10^{-6}}{\sqrt{1 - \dfrac{(0.998c)^2}{c^2}}} \text{ sec} \\
&= \frac{2 \times 10^{-6}}{0.063} \text{ sec} \\
&= 31.7 \times 10^{-6} \text{ sec}
\end{aligned}
$$

almost 16 times greater than when it is at rest with respect to us. In 31.7×10^{-6} sec a meson whose speed is $0.998c$ can travel a distance

$$
\begin{aligned}
y_0 &= vt \\
&= 2.994 \times 10^8 \text{ m/sec} \times 31.7 \times 10^{-6} \text{ sec} \\
&= 9{,}500 \text{ m}
\end{aligned}
$$

the same distance obtained before. The two points of view give identical results.

The relative character of time as well as space has many implications. For example, events that seem to take place simultaneously to one observer may not be simultaneous to another observer in relative motion, and vice versa. Physical theories which require simultaneity in events at different locations must therefore be discarded. The principle of conservation of energy in its elementary form states that the total energy content of the universe is constant, but it does not rule out a process in which a certain amount of energy ΔE vanishes at one point while an equal amount of energy ΔE spontaneously comes into being somewhere else with no actual transport of energy from one place to the other. Because simultaneity is relative, some observers of the process will find energy not being conserved. To rescue conservation of energy in the light of special relativity, then, it is necessary to say that, when energy disappears somewhere and appears elsewhere, it has actually *flowed* from the first location to the second. (There are many ways in which a flow of energy can occur, of course.) Thus energy is conserved *locally* in any arbitrary region of space at any time, not merely when the universe as a whole is considered—a much stronger statement of this principle.

1.7 The Relativity of Mass

We have seen that such fundamental physical quantities as length and time have meaning only when the particular reference frame in which they are measured is specified. Given this frame, we may compute what the values of these quantities are when measured in other reference frames in motion relative to the specified frame by applying the Lorentz transformation equations. An event in time and space—a collision between two bodies, for instance—will have a different appearance in different frames of reference. According to the postulates of special relativity, however, the laws of motion arrived at by observing events of this kind must have the same form in all frames of reference. In the remainder of this chapter we examine how the principles of energy and momentum conservation and the theorem of velocity addition must be modified in the light of this requirement.

We begin by considering an elastic collision (that is, a collision in which kinetic energy is conserved) between two particles A and B, as witnessed by observers in the reference frames S and S′ which are in uniform relative motion. The properties of A and B are identical when determined in reference frames in which they are at rest. The frames S and S′ are oriented as in Fig. 1-10, with S′ moving in the $+x$ direction with respect to S at the velocity v.

Before the collision, particle A has been at rest in frame S and particle B in frame S′. Then, at the same instant, A is thrown in the $+y$ direction at the speed V_A while B is thrown in the $-y'$ direction at the speed V_B', where

1.33 $$V_A = V_B'$$

Hence the behavior of A as seen from S is exactly the same as the behavior of B as seen from S′. When the two particles collide, A rebounds in the $-y$ direction at the speed V_A, while B rebounds in the $+y'$ direction at the speed V_B'. If the particles are thrown from positions Y apart, an observer in S finds that the collision occurs at $y = \frac{1}{2}Y$ and one in S′ finds that it occurs at $y' = \frac{1}{2}Y$. The round-trip time T_0 for A as measured in frame S is therefore

1.34 $$T_0 = \frac{Y}{V_A}$$

and it is the same for B in S′,

$$T_0 = \frac{Y}{V_B'}$$

If momentum is conserved in the S frame, it must be true that

1.35 $$m_A V_A = m_B V_B$$

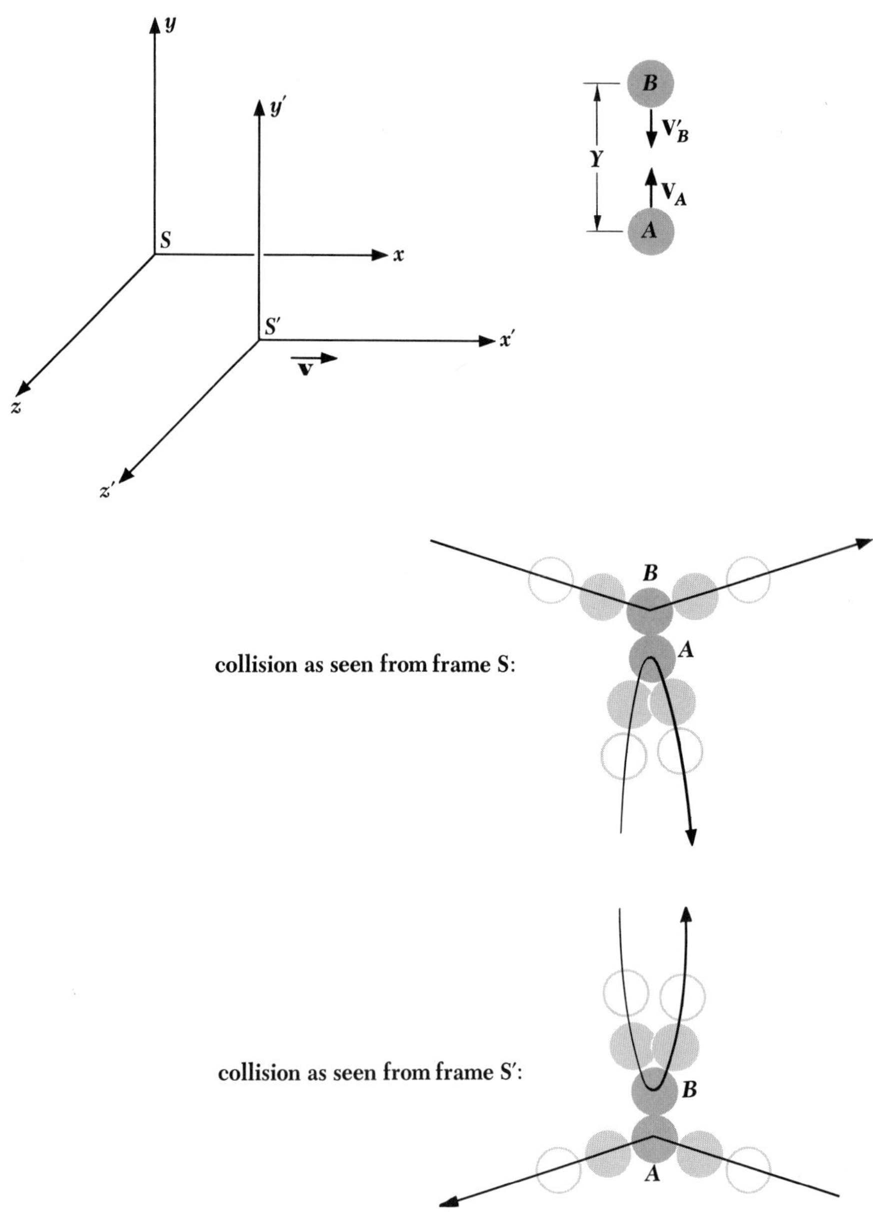

FIGURE 1-10 **An elastic collision as observed in two different frames of reference.**

where m_A and m_B are the masses of A and B, and V_A and V_B their velocities *as measured in the S frame.* In S the speed V_B is found from

$$V_B = \frac{Y}{T} \tag{1.36}$$

where T is the time required for B to make its round trip *as measured in* S. In S′, however, B's trip requires the time T_0, where

1.37 $$T = \frac{T_0}{\sqrt{1 - v^2/c^2}}$$

according to our previous results. Although observers in both frames see the same event, they disagree as to the length of time the particle thrown from the other frame requires to make the collision and return.

Replacing T in Eq. 1.36 with its equivalent in terms of T_0, we have

$$V_B = \frac{Y\sqrt{1 - v^2/c^2}}{T_0}$$

From Eq. 1.34

$$V_A = \frac{Y}{T_0}$$

Inserting these expressions for V_A and V_B in Eq. 1.35, we see that momentum is conserved if

1.38 $$m_A = m_B\sqrt{1 - v^2/c^2}$$

Our original hypothesis was that A and B are identical when at rest with respect to an observer; the difference between m_A and m_B therefore means that measurements of mass, like those of space and time, depend upon the relative speed between an observer and whatever he is observing.

In the above example both A and B are moving in S. In order to obtain a formula giving the mass m of a body measured while in motion in terms of its mass m_0 when measured at rest, we need only consider a similar example in which V_A and V_B' are very small. In this case an observer in S will see B approach A with the velocity v, make a glancing collision (since $V_B' \ll v$), and then continue on. In S

$$m_A = m_0$$

and

$$m_B = m$$

and so

1.39 $$m = \frac{m_0}{\sqrt{1 - v^2/c^2}}$$ **Relativistic mass**

The mass of a body moving at the speed v relative to an observer is larger than its mass when at rest relative to the observer by the factor $1/\sqrt{1 - v^2/c^2}$. This mass increase is reciprocal; to an observer in S′

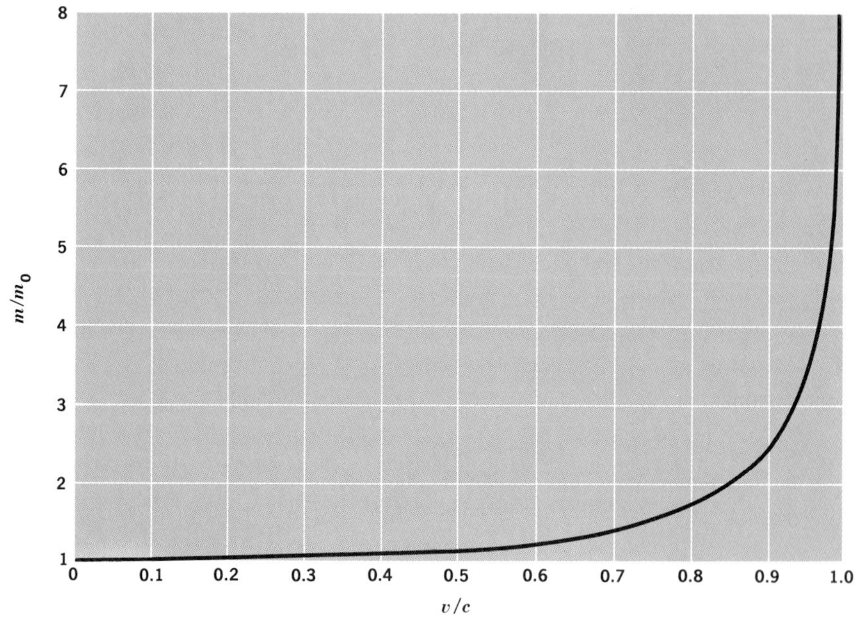

FIGURE 1-11 **The relativity of mass.**

$$m_A = m$$

and

$$m_B = m_0$$

Measured from the earth, a rocket ship in flight is shorter than its twin still on the ground and its mass is greater. To somebody on the rocket ship in flight the ship on the ground also appears shorter and to have a greater mass. (The effect is, of course, unobservably small for actual rocket speeds.) Equation 1.39 is plotted in Fig. 1-11.

Provided that momentum is defined as

1.40 $$mv = \frac{m_0 v}{\sqrt{1 - v^2/c^2}}$$

conservation of momentum is valid in special relativity just as in classical physics. However, Newton's second law of motion is correct only in the form

$$F = \frac{d}{dt}(mv)$$

1.41 $$= \frac{d}{dt}\left[\frac{m_0 v}{\sqrt{1 - v^2/c^2}}\right]$$

This is *not* equivalent to saying that

$$F = ma$$
$$= m\frac{dv}{dt}$$

even with m given by Eq. 1.39, because

$$\frac{d}{dt}(mv) = m\frac{dv}{dt} + v\frac{dm}{dt}$$

and dm/dt does not vanish if the speed of the body varies with time. The resultant force on a body is always equal to the time rate of change of its momentum.

Relativistic mass increases are significant only at speeds approaching that of light. At a speed one-tenth that of light the mass increase amounts to only 0.5 per cent, but this increase is over 100 per cent at a speed nine-tenths that of light. Only atomic particles such as electrons, protons, mesons, and so on have sufficiently high speeds for relativistic effects to be measurable, and in dealing with these particles the "ordinary" laws of physics cannot be used. Historically, the first confirmation of Eq. 1.39 was the discovery by Bücherer in 1908 that the ratio e/m of the electron's charge to its mass is smaller for fast electrons than for slow ones; this equation, like the others of special relativity, has been verified by so many experiments that it is now recognized as one of the basic formulas of physics.

1.8 Mass and Energy

The most famous relationship Einstein obtained from the postulates of special relativity concerns mass and energy. This relationship can be derived directly from the definition of the kinetic energy T of a moving body as the work done in bringing it from rest to its state of motion. That is,

1.42 $$T = \int_0^s F\,ds$$

where F is the component of the applied force in the direction of the displacement ds and s is the distance over which the force acts. Using the relativistic form of the second law of motion

$$F = \frac{d(mv)}{dt}$$

The Čerenkov Effect

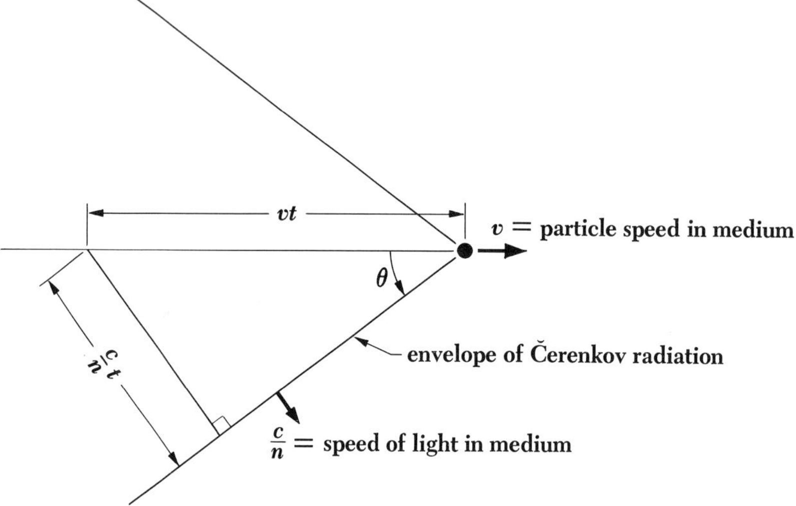

The relativity of mass places a limit on the speed a body can have relative to an observer. As the body's speed v approaches the speed of light c, the ratio v^2/c^2 approaches 1 and the quantity $\sqrt{1 - v^2/c^2}$ approaches 0. Hence the mass m of the body, given by

$$m = \frac{m_0}{\sqrt{1 - v^2/c^2}}$$

approaches infinity as v approaches c. The relative speed v therefore must be less than c. An infinite force would be needed to accelerate a body to a speed at which its mass is infinite, but there are neither infinite forces nor infinite masses in the universe.

The speed of light c in special relativity is the *speed of light in free space*, 3×10^8 m/sec. In all material media, such as water, glass, or air, light travels more slowly than this, and atomic particles are capable of moving with higher speeds in such media than the speed of light *in them*, though never faster than the speed of light in free space. When an electrically charged particle moves through a substance at a speed exceeding that of light in the substance, a cone of light waves is emitted in a process roughly similar to that in which a ship produces a bow wave as it moves through the water at a speed greater than that of water waves. These light waves are known as *Čerenkov radiation.*

The production of Čerenkov radiation is shown above. The medium through which the particle is moving has an index of refraction n, which is defined as

$$n = \frac{\text{speed of light in free space}}{\text{speed of light in medium}}$$

The speed of light in the medium is therefore c/n, which is less than the particle's speed v. As the particle travels through the medium, it creates disturbances along its path from which light waves spread out at the speed c/n. The envelope of these waves is a cone of half angle θ with the particle at its apex, where, from the figure,

$$\begin{aligned}\sin\theta &= \frac{(c/n)t}{vt} \\ &= \frac{c}{nv}\end{aligned}$$

When energetic atomic particles move through such media as glass or transparent plastics, the cone of Čerenkov radiation can be detected and its half angle θ measured, making it possible to determine the speed of the particles. The Čerenkov radiation is visible as a bluish glow when an intense beam of such particles is involved.

Eq. 1.42 becomes

$$\begin{aligned}T &= \int_0^s \frac{d(mv)}{dt}\,ds \\ &= \int_0^{mv} v\,d(mv) \\ &= \int_0^v v\,d\left(\frac{m_0 v}{\sqrt{1-v^2/c^2}}\right)\end{aligned}$$

Integrating by parts ($\int x\,dy = xy - \int y\,dx$),

$$\begin{aligned}T &= \frac{m_0 v^2}{\sqrt{1-v^2/c^2}} - m_0\int_0^v \frac{v\,dv}{\sqrt{1-v^2/c^2}} \\ &= \frac{m_0 v^2}{\sqrt{1-v^2/c^2}} + m_0c^2\sqrt{1-v^2/c^2}\,\Big|_0^v \\ &= \frac{m_0 c^2}{\sqrt{1-v^2/c^2}} - m_0c^2\end{aligned}$$

1.43 $$= mc^2 - m_0c^2$$

Equation 1.43 states that the kinetic energy of a body is equal to the increase in its mass consequent upon its relative motion multiplied by the square of the speed of light.

Equation 1.43 may be rewritten

1.44 $$mc^2 = T + m_0c^2$$

Mass and Energy: Alternative Derivation

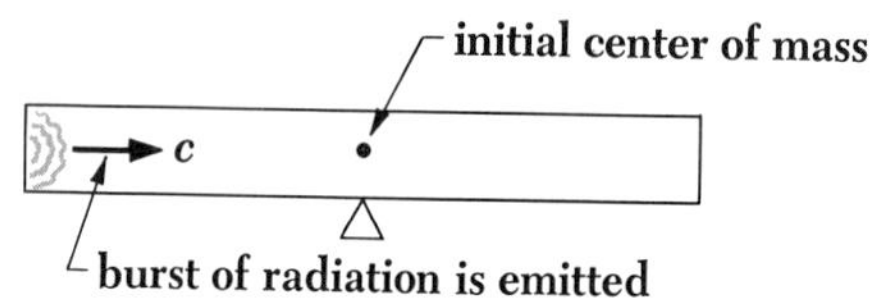

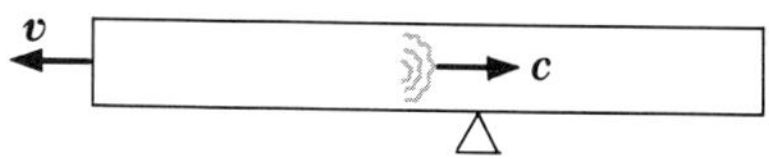

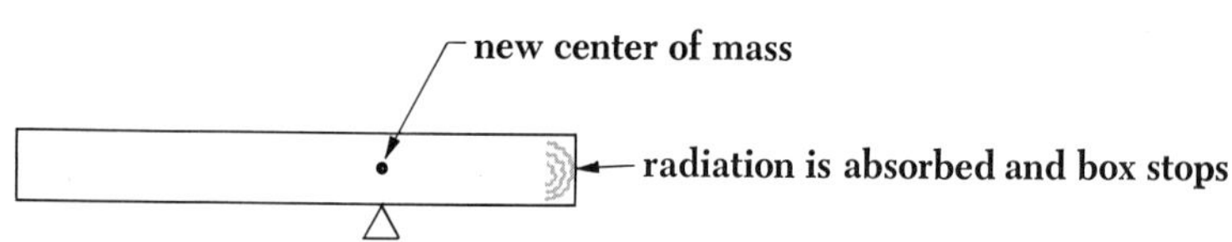

The equivalence of mass and energy can be demonstrated in a number of different ways. An interesting derivation that is somewhat different from the one given in the text, also suggested by Einstein, makes use of the basic notion that the center of mass of an isolated system (one that does not interact with its surroundings) cannot be changed by any process occurring within the system. In this derivation we imagine a closed box from one end of which a burst of electromagnetic radiation is emitted, as above. This radiation carries energy and momentum, and when the emission occurs, the box recoils in order that the total momentum of the system remain constant. When the radiation is absorbed at the opposite end of the box, its momentum cancels the momentum of the box, which then comes to rest. During the time in which the radiation was in transit, the box has moved a distance s. If the center of mass of the system is still to be in the same location in space, the radiation must have transferred mass from the end at which it was emitted to the end at which it was absorbed. We shall compute the amount of mass that must be transferred if the center of mass of the system is to remain unchanged.

For simplicity we shall consider the sides of the box to be massless and its ends to have the mass $\frac{1}{2}M$ each. The center of mass is therefore at the center of the box, a distance $\frac{1}{2}L$ from each end. A burst of electromagnetic radiation that has the energy E carries the momentum E/c and, by hypothesis, has associated with it an amount of mass m. When the radiation is emitted, the box, whose mass is now $M - m$, recoils with the velocity v. From the principle of conservation of momentum,

$$p_{\text{box}} = p_{\text{radiation}}$$

$$(M - m)v = \frac{E}{c}$$

and so the recoil velocity of the box is

$$v = \frac{E}{(M - m)c} \approx \frac{E}{Mc}$$

since m is much smaller than M. The time t during which the box moves is equal to the time required by the radiation to reach the opposite end of the box, a distance L away; this means that $t = L/c$. During the time t the box is displaced to the left by $s = vt = EL/Mc^2$.

After the box has stopped, the mass of its left-hand end is $\frac{1}{2}M - m$ and the mass of its right-hand end is $\frac{1}{2}M + m$ owing to the transfer of the mass m associated with the energy E of the radiation. If the center of mass is to be in the same place it was originally,

$$(\tfrac{1}{2}M - m)(\tfrac{1}{2}L + s) = (\tfrac{1}{2}M + m)(\tfrac{1}{2}L - s)$$

or

$$m = \frac{Ms}{L}$$

Inserting the value of the displacement s,

$$m = \frac{E}{c^2}$$

The mass associated with an amount of energy E is equal to E/c^2.

In the above derivation we assumed that the box is a perfectly rigid body: that the entire box starts to move when the radiation is emitted and the entire box comes to a stop when the radiation is absorbed. Actually, of course, there is no such thing as a rigid body that meets this specification; for example, the radiation, which travels with the speed of light, will arrive at the right-hand end of the box *before* that end begins to move! When the finite speed of elastic waves in the box is taken into account in a more elaborate calculation, however, the same result that $m = E/c^2$ is obtained.

If we interpret mc^2 as the *total energy* E of the body, it follows that, when the body is at rest and $T = 0$, it nevertheless possesses the energy m_0c^2. Accordingly m_0c^2 is called the *rest energy* E_0 of a body whose mass at rest is m_0. Equation 1.44 therefore becomes

1.45 $$E = E_0 + T$$

where

1.46 $$E = mc^2$$

and

1.47 $$E_0 = m_0c^2$$ **Rest energy**

In addition to its kinetic, potential, electromagnetic, thermal, and other familiar guises, then, energy can manifest itself as mass. The conversion factor between the unit of mass (kg) and the unit of energy (joule) is c^2, so 1 kg of matter has an energy content of 9×10^{16} joules. Even a minute bit of matter represents a vast amount of energy, and, in fact, the conversion of matter into energy is the source of the power liberated in all of the exothermic reactions of physics and chemistry.

Since mass and energy are not independent entities, the separate conservation principles of energy and mass are properly a single one, the principle of conservation of mass energy. Mass *can* be created or destroyed, but only if an equivalent amount of energy simultaneously vanishes or comes into being, and vice versa.

When the relative speed v is small compared with c, the formula for kinetic energy must reduce to the familiar $\frac{1}{2}m_0v^2$, which has been verified by experiment at low speeds. Let us see whether this is true. As we saw earlier, the binomial theorem of algebra tells us that if some quantity x is much smaller than 1,

$$(1 \pm x)^n \approx 1 \pm nx$$

The relativistic formula for kinetic energy is

$$\begin{aligned} T &= mc^2 - m_0c^2 \\ &= \frac{m_0c^2}{\sqrt{1 - v^2/c^2}} - m_0c^2 \end{aligned}$$

Expanding the first term of this formula with the help of the binomial theorem, with $v^2/c^2 \ll 1$ since v is much less than c,

$$\begin{aligned} T &= (1 + \tfrac{1}{2}v^2/c^2)m_0c^2 - m_0c^2 \\ &= \tfrac{1}{2}m_0v^2 \end{aligned}$$

Hence at low speeds the relativistic expression for the kinetic energy of a moving particle reduces to the classical one. The total energy of such a particle is

$$E = m_0c^2 + \tfrac{1}{2}m_0v^2$$

In the foregoing calculation relativity has once again met an important test; it has yielded exactly the same results as those of ordinary mechanics at low speeds, where we know by experience that the latter are perfectly valid. It is nevertheless important to keep in mind that, so far as is known, the correct formulation of mechanics has its basis in relativity, with classical mechanics no more than an approximation correct only under certain circumstances.

1.9 Some Relativistic Formulas

It is often convenient to express several of the relativistic formulas obtained above in forms somewhat different from their original ones. The new equations are so easy to derive that we shall simply state them without proof:

1.48 $$E = \sqrt{m_0{}^2c^4 + p^2c^2}$$

1.49 $$p = m_0c\sqrt{\frac{1}{1 - v^2/c^2} - 1}$$

1.50 $$T = m_0c^2\left[\frac{1}{\sqrt{1 - v^2/c^2}} - 1\right]$$

1.51 $$\frac{v}{c} = \sqrt{1 - \frac{1}{[1 + (T/m_0c^2)]^2}}$$

1.52 $$\frac{1}{\sqrt{1 - v^2/c^2}} = \sqrt{1 + \frac{p^2}{m_0{}^2c^2}}$$

1.53 $$= 1 + \frac{T}{m_0c^2}$$

These formulas are particularly useful in nuclear and elementary-particle physics, where the kinetic energies of moving particles are customarily specified, rather than their velocities. Equation 1.51, for instance, permits us to find v/c directly from T/m_0c^2, the ratio between the kinetic and rest energies of a particle. Table 1.1 is a list of the rest energies of the most stable elementary particles expressed in millions of electron volts (Mev), where 1 Mev = 1.6×10^{-13} joule; the electron volt is the usual energy unit in atomic and nuclear physics.

From Eq. 1.48 we see that, when $E \gg m_0c^2$,

1.54 $$E \approx pc$$

Hence, if we establish as our momentum unit the Mev/c, where, since 1 Mev = 1 million ev = 10^6 ev,

$$1 \text{ Mev}/c = \frac{10^6 \text{ ev} \times 1.6 \times 10^{-19} \text{ joule/ev}}{3 \times 10^8 \text{ m/sec}}$$
$$= 5.3 \times 10^{-22} \text{ kg-m/sec}$$

the energy and momentum of a particle will be numerically the same when $E \gg m_0c^2$.

Table 1.1

REST MASSES OF THE MOST STABLE ELEMENTARY PARTICLES

Particle	Symbol°	Rest mass, Mev
Photon	$\gamma(\overline{\gamma} \equiv \gamma)$	0
Neutrino	$\nu_e, \nu_\mu(\overline{\nu}_e, \overline{\nu}_\mu)$	0
Electron	$e(e^+)$	0.51
Mu meson	$\mu^-(\mu^+)$	106
Pi meson	$\pi^+(\pi^-)$	140
	$\pi^0(\overline{\pi^0} \equiv \pi^0)$	135
K meson	$K^+(K^-)$	494
	$K^0(\overline{K^0})$	498
Eta meson	$\eta^0(\overline{\eta^0} \equiv \eta^0)$	548
Proton	$p(\overline{p})$	938.2
Neutron	$n(\overline{n})$	939.5
Λ Hyperon	$\Lambda(\overline{\Lambda})$	1115
Σ Hyperon	$\Sigma^+(\overline{\Sigma^-})$	1192
	$\Sigma^0(\overline{\Sigma^0})$	1194
	$\Sigma^-(\overline{\Sigma^+})$	1197
Ξ Hyperon	$\Xi^0(\overline{\Xi^0})$	1310
	$\Xi^-(\overline{\Xi^+})$	1320
Ω Hyperon	$\Omega^-(\overline{\Omega^+})$	1676

° Antiparticle symbol in parenthesis.

1.10 Velocity Addition

One of the postulates of special relativity states that the speed of light c in free space has the same value for all observers, regardless of their relative motion. But "common sense" tells us that, if we throw a ball forward at 50 ft/sec from a car moving at 80 ft/sec, the ball's speed relative to the ground is 130 ft/sec, the sum of the two speeds. Hence we would expect that a ray of light emitted in a frame of reference S′ in the direction of its motion at velocity v relative to another frame S will have a speed of $c + v$ as measured in S, contradicting the above postulate. "Common sense" is no more reliable as a guide in science than it is elsewhere, and we must turn to the Lorentz transformation equations for the correct scheme of velocity addition.

Let us consider something moving relative to both S and S′. An observer in S measures its three velocity components to be

$$V_x = \frac{dx}{dt}$$

$$V_y = \frac{dy}{dt}$$

$$V_z = \frac{dz}{dt}$$

while to an observer in S' they are

$$V_x' = \frac{dx'}{dt'}$$

$$V_y' = \frac{dy'}{dt'}$$

$$V_z' = \frac{dz'}{dt'}$$

By differentiating the Lorentz transformation equations for x', y', z', and t', we obtain

$$dx' = \frac{dx - v\,dt}{\sqrt{1 - v^2/c^2}}$$

$$dy' = dy$$

$$dz' = dz$$

$$dt' = \frac{dt - \dfrac{v\,dx}{c^2}}{\sqrt{1 - v^2/c^2}}$$

and so

$$V_x' = \frac{dx'}{dt'}$$

$$= \frac{dx - v\,dt}{dt - \dfrac{v\,dx}{c^2}}$$

$$= \frac{\dfrac{dx}{dt} - v}{1 - \dfrac{v}{c^2}\dfrac{dx}{dt}}$$

1.55 $$= \frac{V_x - v}{1 - \dfrac{vV_x}{c^2}}$$ **Relativistic velocity transformation**

$$V_y' = \frac{dy'}{dt'}$$

$$= \frac{dy\sqrt{1 - v^2/c^2}}{dt - \dfrac{v\,dx}{c^2}}$$

$$= \frac{\dfrac{dy}{dt}\sqrt{1 - v^2/c^2}}{1 - \dfrac{v}{c^2}\dfrac{dx}{dt}}$$

1.56 $$= \frac{V_y\sqrt{1 - v^2/c^2}}{1 - \dfrac{vV_x}{c^2}}$$ **Relativistic velocity transformation** and, similarly,

$$V_z' = \frac{V_z\sqrt{1 - v^2/c^2}}{1 - \dfrac{vV_x}{c^2}} \tag{1.57}$$

Relativistic velocity transformation

Equations 1.55 to 1.57 constitute the relativistic velocity transformation. When v is small compared with c, these equations reduce to the classical equations 1.8 to 1.10,

$$\begin{aligned} V_x' &= V_x - v \\ V_y' &= V_y \\ V_z' &= V_z \end{aligned}$$

To express V_x, V_y, V_z in terms of V_x', V_y', V_z', we need only replace v with $-v$ and exchange primed and unprimed quantities in Eqs. 1.55 to 1.59. Hence

$$V_x = \frac{V_x' + v}{1 + \dfrac{vV_x'}{c^2}} \tag{1.58}$$

Inverse velocity transformation

$$V_y = \frac{V_y'\sqrt{1 - v^2/c^2}}{1 + \dfrac{vV_x'}{c^2}} \tag{1.59}$$

Inverse velocity transformation

$$V_z = \frac{V_z'\sqrt{1 - v^2/c^2}}{1 + \dfrac{vV_x'}{c^2}} \tag{1.60}$$

Inverse velocity transformation

If $V_x' = c$, that is, if a ray of light is emitted in the moving reference frame S′ in its direction of motion relative to S, an observer in frame S will measure the velocity

$$\begin{aligned} V_x &= \frac{V_x' + v}{1 + \dfrac{vV_x'}{c^2}} \\ &= \frac{c + v}{1 + \dfrac{vc}{c^2}} \\ &= \frac{c(c + v)}{c + v} \\ &= c \end{aligned}$$

Both observers determine the same value for the speed of light, as they must according to the first postulate of relativity.

The relativistic velocity transformation has other peculiar consequences. For instance, we might imagine wishing to pass a space ship whose speed with respect to the earth is $0.9c$ at a relative speed of $0.5c$. According to conventional mechanics our required speed relative to the earth would have to be $1.4c$, more than the velocity of light. According to Eq. 1.58, however, with $V_x' = 0.5c$ and $v = 0.9c$, the necessary speed is only

$$V_x = \frac{V_x' + v}{1 + \dfrac{vV_x'}{c^2}}$$

$$= \frac{0.5c + 0.9c}{1 + \dfrac{(0.9c)(0.5c)}{c^2}}$$

$$= 0.9655c$$

which is less than c. We need go less than 10 per cent faster than a space ship traveling at $0.9c$ in order to pass it at a relative speed of $0.5c$.

Problems

1. In what sense is the second postulate of special relativity a corollary of the first?

2. Many years before the special theory of relativity, the Lorentz-FitzGerald contraction was proposed in order to account for the negative result of the Michelson-Morley experiment. Show that it does so.

3. Starting from the Lorentz transformation equations 1.19 to 1.22, prove that the inverse transformation of Eqs. 1.23 to 1.26 is correct.

4. A meter stick is projected into space at so great a speed that its length appears contracted to only 50 cm. How fast is it going in miles per second?

5. A rocket ship is 100 m long on the ground. When it is in flight, its length is 99 m to an observer on the ground. What is its speed?

6. A rocket ship leaves the earth at a speed of $0.98c$. How much time does it take for the minute hand of a clock in the ship to make a complete revolution as measured by an observer on the earth?

7. An airplane circles the earth at 300 m/sec. How many years must elapse before a clock in the airplane and one on the ground differ by 1 sec?

8. A man leaves the earth in a rocket ship that makes a round trip to the nearest star, 4 light-years distant, at a speed of $0.9c$. How much younger is he upon his return than his twin brother who remained behind? (A light-year is the distance light travels in a year. It is equal to 9.46×10^{15} m.)

9. A certain particle has a lifetime of 10^{-7} sec when measured at rest. How far does it go before decaying if its speed is $0.99c$ when it is created?

10. Why must the derivation of the Doppler effect in light waves be different from that for sound waves?

11. A man has a mass of 100 kg on the ground. When he is in a rocket ship in flight, his mass is 101 kg as determined by an observer on the ground. What is the speed of the rocket ship?

12. How fast must an electron move in order that its mass equal the rest mass of the proton?

13. Find the speed of a 0.1-Mev electron according to classical and relativistic mechanics.

14. It is possible for the electron beam in a television picture tube to move across the screen at a speed faster than the speed of light. Why does this not contradict special relativity?

15. A beam of 0.8-Mev electrons enters a transparent material whose index of refraction is 1.4. Find the angle between the beam and the emitted Čerenkov radiation, which is the complement of θ on p. 28.

16. What is the minimum speed a charged particle must have if its passage through a medium whose index of refraction is 1.33 is to be accompanied by the production of Čerenkov radiation?

17. A beam of protons of unknown energy moves through a transparent material whose index of refraction is 1.8. Čerenkov radiation is emitted at an angle of 11° with the beam. Find the kinetic energy of the protons in Mev. (The above angle is the complement of θ on p. 28.)

18. How much mass does a proton gain when it is accelerated to a kinetic energy of 500 Mev?

19. How much mass does an electron gain when it is accelerated to a kinetic energy of 500 Mev?

20. The total energy of a particle is exactly twice its rest energy. Find its speed.

21. How much energy per kilogram of rest mass is required to accelerate a rocket ship to a speed of $0.98c$?

22. An electron has an initial speed of 1.4×10^8 m/sec. How much additional energy must be imparted to it for its speed to double?

23. A certain quantity of ice at 0°C melts into water at 0°C and in so doing gains 1 kg of mass. What was its initial mass?

24. Find the mass and speed of a 2-Mev electron.

25. What is the mass of a 1-Bev electron in terms of its rest mass?

26. Express the energy equivalent of 1 g of mass in kilowatt hours.

27. How much work must be done in order to increase the speed of an electron from 1.2×10^8 m/sec to 2.4×10^8 m/sec?

28. Prove that $\frac{1}{2}mv^2$, where $m = m_0/\sqrt{1 - v^2/c^2}$, does *not* equal the kinetic energy of a particle moving at relativistic speeds.

29. Express the relativistic form of the second law of motion, $F = d(mv)/dt$, in terms of the rest mass and velocity of a particle.

30. A particle has a kinetic energy of 62 Mev and a momentum of 335 Mev/c. Find its mass and speed.

31. A proton has a kinetic energy of m_0c^2. Find its momentum in units of Mev/c.

32. An observer moving in the $+x$ direction at a speed (in the laboratory system) of 2.9×10^8 m/sec finds the speed of an object moving in the $-x$ direction to be 2.998×10^8 m/sec. What is the speed of the object in the laboratory system?

33. A man on the moon observes two space ships coming toward him from opposite directions at speeds of $0.8c$ and $0.9c$ respectively. What is the relative speed of the two space ships as measured by an observer on either one?

34. In 1903 Trouton and Noble performed an interesting experiment to detect the existence of an absolute frame of reference. What they did, in essence, was to place a charge $+q$ at one end of a rod L long and another charge $-q$ at the opposite end. The rod was suspended from its center with a tiny fiber so it could rotate in a horizontal plane. If it is supposed that all electromagnetic phenomena take place in a stationary frame of reference (the "ether"), then the charges are moving through the ether with the speed v, and each charge gives rise to a magnetic field. If the angle between the rod and the direction of v is θ, show that the torque on the rod produced by the action of these magnetic fields on the moving charges is $(q^2v^2 \sin 2\theta)/8\pi\varepsilon_0 Lc^2$.

2 THE PARTICLE PROPERTIES OF WAVES

In our everyday experience there is nothing mysterious or ambiguous about the concepts of *particle* and *wave*. A stone dropped into a lake and the ripples that spread out from its point of impact apparently have in common only the ability to carry energy and momentum from one place to another. Classical physics, which mirrors the "physical reality" of our sense impressions, treats particles and waves as separate components of that reality. The mechanics of particles and the optics of waves are traditionally independent disciplines, each with its own chain of experiments and hypotheses. But the physical reality we perceive arises from phenomena that occur in the microscopic world of atoms and molecules, electrons and nuclei, and in this world there are neither particles nor waves in our sense of these terms. We regard electrons as particles because they possess charge and mass and behave according to the laws of particle mechanics in such familiar devices as television picture tubes. We shall see, however, that there is as much evidence in favor of interpreting a moving electron as a wave manifestation as there is in favor of interpreting it as a particle manifestation. We regard electromagnetic waves as waves because under suitable circumstances they exhibit diffraction, interference, and polarization. Similarly, we shall see that under other circumstances electromagnetic waves behave as though they consist of streams of particles. Together with special relativity, the wave-particle duality is central to an understanding of modern physics, and in this book there are few arguments that do not draw upon these fundamental principles.

2.1 The Photoelectric Effect

Late in the nineteenth century a series of experiments revealed that electrons are emitted from a metal surface when light of sufficiently high frequency (ultraviolet light is required for all but the alkali metals) falls upon it. This phenomenon is known as the *photoelectric effect*. Figure 2-1 illustrates the type of apparatus that was employed in the more precise of these experiments.

FIGURE 2-1 Experimental observation of the photoelectric effect.

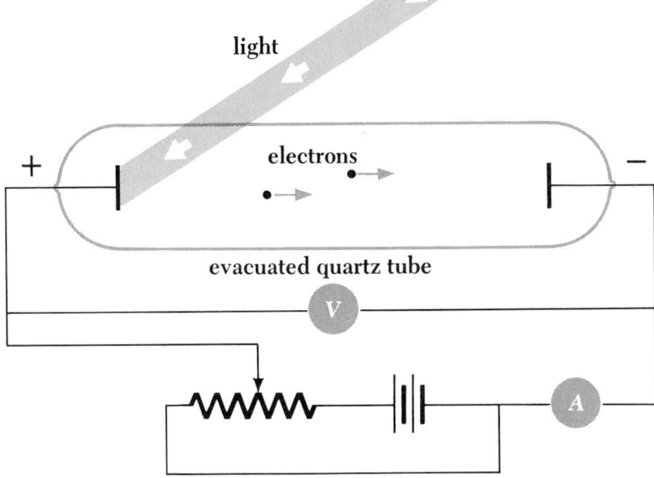

An evacuated tube contains two electrodes connected to an external circuit like that shown schematically, with the metal plate whose surface is to be irradiated as the anode. Some of the photoelectrons that emerge from the irradiated surface have sufficient energy to reach the cathode despite its negative polarity, and they constitute the current that is measured by the ammeter in the circuit. As the retarding potential V is increased, fewer and fewer electrons get to the cathode and the current drops. Ultimately, when V equals or exceeds a certain value, of the order of a few volts, no further electrons strike the cathode and the current ceases.

The existence of the photoelectric effect ought not to be surprising; after all, light waves carry energy, and some of the energy absorbed by the metal may somehow concentrate on individual electrons and reappear as kinetic energy. When we look more closely at the data, however, we find that the photoelectric effect can hardly be interpreted so simply.

One of the features of the photoelectric effect that particularly puzzled its discoverers is that the energy distribution in the emitted electrons (called *photoelectrons*) is independent of the intensity of the light. A strong light beam yields more photoelectrons than a weak one of the same frequency, but the average electron energy is the same. Also, within the limits of experimental accuracy (about 3×10^{-9} sec), there is no time lag between the arrival of light at a metal surface and the emission of photoelectrons. These observations cannot be understood from the electromagnetic theory of light. Let us consider violet light falling on a sodium surface, in an apparatus like that of Fig. 2-1. There will be a detectable photoelectric current when 10^{-6} watt/m^2 of electromagnetic energy is absorbed by the surface (a more intense beam than this is required, of course, since sodium is a good reflector of light). Now there are about 10^{19} atoms in a layer of sodium 1 atom thick and 1 m^2 in area, so that, if we assume that the incident light is absorbed in the 10 uppermost layers of sodium atoms, the 10^{-6} watt/m^2 is distributed among 10^{20} atoms. Hence each atom receives energy at the average rate of 10^{-26} watt, which is less than 10^{-7} ev/sec. It should therefore take more than 10^7 sec, or almost a year, for any single electron to accumulate the 1 ev or so of energy that the photoelectrons are found to have! In the maximum possible time of 3×10^{-9} sec, an average electron, according to electromagnetic theory, will have gained only 3×10^{-16} ev. Even if we call upon some kind of resonance process to explain why some electrons acquire more energy than others, the fortunate electrons will still have no more than 10^{-10} of the observed energy.

Equally odd from the point of view of the wave theory is the fact that the photoelectron energy depends upon the *frequency* of the light employed. At frequencies below a certain critical frequency characteristic of each particular metal, no electrons whatever are emitted. Above this threshold frequency the photoelectrons have a range of energies from 0 to a certain maximum value, and this maximum energy increases linearly with increasing frequency. High frequencies result in high-maximum photoelectron energies, low frequencies in low-maximum photoelectron energies. Thus a faint blue light produces electrons with more energy than those produced by a bright red light, although the latter yields a greater number of them.

Figure 2-2 is a plot of maximum photoelectron energy T_{max} versus the frequency ν of the incident light in a particular experiment. It is clear that the relationship between T_{max} and the frequency ν involves a proportionality, which we can express in the form

$$\begin{aligned} T_{max} &= h(\nu - \nu_0) \\ &= h\nu - h\nu_0 \end{aligned} \quad (2.1)$$

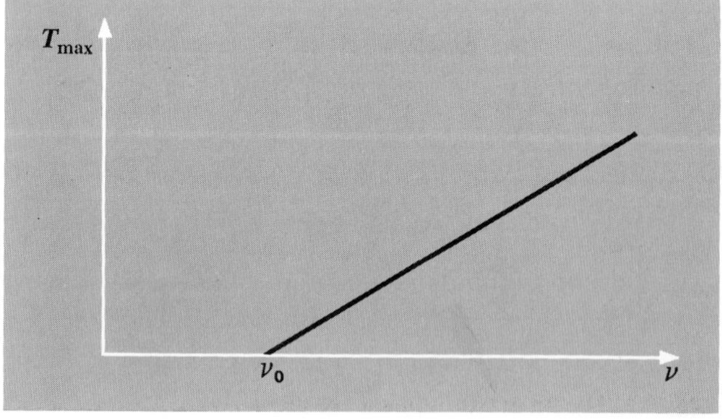

FIGURE 2-2 The maximum photoelectron energy as a function of the frequency of the incident light.

where ν_0 is the threshold frequency below which no photoemission occurs and h is a constant. Significantly, the value of h,

$$h = 6.63 \times 10^{-34} \text{ joule-sec}$$

is *always the same,* although ν_0 varies with the particular metal being illuminated.

2.2 The Quantum Theory of Light

The electromagnetic theory of light accounts so well for such a variety of phenomena that it must contain some measure of truth. Yet this well-founded theory is completely at odds with the photoelectric effect. In 1905 Albert Einstein found that the paradox presented by the photoelectric effect could be understood only by taking seriously a notion proposed five years earlier by the German theoretical physicist Max Planck. Planck was seeking to explain the characteristics of the radiation emitted by bodies hot enough to be luminous, a problem notorious at the time for its resistance to solution. Planck was able to derive a formula for the spectrum of this radiation (that is, the relative brightness of the various colors present) as a function of the temperature of the body that was in agreement with experiment provided he assumed that the radiation is emitted *discontinuously* as little bursts of energy. These bursts of energy are called *quanta.* Planck found that the quanta associated with a particular frequency ν of light all have the same energy and that this energy E is directly proportional to ν. That is,

2.2 $$E = h\nu$$ **Quantum energy**

where h, today known as *Planck's constant,* has the value

$$h = 6.63 \times 10^{-34} \text{ joule-sec}$$

We shall examine some of the details of this interesting problem and its solution in Chap. 10.

While he had to assume that the electromagnetic energy radiated by a hot object emerges intermittently, Planck did not doubt that it propagates continuously through space as electromagnetic waves. Einstein proposed that light not only is emitted a quantum at a time, but also propagates as individual quanta. In terms of this hypothesis the photoelectric effect can be readily explained. The empirical formula Eq. 2.1 may be rewritten

2.3 $$h\nu = T_{max} + h\nu_0$$ **Photoelectric effect**

Einstein's proposal means that the three terms of Eq. 2.3 are to be interpreted as follows:

$$h\nu = \text{the energy content of each quantum of the incident light}$$

$$T_{max} = \text{the maximum photoelectron energy}$$

$$h\nu_0 = \text{the minimum energy needed to dislodge an electron from the metal surface being illuminated}$$

There must be a minimum energy required by an electron in order to escape from a metal surface, or else electrons would pour out even in the absence of light. The energy $h\nu_0$ characteristic of a particular surface is called its *work function.* Hence Eq. 2.3 states that

$$\text{Quantum energy} = \text{maximum electron energy} + \text{work function of surface}$$

It is easy to see why not all photoelectrons have the same energy, but emerge with all energies up to T_{max}: $h\nu_0$ is the work that must be done to take an electron through the metal surface from just beneath it, and more work is required when the electron originates deeper in the metal.

The validity of this interpretation of the photoelectric effect is confirmed by studies of *thermionic emission.* Long ago it became known that the presence of a very hot object increases the electrical conductivity of the surrounding air, and late in the nineteenth century the reason for this phenomenon was found to be the emission of electrons from such an object. Thermionic emission makes possible the operation of the vacuum tubes so widely used in electronics, where metal filaments or specially coated cathodes at high temperature supply dense streams of electrons. The emitted electrons evidently obtain their energy from the thermal agitation of the particles constituting the metal, and we should expect that the electrons must acquire a certain minimum energy in order to escape. This minimum energy can be determined for many surfaces, and it is always almost identical to the photoelectric work function for the same surfaces. In photoelectric emission, photons of light provide the energy required by an electron to escape, while in thermionic emission heat does so: in both cases the physical processes involved in the emergence of an electron from a metal surface are the same.

Let us apply Eq. 2.3 to a specific situation. The work function of potassium is 2.0 ev. When ultraviolet light of wavelength 3,500 A (1 A = 1 Angstrom unit = 10^{-10} m) falls on a potassium surface, what is the maximum energy in electron volts of the photoelectrons? From Eq. 2.3,

$$T_{max} = h\nu - h\nu_0$$

Since $h\nu_0$ is already expressed in electron volts, we need only compute the quantum energy $h\nu$ of 3,500-A light. This is

$$
\begin{aligned}
h\nu &= \frac{hc}{\lambda} \\
&= \frac{6.63 \times 10^{-34} \text{ joule-sec} \times 3 \times 10^{8} \text{ m/sec} \times 10^{10} \text{ A/m}}{3{,}500 \text{ A}} \\
&= 5.7 \times 10^{-19} \text{ joule}
\end{aligned}
$$

To convert this energy from joules to electron volts, we recall that

$$1 \text{ ev} = 1.6 \times 10^{-19} \text{ joule}$$

and so

$$
\begin{aligned}
h\nu &= \frac{5.7 \times 10^{-19} \text{ joule}}{1.6 \times 10^{-19} \text{ joule/ev}} \\
&= 3.6 \text{ ev}
\end{aligned}
$$

Hence the maximum photoelectron energy is

$$
\begin{aligned}
T_{\max} &= h\nu - h\nu_0 \\
&= 3.6 \text{ ev} - 2.0 \text{ ev} \\
&= 1.6 \text{ ev}
\end{aligned}
$$

The view that light propagates as a series of little packets of energy (sometimes called photons) is directly opposed to the wave theory of light. The latter, which provides the sole means of explaining a host of optical effects—notably diffraction and interference—is one of the most securely established of physical theories. Planck's suggestion that a hot object emits light in separate quanta was not incompatible with the propagation of light as a wave. Einstein's suggestion in 1905 that light travels through space in the form of distinct photons, however, elicited incredulity from his contemporaries. According to the wave theory, light waves spread out from a source in the way ripples spread out on the surface of a lake when a stone falls into it. The energy carried by the light, in this analogy, is distributed continuously throughout the wave pattern. According to the quantum theory, on the other hand, light spreads out from a source as a series of localized concentrations of energy, each sufficiently small to be capable of absorption by a single electron. Curiously, the quantum theory of light, which treats it strictly as a particle phenomenon, explicitly involves the light frequency ν, strictly a wave concept.

The quantum theory of light is strikingly successful in explaining the photoelectric effect. It predicts correctly that the maximum photoelectron energy should depend upon the frequency of the incident light and not upon its intensity, contrary to what the wave theory suggests, and it is able to explain why even the feeblest light can lead to the immediate emission of photoelectrons, again contrary to the wave theory. The wave theory can give no reason why there should be a threshold frequency such that, when light of lower frequency is employed, no photoelectrons are observed no matter how strong the light beam, something that follows naturally from the quantum theory.

Which theory are we to believe? A great many physical hypotheses have had to be altered or discarded when they were found to disagree with experiment, but never before have we had to devise two totally different theories to account for a single physical phenomenon. The situation here is fundamentally different from what it is, say, in the case of relativistic versus Newtonian mechanics, where the latter turns out to be an approximation of the former. There is no way of deriving the quantum theory of light from the wave theory of light or vice versa.

In a specific event light exhibits *either* a wave or a particle nature, never both simultaneously. The same light beam that is diffracted by a grating can cause the emission of photoelectrons from a suitable surface, but these processes occur independently. **The wave theory of light and the quantum theory of light complement each other.** Electromagnetic waves provide the sole possible explanation for certain experiments involving light and optical phenomena, while photons provide the sole possible explanation for all of the other experiments in this field. We have no alternative to regarding light as something that manifests itself as a stream of discrete photons on occasion and as a wave train the rest of the time. The "true nature" of light is no longer a meaningful concept, and we must accept both wave and quantum theories, contradictions and all, as the closest we can get to a complete description of light.

2.3 X Rays

The photoelectric effect provides convincing evidence that photons of light can transfer energy to electrons. Is the inverse process also possible? That is, can part or all of the kinetic energy of a moving electron be converted into a photon? As it happens, the inverse photoelectric effect not only does occur, but had been discovered (though not at all understood) prior to the theoretical work of Planck and Einstein.

In 1895 Wilhelm Roentgen made the classic observation that a highly penetrating radiation of unknown nature is produced when fast electrons impinge on matter. These *X rays* were soon found to travel in straight lines, even through electric and magnetic fields, to pass readily through opaque materials, to cause phosphorescent substances to glow, and to expose photographic plates. The faster the original electrons, the more penetrating the resulting X rays, and the greater the number of electrons, the greater the intensity of the X-ray beam.

X-ray Polarization

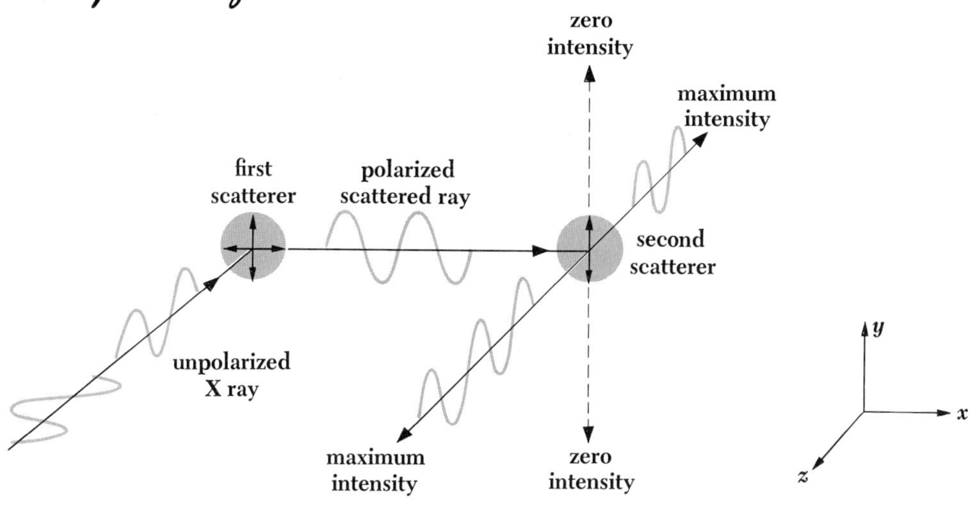

The wave nature of X rays was first established in 1906 by Barkla, who was able to exhibit their polarization. Barkla's experimental arrangement is sketched above. We shall analyze this classic experiment under the assumption that X rays are electromagnetic waves. At the left a beam of unpolarized X rays heading in the $-z$ direction impinges on a small block of carbon. These X rays are scattered by the carbon; this means that electrons in the carbon atoms are set in vibration by the electric vectors of the X rays and then reradiate. Because the electric vector in an electromagnetic wave is perpendicular to its direction of propagation, the initial beam of X rays contains electric vectors that lie in the xy plane only. The target electrons therefore are induced to vibrate in the xy plane. A scattered X ray that proceeds in the $+x$ direction can have an electric vector in the y direction only, and so it is plane-polarized. To demonstrate this polarization, another carbon block is placed in the path of the ray, as at the right. The electrons in this block are restricted to vibrate in the y direction and therefore reradiate X rays that propagate in the xz plane exclusively, and not at all in the y direction. The observed absence of scattered X rays outside the xz plane confirms the wave character of X rays.

Not long after their discovery it began to be suspected that X rays are electromagnetic waves. After all, electromagnetic theory predicts that an accelerated electric charge will radiate electromagnetic waves, and a rapidly moving electron suddenly brought to rest is certainly accelerated. Radiation produced under these circumstances is given the German name *bremsstrahlung* ("braking radiation"). The absence of any perceptible X-ray refraction in the early work could be attributed to very short wavelengths, below those in the ultraviolet range, since the refractive index of a substance decreases to unity (corresponding to straight-line propagation) with decreasing wavelength. Subsequently Barkla was able to show that X rays that have been scattered upon striking some material are polarized, further supporting the wave hypothesis.

In 1912 a method was devised for actually measuring the wavelengths of X rays. A diffraction experiment had been recognized as ideal, but, as we recall from physical optics, the spacing between adjacent lines on a diffraction grating must be of the same order of magnitude as the wavelength of the light for satisfactory results, and gratings cannot be ruled with the minute spacing required by X rays. In 1912, however, Max von Laue recognized that the wavelengths hypothesized for X rays were about the same order of magnitude as the spacing between adjacent atoms in crystals, which is about 1 A. He therefore proposed that crystals be used to diffract X rays, with their regular lattices acting as a kind of three-dimensional grating. Suitable experiments were performed in the next year, and the wave nature of X rays was successfully demonstrated. In these experiments wavelengths from 1.3×10^{-11} m to 4.8×10^{-11} m were found, 0.13 A to 0.48 A, 10^{-4} of those in visible light, and hence having quanta 10^4 times as energetic. We shall consider X-ray diffraction in Sec. 2.4.

For purposes of classification, electromagnetic radiations with wavelengths in the approximate interval from 10^{-11} to 10^{-8} m (0.1 to 100 A) are today considered as X rays.

Figure 2-3 is a diagram of an X-ray tube. A cathode, heated by an adjacent filament through which an electric current is passed, supplies electrons copiously by thermionic emission. The high potential difference V

FIGURE 2-3 An X-ray tube.

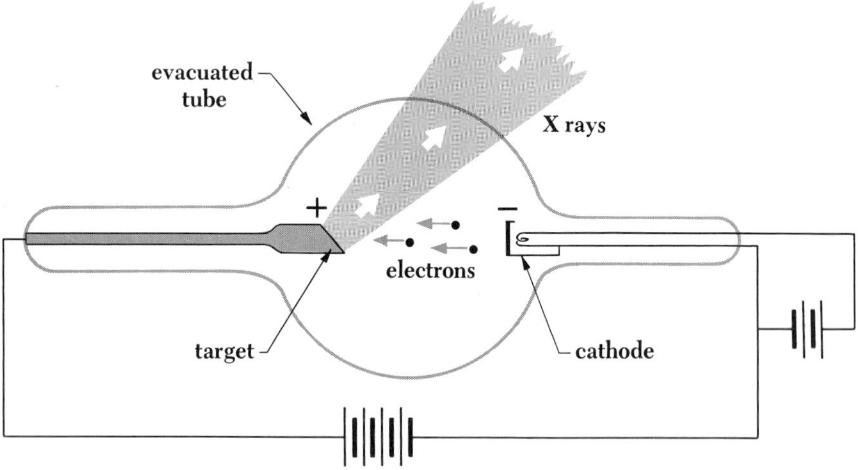

maintained between the cathode and a metallic target accelerates the electrons toward the latter. The face of the target is at an angle relative to the electron beam, and the X rays that emerge from the target pass through the side of the tube. The tube is evacuated to permit the electrons to get to the target unimpeded.

As we said earlier, classical electromagnetic theory predicts the production of *bremsstrahlung* when electrons are accelerated, thereby apparently accounting for the X rays emitted when fast electrons are stopped by the target of an X-ray tube. However, the agreement between the classical theory and the experimental data is not satisfactory in certain important respects. Figures 2-4 and 2-5 show the X-ray spectra that result when tungsten and molybdenum targets are bombarded by electrons at several different accelerating potentials. The curves exhibit two distinctive features not accountable in terms of electromagnetic theory:

1. In the case of molybdenum there are pronounced intensity peaks at certain wavelengths, indicating the enhanced production of X rays. These peaks occur at various specific wavelengths for each target material and originate in rearrangements of the electron structures of the target atoms after having been disturbed by the bombarding electrons. The important thing to note at this point is the production of X rays of specific wavelengths, a decidedly nonclassical effect, in addition to the production of a continuous X-ray spectrum.

2. The X rays produced at a given accelerating potential V vary in wavelength, but none has a wavelength shorter than a certain value λ_{min}. Increasing V decreases λ_{min}. At a particular V, λ_{min} is the *same* for both the

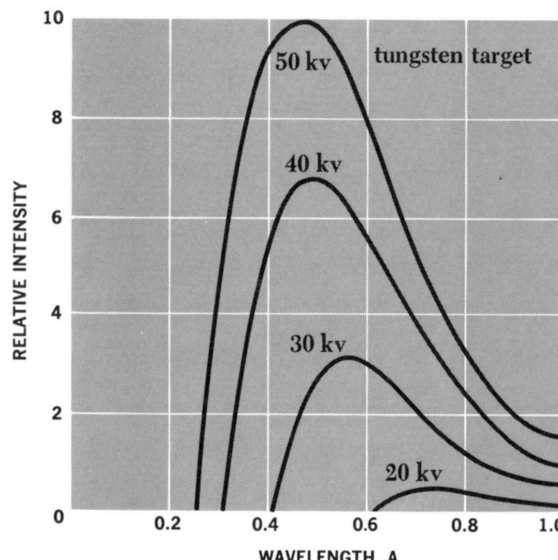

FIGURE 2-4 X-ray spectra of tungsten at various accelerating potentials.

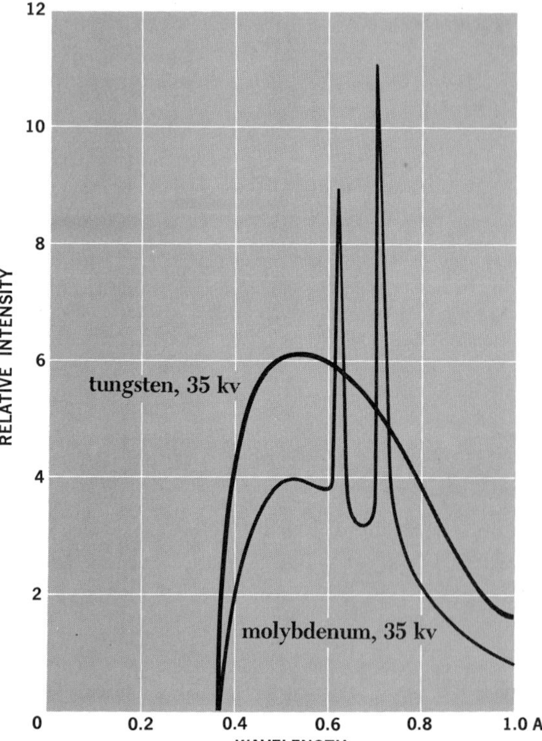

FIGURE 2-5 X-ray spectra of tungsten and molybdenum at 35 kv accelerating potential.

tungsten and molybdenum targets. Duane and Hunt have found that λ_{min} is inversely proportional to V; their precise relationship is

$$\lambda_{min} = \frac{1.24 \times 10^{-6} \text{ volt-m}}{V} \tag{2.4}$$

X-ray production

The second observation is readily understood in terms of the quantum theory of radiation. Most of the electrons incident upon the target lose their kinetic energy gradually in numerous collisions, their energy going simply into heat. (This is the reason that the targets in X-ray tubes are normally of high-melting-point metals, and an efficient means of cooling the target is often employed.) A few electrons, though, lose most or all of their energy in single collisions with target atoms; this is the energy that is evolved as X rays. X-ray production, then, except for the peaks mentioned in observation 1 above, represents an inverse photoelectric effect. Instead of photon energy being transformed into electron kinetic energy, electron kinetic energy is being transformed into photon energy. A short wavelength means a high frequency, and a high frequency means a high photon energy $h\nu$. It is therefore logical to interpret the short wavelength limit λ_{min} of Eq. 2.4 as corresponding to a maximum photon energy $h\nu_{max}$, where

2.5 $$h\nu_{max} = \frac{hc}{\lambda_{min}}$$

Since work functions are only several electron volts while the accelerating potentials in X-ray tubes are tens or hundreds of thousands of volts, we may assume that the kinetic energy T of the bombarding electrons is

2.6 $$T = eV$$

When the entire kinetic energy of an electron is lost to create a single photon, then

2.7 $$h\nu_{max} = T$$

Substituting Eqs. 2.5 and 2.6 into 2.7, we see that

$$h\nu_{max} = T$$

$$\frac{hc}{\lambda_{min}} = eV$$

$$\lambda_{min} = \frac{hc}{eV}$$

$$= \frac{6.63 \times 10^{-34}\ \text{joule-sec} \times 3 \times 10^{8}\ \text{m/sec}}{1.6 \times 10^{-19}\ \text{coulomb} \times V}$$

$$= \frac{1.24 \times 10^{-6}}{V}\ \text{volt-m}$$

which is just the experimental relationship of Eq. 2.4. It is therefore correct to regard X-ray production as the inverse of the photoelectric effect.

A conventional X-ray machine might have an accelerating potential of 50,000 volts. To find the shortest wavelength present in its radiation, we use Eq. 2.4, with the result that

$$\lambda_{min} = \frac{1.24 \times 10^{-6}\ \text{volt-m}}{5 \times 10^{4}\ \text{volts}}$$

$$= 2.5 \times 10^{-11}\ \text{m}$$

$$= 0.25\ \text{A}$$

This wavelength corresponds to the frequency

$$\nu_{max} = \frac{c}{\lambda_{min}}$$

$$= \frac{3 \times 10^{8}\ \text{m/sec}}{2.5 \times 10^{-11}\ \text{m}}$$

$$= 1.2 \times 10^{19}\ \text{sec}^{-1}$$

2.4 X-ray Diffraction

Let us now return to the question of how X rays may be demonstrated to consist of electromagnetic waves. A crystal consists of a regular array of atoms each of which is able to scatter any electromagnetic waves that happen to strike it. A monochromatic beam of such waves that falls upon a crystal will be scattered in all directions within it, but, owing to the regular arrangement of the atoms, in certain directions the scattered waves will constructively interfere with one another while in others they will destructively interfere. The atoms in a crystal may be thought of as defining families of parallel planes, as in Fig. 2-6, with each family having a characteristic separation between its component planes. This analysis was suggested in 1913 by W. L. Bragg, in honor of whom the above planes are called *Bragg planes.* The conditions that must be fulfilled for radiation scattered by crystal atoms to undergo constructive interference may be obtained from a diagram like that in Fig. 2-7. A beam containing X rays of wavelength λ is incident upon a crystal at an angle θ with a family of Bragg planes whose spacing is d. The beam goes past atom A in the first plane and atom B in the next, and each of them scatters part of the beam in random directions. Constructive interference takes place only between those scattered rays that are parallel and whose paths differ by exactly λ, 2λ, 3λ, and so on. That is, the path difference must be $n\lambda$, where n is an integer. The only rays scattered by A and B for which this is true are those labeled I and II in Fig. 2-7. The first condition upon I and II is that their common scattering angle be equal to the angle of incidence θ of the original beam. (This condition is the same as that for ordinary optical reflection: angle of incidence = angle of reflection. For this reason X-ray scattering from atomic planes in a crystal is commonly, though incorrectly, termed *Bragg reflection.*) The second condition is that

$$2d \sin \theta = n\lambda \qquad n = 1, 2, 3, \ldots \tag{2.8}$$

since ray II must travel the distance $2d \sin \theta$ farther than ray I. The integer n is the *order* of the scattered beam.

The schematic design of an X-ray spectrometer based upon Bragg's analysis is shown in Fig. 2-8. A collimated beam of X rays falls upon a crystal at an angle θ, and a detector is placed so that it records those rays whose scattering angle is also θ. Any X rays reaching the detector therefore obey the first Bragg condition. As θ is varied, the detector will record intensity peaks corresponding to the orders predicted by Eq. 2.8. If the spacing d between adjacent Bragg planes in the crystal is known, the X-ray wavelength λ may be calculated.

How can we determine the value of d? This is a simple task in the case of crystals whose atoms are arranged in cubic lattices similar to that of rock

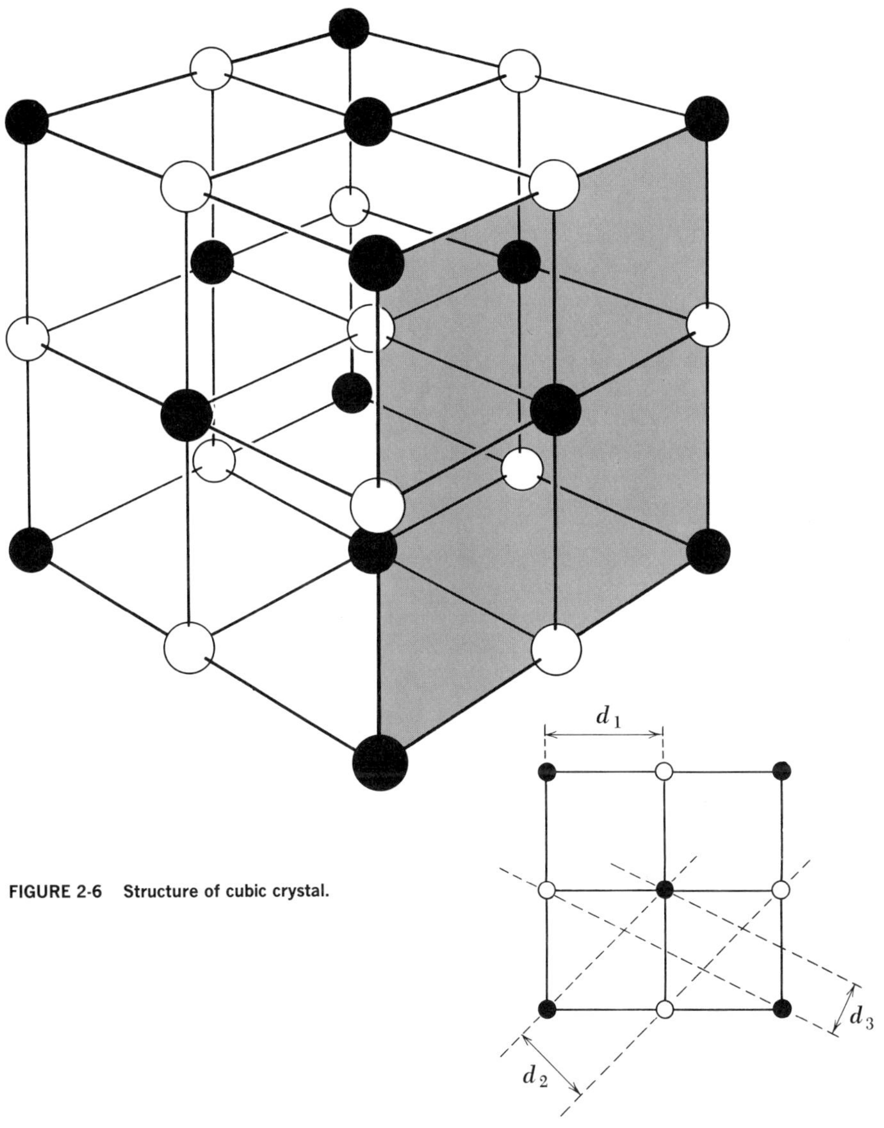

FIGURE 2-6 Structure of cubic crystal.

salt (NaCl), illustrated in Fig. 2-6. As an example let us compute the separation of adjacent atoms in a crystal of NaCl. The molecular weight of NaCl is 58.5, which means that there are 58.5 g of NaCl per mole. Since there are $N_0 = 6.03 \times 10^{23}$ molecules in a mole of any substance (N_0 is Avogadro's number), the mass of each NaCl molecule is

$$m_{\mathrm{NaCl}} = 58.5 \text{ g/mole} \times \frac{1}{6.03 \times 10^{23} \text{ molecules/mole}}$$

$$= 9.71 \times 10^{-23} \text{ g/molecule}$$

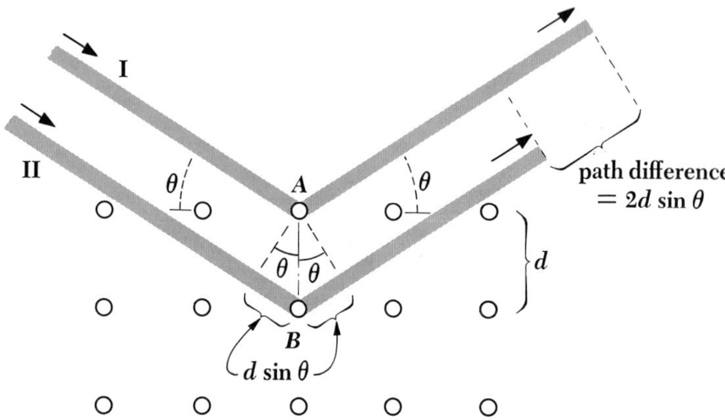

FIGURE 2-7 X-ray scattering from cubic crystal.

Crystalline NaCl has a density of 2.16 g/cm^3, and so, taking into account the presence of two atoms in each NaCl molecule, the number of atoms in 1 cm^3 of NaCl is

$$n = 2 \text{ atoms/molecule} \times 2.16 \text{ g/cm}^3 \times \frac{1}{9.71 \times 10^{-23} \text{ g/molecule}}$$

$$= 4.45 \times 10^{22} \text{ atoms/cm}^3$$

If d is the distance between adjacent atoms in a crystal, there are $1/d$

FIGURE 2-8 X-ray spectrometer.

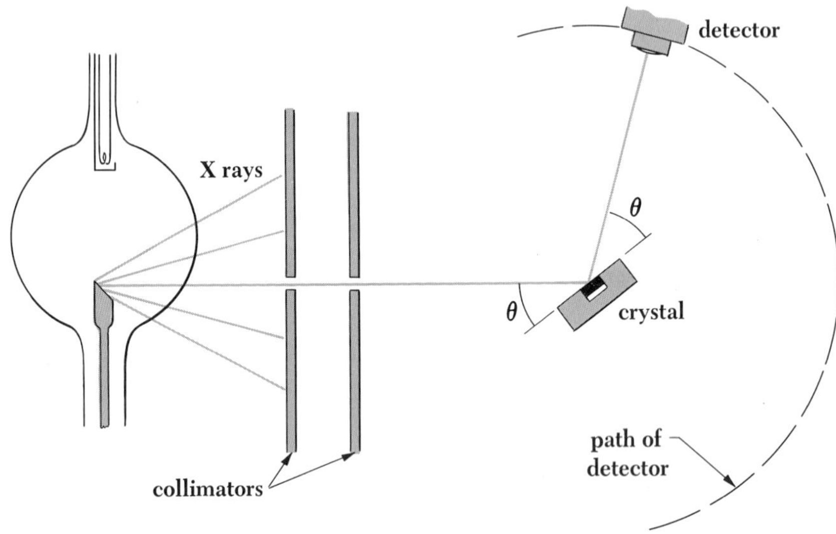

atoms/cm along any of the crystal axes and $1/d^3$ atoms/cm^3 in the entire crystal. Hence

$$\frac{1}{d^3} = n$$

and

$$\begin{aligned} d &= (4.45 \times 10^{22})^{-1/3} \text{ cm} \\ &= 2.82 \times 10^{-8} \text{ cm} \\ &= 2.82 \text{ A} \end{aligned}$$

2.5 The Compton Effect

The quantum theory of light postulates that photons behave like particles except for the absence of any rest mass. If this is true, then it should be possible for us to treat collisions between photons and, say, electrons in the same manner as billiard-ball collisions are treated in elementary mechanics. Figure 2-9 shows how such a collision might be represented, with an X-ray photon striking an electron (assumed to be initially at rest in the laboratory coordinate system) and being scattered away from its original direction of motion while the electron receives an impulse and begins to move. In the collision the photon may be regarded as having lost an amount of energy that is the same as the kinetic energy T gained by the electron, though actually separate photons are involved. If the initial photon has the frequency ν associated with it, the scattered photon has the lower frequency ν', where

$$\text{Loss in photon energy} = \text{gain in electron energy}$$

2.9 $$h\nu - h\nu' = T$$

From the previous chapter we recall that

$$E = \sqrt{m_0{}^2c^4 + p^2c^2}$$

so that, since the photon has no rest mass, its total energy is

$$E = pc$$

Since

$$E = h\nu$$

for a photon, its momentum is

$$p = \frac{E}{c}$$

2.10 $$= \frac{h\nu}{c}$$

Momentum, unlike energy, is a vector quantity, incorporating direction as well as magnitude, and in the collision momentum must be conserved in each

FIGURE 2-9 The Compton effect.

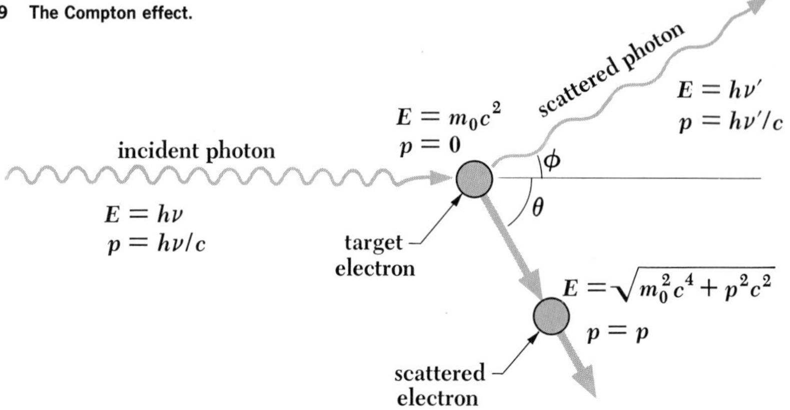

of two mutually perpendicular directions. (When more than two bodies participate in a collision, of course, momentum must be conserved in each of three mutually perpendicular directions.) The directions we choose here are that of the original photon and one perpendicular to it in the plane containing the electron and the scattered photon (Fig. 2-9). The initial photon momentum is $h\nu/c$, the scattered photon momentum is $h\nu'/c$, and the initial and final electron momenta are respectively 0 and p. In the original photon direction

Initial momentum = final momentum

2.11 $$\frac{h\nu}{c} + 0 = \frac{h\nu'}{c} \cos \phi + p \cos \theta$$

and perpendicular to this direction

Initial momentum = final momentum

2.12 $$0 = \frac{h\nu'}{c} \sin \phi - p \sin \theta$$

The angle ϕ is that between the directions of the initial and scattered photons, and θ is that between the directions of the initial photon and the recoil electron. From Eqs. 2.9, 2.11, and 2.12 we shall now obtain a formula relating the wavelength difference between initial and scattered photons with the angle ϕ between their directions, both of which are readily measurable quantities.

The first step is to multiply Eqs. 2.11 and 2.12 by c and rewrite them as

$$pc \cos \theta = h\nu - h\nu' \cos \phi$$
$$pc \sin \theta = h\nu' \sin \phi$$

By squaring each of these equations and adding the new ones together, the angle θ is eliminated, leaving

2.13 $$p^2c^2 = (h\nu)^2 - 2(h\nu)(h\nu')\cos\phi + (h\nu')^2$$

By equating the two expressions for the total energy of a particle

$$E = T + m_0c^2$$
$$E = \sqrt{m_0{}^2c^4 + p^2c^2}$$

from the previous chapter, we have

$$(T + m_0c^2)^2 = m_0{}^2c^4 + p^2c^2$$
$$p^2c^2 = T^2 + 2m_0c^2T$$

Since

$$T = h\nu - h\nu'$$

we have

2.14 $$p^2c^2 = (h\nu)^2 - 2(h\nu)(h\nu') + (h\nu')^2 + 2m_0c^2(h\nu - h\nu')$$

Substituting this value of p^2c^2 in Eq. 2.13, we finally obtain

2.15 $$2m_0c^2(h\nu - h\nu') = 2(h\nu)(h\nu')(1 - \cos\phi)$$

It is more convenient to express this relationship in terms of wavelength rather than frequency. Dividing Eq. 2.15 by $2h^2c^2$,

$$\frac{m_0c}{h}\left(\frac{\nu}{c} - \frac{\nu'}{c}\right) = \frac{\nu}{c}\frac{\nu'}{c}(1 - \cos\phi)$$

and so, since $\nu/c = 1/\lambda$ and $\nu'/c = 1/\lambda'$,

$$\frac{m_0c}{h}\left(\frac{1}{\lambda} - \frac{1}{\lambda'}\right) = \frac{(1 - \cos\phi)}{\lambda\lambda'}$$

2.16 $$\lambda' - \lambda = \frac{h}{m_0c}(1 - \cos\phi)$$ **Compton effect**

Equation 2.16 was derived by Arthur H. Compton in the early 1920s, and the phenomenon it describes, which he was the first to observe, is known as the *Compton effect*. It constitutes very strong evidence in support of the quantum theory of radiation.

Equation 2.16 gives the change in wavelength expected for a photon that is scattered through the angle ϕ by a particle of rest mass m_0; it is independent of the wavelength λ of the incident photon. The quantity h/m_0c is called the *Compton wavelength* of the scattering particle, which for an electron is 0.024 A (2.4×10^{-12} m). From Eq. 2.16 we note that the greatest wavelength change that can occur will take place for $\phi = 180°$, when the wavelength change will be twice the Compton wavelength h/m_0c. Because the Compton wavelength of an electron is 0.024 A, while it is considerably less for other particles owing to their larger rest masses, the maximum wavelength change in the Compton effect is 0.048 A. Changes of this magnitude or less are readily observable only in X rays, since the shift in wavelength for

FIGURE 2-10 Experimental demonstration of the Compton effect.

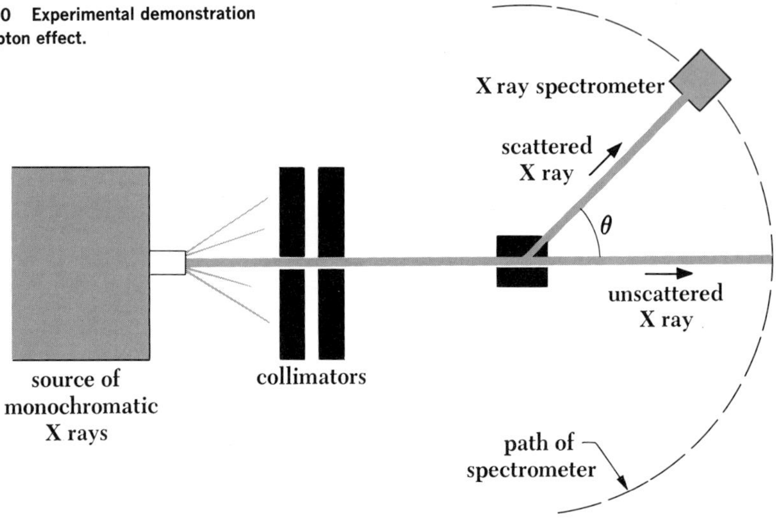

visible light is less than 0.01 per cent of the initial wavelength, while for X rays of $\lambda = 1$ A it is several per cent.

The experimental demonstration of the Compton effect is straightforward. As in Fig. 2-10, a beam of X rays of a single, known wavelength is directed at a target, and the wavelengths of the scattered X rays are determined at various angles θ. The results, shown in Fig. 2-11 (on page 58), exhibit the wavelength shift predicted by Eq. 2.16, but at each angle the scattered X rays also include a substantial proportion having the initial wavelength. This is not hard to understand. In deriving Eq. 2.16, we assumed that the scattering particle is free to move at will, a reasonable assumption since many of the electrons in matter are only loosely bound to their parent atoms. Other electrons, however, are very tightly bound and, when struck by a photon, the entire atom recoils instead of the single electron. In this event the value of m_0 to use in Eq. 2.16 is that of the entire atom, usually tens of thousands of times greater than that of an electron, with the resulting Compton shift accordingly so minute as to be undetectable.

Problems

1. The threshold wavelength for photoelectric emission in tungsten is 2,300 A. What wavelength of light must be used in order for electrons with a maximum energy of 1.5 ev to be ejected?

2. The threshold frequency for photoelectric emission in copper is 1.1×10^{15} sec^{-1}. Find the maximum energy of the photoelectrons (in joules and electron volts) when light of frequency 1.5×10^{15} sec^{-1} is directed on a copper surface.

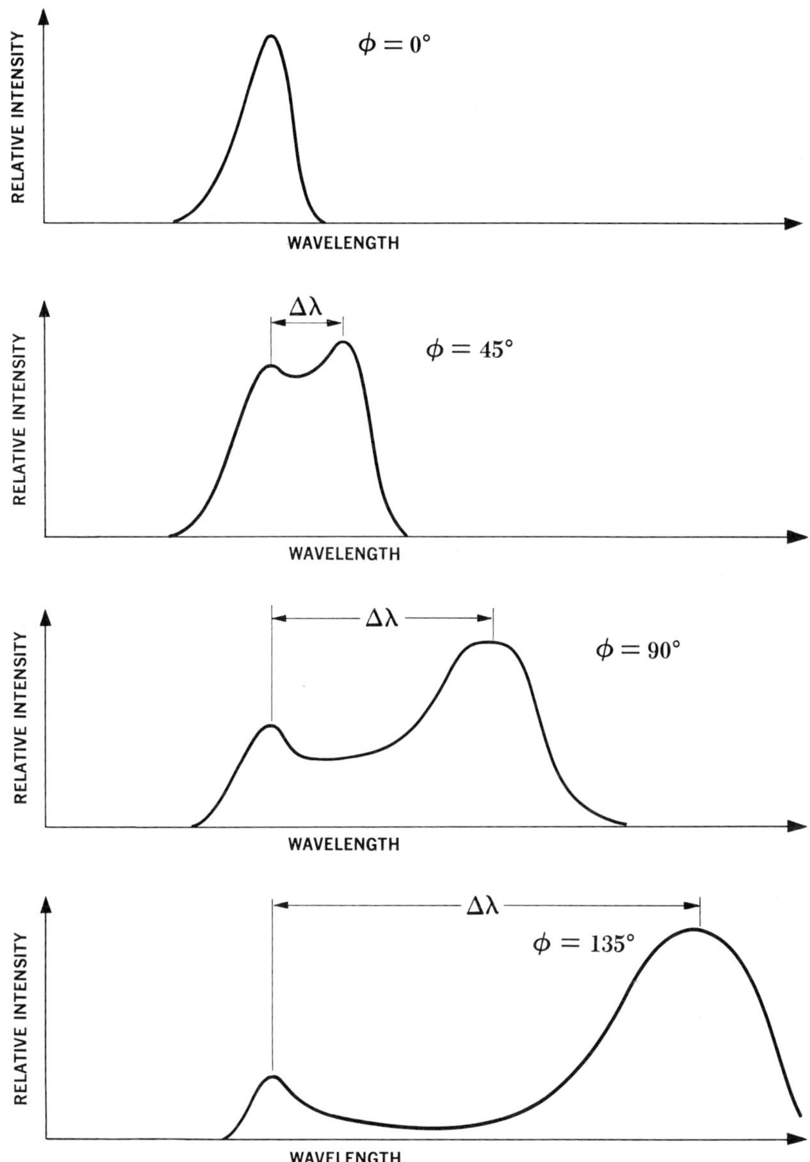

FIGURE 2-11 Compton scattering.

3. The work function of sodium is 2.3 ev. What is the maximum wavelength of light that will cause photoelectrons to be emitted from sodium? What will the maximum kinetic energy of the photoelectrons be if 2,000 A light falls on a sodium surface?

4. Find the wavelength and frequency of a 100-Mev photon.

5. Find the energy of a 7,000-A photon.

6. Under favorable circumstances the human eye can detect 10^{-18} joule of electromagnetic energy. How many 6,000-A photons does this represent?

7. Find the wavelength of a 5×10^{-19} joule photon.

8. What is the wavelength of the X rays emitted when 100-kev electrons strike a target? What is their frequency?

9. An X-ray machine produces 0.1-A X rays. What accelerating voltage does it employ?

10. The distance between adjacent atomic planes in calcite is 3×10^{-8} cm. What is the smallest angle between these planes and an incident beam of 0.3-A X rays at which these X rays can be detected?

11. A potassium chloride crystal has a density of 1.98×10^{3} kg/m^3. The molecular weight of KCl is 74.55. Find the distance between adjacent atoms.

12. How much energy must a photon have if it is to have the momentum of a 10-Mev proton?

13. What is the frequency of an X-ray photon whose momentum is 1.1×10^{-23} kg-m/sec?

14. Prove that it is impossible for a photon to give up all of its energy and momentum to a free electron, so that the photoelectric effect can take place only when photons strike bound electrons.

15. A beam of X rays is scattered by free electrons. At 45° from the beam direction the scattered X rays have a wavelength of 0.022 A. What is the wavelength of the X rays in the direct beam?

16. An X-ray photon whose initial frequency was 1.5×10^{19} sec^{-1} emerges from a collision with an electron with a frequency of 1.2×10^{19} sec^{-1}. How much kinetic energy was imparted to the electron?

17. An X-ray photon of initial frequency 3×10^{19} sec^{-1} collides with an electron and is scattered through 90°. Find its new frequency.

18. Find the energy of an X-ray photon which can impart a maximum energy of 50 kev to an electron.

19. A monochromatic X-ray beam whose wavelength is 0.558 A is scattered through 46°. Find the wavelength of the scattered beam.

THE WAVE PROPERTIES OF PARTICLES 3

In retrospect it may seem odd that two decades passed between the discovery in 1905 of the particle properties of waves and the speculation in 1924 that the converse might also be true. It is one thing, however, to suggest a revolutionary hypothesis to explain otherwise mysterious data and quite another to advance an equally revolutionary hypothesis in the absence of a strong experimental mandate. The latter is just what Louis de Broglie did in 1924 when he proposed that matter possesses wave as well as particle characteristics. So different was the intellectual climate at the time from that prevailing at the turn of the century that de Broglie's notion received immediate and respectful attention, whereas the earlier quantum theory of light of Planck and Einstein created hardly any stir despite its striking empirical support. Although the existence of de Broglie waves was not demonstrated until 1927, the duality principle they represent provided the starting point for Schrödinger's successful development of quantum mechanics in the previous year.

3.1 De Broglie Waves

A photon of light of frequency ν has the momentum

$$p = \frac{h\nu}{c}$$

which can be expressed in terms of wavelength λ as

$$p = \frac{h}{\lambda}$$

since $\lambda\nu = c$. The wavelength of a photon is therefore specified by its momentum according to the relation

$$\lambda = \frac{h}{p} \tag{3.1}$$

Drawing upon an intuitive expectation that nature is symmetric, de Broglie asserted that Eq. 3.1 is a completely general formula that applies to material particles as well as to photons. The momentum of a particle of mass m and velocity v is

$$p = mv$$

and consequently its *de Broglie wavelength* is

3.2 $$\lambda = \frac{h}{mv}$$ **De Broglie waves**

The greater the particle's momentum, the shorter its wavelength. In Eq. 3.2 m is the relativistic mass

$$m = \frac{m_0}{\sqrt{1 - v^2/c^2}}$$

Equation 3.2 has been amply verified by experiments involving the diffraction of fast electrons by crystals, experiments analogous to those that showed X rays to be electromagnetic waves. Before we consider these experiments, it is appropriate to look into the question of what kind of wave phenomenon is involved in the matter waves of de Broglie. In a light wave the electromagnetic field varies in space and time, in a sound wave pressure varies in space and time; what is it whose variations constitute de Broglie waves?

3.2 Wave Function

The variable quantity characterizing de Broglie waves is called the *wave function*, denoted by the symbol Ψ (the Greek letter *psi*). **The value of the wave function associated with a moving body at the particular point x, y, z in space at the time t is related to the likelihood of finding the body there at the time.** Ψ itself, however, has no direct physical significance. There is a simple reason why Ψ cannot be interpreted in terms of an experiment. The probability P that something be somewhere at a given time can have any value between two limits: 0, corresponding to the certainty of its absence, and 1, corresponding to the certainty of its presence. (A probability of 0.2, for instance, signifies a 20 per cent chance of finding the body.) But the amplitude of any wave may be negative as well as positive, and a negative probability is meaningless. Hence Ψ itself cannot be an observable quantity.

This objection does not apply to Ψ^2, the square of the wave function. For this and other reasons Ψ^2 is known as *probability density*. **The probability of experimentally finding the body described by the wave function Ψ at the point x, y, z at the time t is proportional to the value of Ψ^2 there at t.** A large value of Ψ^2 means the strong possibility of the body's presence, while a small value of Ψ^2 means the slight possibility of its presence. As long as Ψ^2 is not actually 0 somewhere, however, there is a definite chance, however small, of detecting it there. This interpretation was first made by Max Born in 1926.

Alternatively, if an experiment involves a great many identical bodies all described by the same wave function Ψ, the *actual density* of bodies at x, y, z at the time t is proportional to the corresponding value of Ψ^2.

While the wavelength of the de Broglie waves associated with a moving body is given by the simple formula

$$\lambda = \frac{h}{mv}$$

determining their amplitude Ψ as a function of position and time usually presents a formidable problem. We shall discuss the calculation of Ψ in Chap. 6 and then go on to apply the ideas developed there to the structure of the atom in Chap. 7. Until then we shall assume that we have whatever knowledge of Ψ is required by the situation at hand.

In the event that a wave function Ψ is complex, with both real and imaginary parts, the probability density is given by the product $\Psi\Psi^*$ of Ψ and its *complex conjugate* Ψ^*. The complex conjugate of any function is obtained by replacing i by $-i$ wherever it appears in the function. (The symbol i represents $\sqrt{-1}$.) Hence $\Psi\Psi^*$ is always a positive quantity.

3.3 De Broglie Wave Velocity

With what velocity do de Broglie waves travel? Since we associate a de Broglie wave with a moving body, it is reasonable to expect that this wave travels at the same velocity v as that of the body. If we call the de Broglie wave velocity w, we may apply the usual formula

$$w = \nu\lambda$$

to determine the value of w. The wavelength λ is just the de Broglie wavelength

$$\lambda = \frac{h}{mv}$$

We shall take the frequency ν to be that specified by the quantum equation

$$E = h\nu$$

Hence

$$\nu = \frac{E}{h}$$

or, since

$$E = mc^2$$

we have

$$\nu = \frac{mc^2}{h}$$

The de Broglie wave velocity is therefore

3.3 $$\begin{aligned} w &= \nu\lambda \\ &= \frac{mc^2}{h}\frac{h}{mv} \\ &= \frac{c^2}{v} \end{aligned}$$

Since the particle velocity v cannot equal or exceed the velocity of light c, the de Broglie wave velocity w is always greater than c! Clearly v and w are never equal for a moving body. In order to understand this unexpected result, we shall digress briefly to consider the notions of *wave velocity* and *group velocity*. (Wave velocity is sometimes called *phase velocity*.)

Let us begin by reviewing how waves are described mathematically. For clarity we shall consider a string stretched along the x axis whose vibrations are in the y direction, as in Fig. 3-1. If we choose $t = 0$ when the displacement y of the string at $x = 0$ is a maximum, its displacement at any future time t at the same place is given by the formula

3.4 $$y = A \cos 2\pi\nu t$$

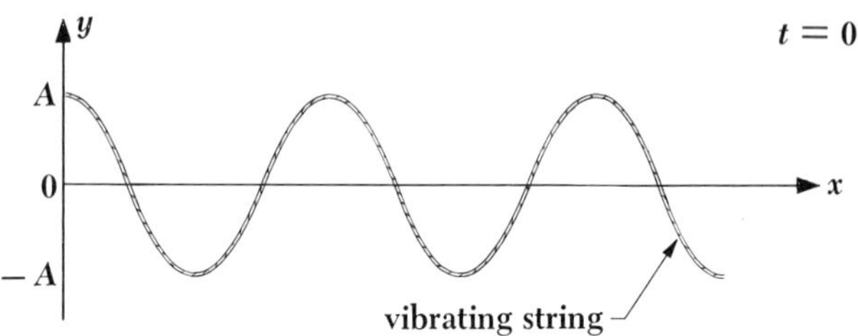

FIGURE 3-1 Wave motion.

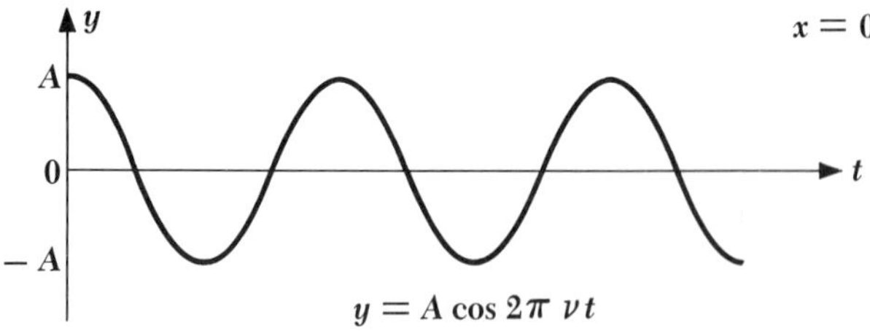

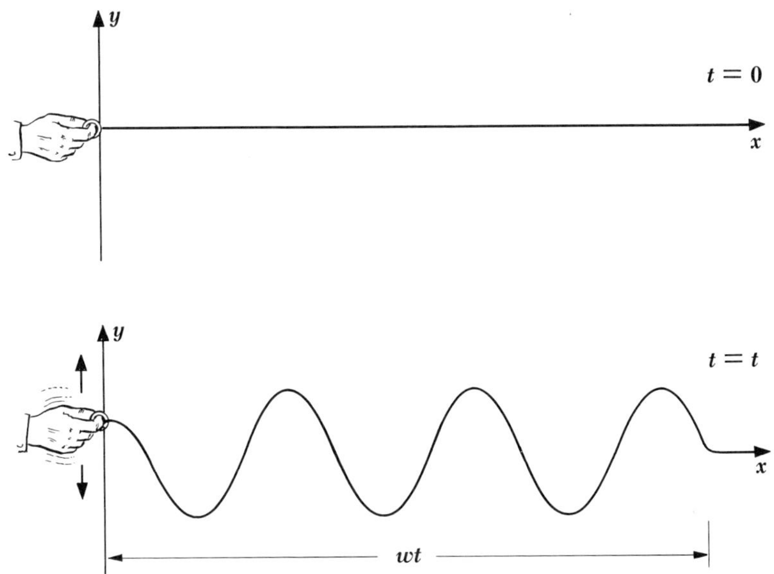

FIGURE 3-2 Wave propagation.

where A is the amplitude of the vibrations (that is, their maximum displacement on either side of the x axis) and ν their frequency.

Equation 3.4 tells us what the displacement of a single point on the string is as a function of time t. A complete description of wave motion in a stretched string, however, should tell us what y is at *any* point on the string at *any* time. What we want is a formula giving y as a function of both x and t. To obtain such a formula, let us imagine that we shake the string at $x = 0$ when $t = 0$, so that a wave starts to travel down the string in the $+x$ direction (Fig. 3-2). This wave has some speed w that depends upon the properties of the string. The wave travels the distance $x = wt$ in the time t; hence the time interval between the formation of the wave at $x = 0$ and its arrival at the point x is x/w. Accordingly the displacement y of the string at x at any time t is exactly the same as the value of y at $x = 0$ *at the earlier time* $t - x/w$. By simply replacing t in Eq. 3.4 with $t - x/w$, then, we have the desired formula giving y in terms of both x and t:

3.5 $$y = A \cos 2\pi\nu\left(t - \frac{x}{w}\right)$$

As a check, we note that Eq. 3.5 reduces to Eq. 3.4 at $x = 0$.

Equation 3.5 may be rewritten

$$y = A \cos 2\pi\left(\nu t - \frac{\nu x}{w}\right)$$

Since

$$w = \nu\lambda$$

we have

3.6 $$y = A \cos 2\pi \left(\nu t - \frac{x}{\lambda}\right)$$

Equation 3.6 is often more convenient to apply than Eq. 3.5.

Perhaps the most widely used description of a wave, however, is still another form of Eq. 3.5. We define the quantities *angular frequency* ω and *propagation constant* k by the formulas

3.7 $$\omega = 2\pi\nu$$

3.8 $$k = \frac{2\pi}{\lambda}$$

3.9 $$= \frac{\omega}{w}$$

Angular frequency gets its name from uniform circular motion, where a particle that moves around a circle ν times per second sweeps out $2\pi\nu$ radians per second. The propagation constant is equal to the number of radians corresponding to a wave train 1 m long, since there are 2π radians in one complete wave. In terms of ω and k, Eq. 3.5 becomes

3.10 $$y = A \cos (\omega t - kx)$$

In three dimensions k becomes a vector $\mathbf{k}$ normal to the wave fronts and x is replaced by the radius vector $\mathbf{r}$. The scalar product $\mathbf{k} \cdot \mathbf{r}$ is then used instead of kx in Eq. 3.10.

3.4 Wave and Group Velocities

The amplitude of the de Broglie waves that correspond to a moving body reflects the probability that it be found at a particular place at a particular time. It is clear that de Broglie waves cannot be represented simply by a formula resembling Eq. 3.10, which describes an indefinite series of waves all with the same amplitude A. Instead, we expect the wave representation of a moving body to correspond to a *wave packet,* or *wave group,* like that shown in Fig. 3-3, whose constituent waves have amplitudes which vary with the likelihood of detecting the body.

A familiar example of how wave groups come into being is the case of *beats.* When two sound waves of the same amplitude, but of slightly different frequencies, are produced simultaneously, the sound we hear has a fre-

FIGURE 3-3 A wave group.

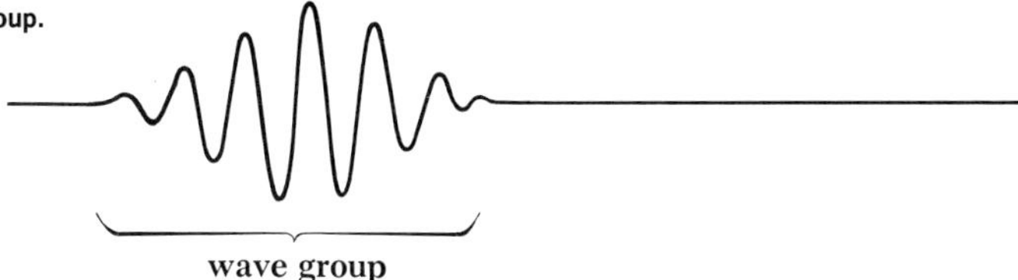

quency equal to the average of the two original frequencies and its amplitude rises and falls periodically. The amplitude fluctuations occur a number of times per second equal to the difference between the two original frequencies. If the original sounds have frequencies of, say, 440 and 442 cycles/sec, we will hear a fluctuating sound of frequency 441 cycles/sec with two loudness peaks, called *beats*, per second. The production of beats is illustrated in Fig. 3-4.

A way of mathematically describing a wave group, then, is in terms of a series of individual waves, differing slightly in wavelength, whose interference with one another results in the variation in amplitudes that defines the group shape. If the speeds of the waves are the same, the speed with which the wave group travels is identical with the common wave speed. However, if the wave speed varies with wavelength, the different individual waves do not proceed together, and the wave group has a speed different from that of the waves that compose it.

FIGURE 3-4 The production of beats.

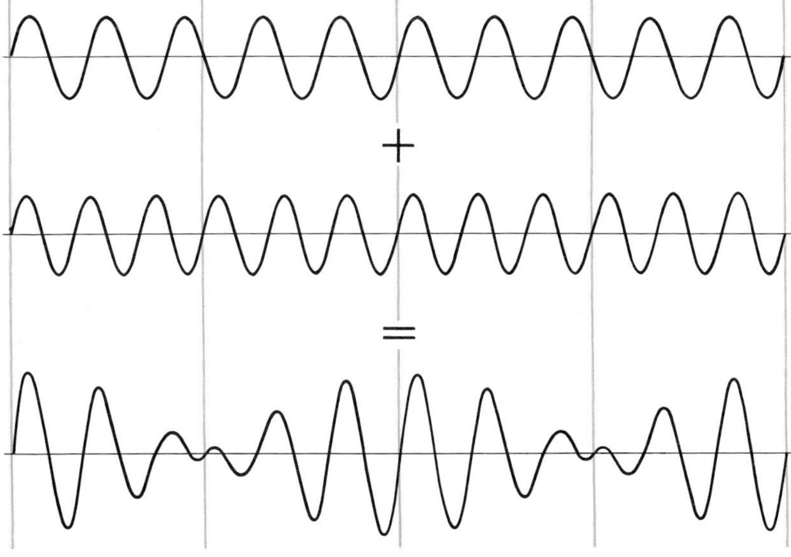

It is not difficult to compute the speed u with which a wave group travels. Let us suppose that a wave group arises from the combination of two waves with the same amplitude A but differing by an amount $d\omega$ in angular frequency and an amount dk in propagation constant. We may represent the original waves by the formulas

$$y_1 = A \cos (\omega t - kx)$$
$$y_2 = A \cos [(\omega + d\omega)t - (k + dk)x]$$

The resultant displacement y at any time t and any position x is the sum of y_1 and y_2. With the help of the identity

$$\cos \alpha + \cos \beta = 2 \cos \tfrac{1}{2}(\alpha + \beta) \cos \tfrac{1}{2}(\alpha - \beta)$$

and the relation

$$\cos (-\theta) = \cos \theta$$

we find that

$$\begin{aligned} y &= y_1 + y_2 \\ &= 2A \cos \tfrac{1}{2}[(2\omega + d\omega)t - (2k + dk)x] \cos \tfrac{1}{2}(d\omega\, t - dk\, x) \end{aligned}$$

Since $d\omega$ and dk are small compared with ω and k respectively,

$$2\omega + d\omega \approx 2\omega$$
$$2k + dk \approx 2k$$

and

3.11 $$y = 2A \cos (\omega t - kx) \cos \left(\frac{d\omega}{2} t - \frac{dk}{2} x\right)$$

Equation 3.11 represents a wave of angular frequency ω and propagation constant k that has superimposed upon it a modulation of angular frequency $\tfrac{1}{2}d\omega$ and propagation constant $\tfrac{1}{2}dk$. The effect of the modulation is to produce successive wave groups, as in Fig. 3-4. The wave velocity w is

3.12 $$w = \frac{\omega}{k}$$ **Wave velocity**

while the velocity u of the wave groups is

3.13 $$u = \frac{d\omega}{dk}$$ **Group velocity**

In general, depending upon the manner in which wave velocity varies with wavelength in a particular medium, the group velocity may be greater than or less than the wave velocity. If the wave velocity w is the same for all wavelengths, the group and wave velocities are the same.

The angular frequency and propagation constant of the de Broglie waves associated with a body of rest mass m_0 moving with the velocity v are

3.14 $$\omega = 2\pi\nu = \frac{2\pi mc^2}{h} = \frac{2\pi m_0 c^2}{h\sqrt{1 - v^2/c^2}}$$

and

3.15 $$k = \frac{2\pi}{\lambda} = \frac{2\pi mv}{h} = \frac{2\pi m_0 v}{h\sqrt{1 - v^2/c^2}}$$

Both ω and k are functions of the velocity v. The wave velocity w is, as we found earlier,

$$w = \frac{\omega}{k} = \frac{c^2}{v}$$

which exceeds both the velocity of the body v and the velocity of light c, since $v < c$.

The group velocity u of the de Broglie waves associated with the body is

$$u = \frac{d\omega}{dk} = \frac{d\omega/dv}{dk/dv}$$

Now

$$\frac{d\omega}{dv} = \frac{2\pi m_0 v}{h(1 - v^2/c^2)^{3/2}}$$

and

$$\frac{dk}{dv} = \frac{2\pi m_0}{h(1 - v^2/c^2)^{3/2}}$$

and so the group velocity is

3.16 $$u = v$$

The de Broglie wave group associated with a moving body travels with the same velocity as the body. The wave velocity w of the de Broglie waves evidently has no simple physical significance in itself.

3.5 The Diffraction of Particles

A wave manifestation having no analog in the behavior of Newtonian particles is diffraction. In 1927 Davisson and Germer in the United States and G. P. Thomson in England independently confirmed de Broglie's hypothesis by demonstrating that electrons exhibit diffraction when they are scattered from crystals whose atoms are spaced appropriately. We shall consider the experiment of Davisson and Germer because its interpretation is more direct.

Davisson and Germer were studying the scattering of electrons from a solid, using an apparatus like that sketched in Fig. 3-5. The energy of the electrons in the primary beam, the angle at which they are incident upon the target, and the position of the detector can all be varied. Classical physics predicts that the scattered electrons will emerge in all directions with only a moderate dependence of their intensity upon scattering angle and even less upon the energy of the primary electrons. Using a block of nickel as the target, Davisson and Germer verified these predictions.

In the midst of their work there occurred an accident that allowed air to enter their apparatus and oxidize the metal surface. To reduce the oxide to pure nickel, the target was baked in a high-temperature oven. After this treatment, the target was returned to the apparatus and the measurements resumed. Now the results were very different from what had been found before the accident: instead of a continuous variation of scattered electron intensity with angle, distinct maxima and minima were observed whose positions depended upon the electron energy! Typical polar graphs of electron intensity after the accident are shown in Fig. 3-6; the method of plotting is such that the intensity at any angle is proportional to the distance of the curve at that angle from the point of scattering.

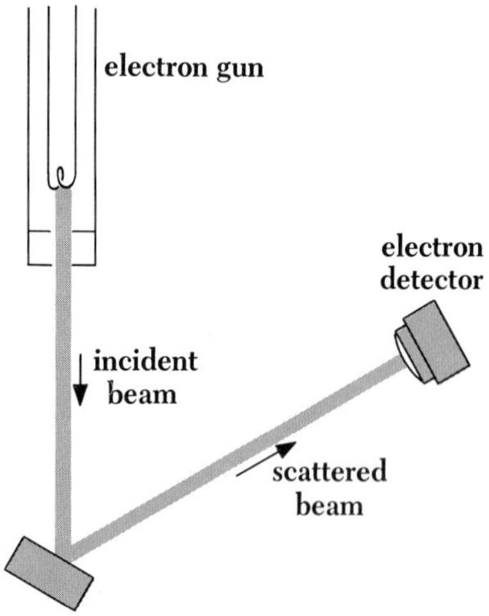

electron gun

electron detector

incident beam

scattered beam

FIGURE 3-5 The Davisson-Germer experiment.

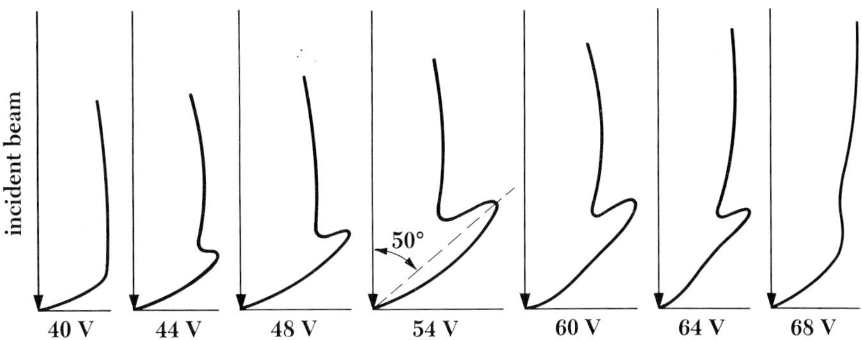

FIGURE 3-6 Results of the Davisson-Germer experiment.

Two questions come to mind immediately: what is the reason for this new effect, and why did it not appear until after the nickel target was baked?

De Broglie's hypothesis suggested the interpretation that electron waves were being diffracted by the target, much as X rays are diffracted by Bragg reflection from crystal planes. This interpretation received support when it was realized that the effect of heating a block of nickel at high temperature is to cause the many small individual crystals of which it is normally composed to form into a single large crystal, all of whose atoms are arranged in a regular lattice.

Let us see whether we can verify that de Broglie waves are responsible for the findings of Davisson and Germer. In a particular determination, a beam of 54-ev electrons was directed perpendicularly at the nickel target, and a sharp maximum in the electron distribution occurred at an angle of 50° with the original beam. The angles of incidence and scattering relative to the family of Bragg planes shown in Fig. 3-7 will both be 65°. The spacing of the planes in this family, which can be measured by X-ray diffraction, is 0.91 A. The Bragg equation for maxima in the diffraction pattern is

$$n\lambda = 2d \sin \theta$$

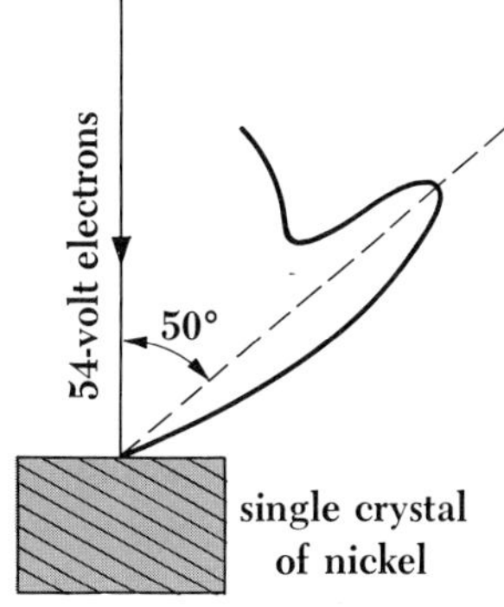

FIGURE 3-7 The diffraction of de Broglie waves by the target is responsible for the results of Davisson and Germer.

Here $d = 0.91$ A and $\theta = 65°$; assuming that $n = 1$, the de Broglie wavelength λ of the diffracted electrons is

$$\begin{aligned} n\lambda &= 2d \sin \theta \\ &= 2 \times 0.91 \text{ A} \times \sin 65° \\ &= 1.65 \text{ A} \end{aligned}$$

Now we use de Broglie's formula

$$\lambda = \frac{h}{mv}$$

to calculate the expected wavelength of the electrons. The electron kinetic energy of 54 ev is small compared with its rest energy m_0c^2 of 5.1×10^5 ev, and so we can ignore relativistic considerations. Since

$$T = \tfrac{1}{2}mv^2$$

the electron momentum mv is

$$\begin{aligned} mv &= \sqrt{2mT} \\ &= \sqrt{2 \times 9.1 \times 10^{-31} \text{ kg} \times 54 \text{ ev} \times 1.6 \times 10^{-19} \text{ joule/ev}} \\ &= 4.0 \times 10^{-24} \text{ kg-m/sec} \end{aligned}$$

The electron wavelength is therefore

$$\begin{aligned} \lambda &= \frac{h}{mv} \\ &= \frac{6.63 \times 10^{-34} \text{ joule-sec}}{4.0 \times 10^{-24} \text{ kg-m/sec}} \\ &= 1.66 \times 10^{-10} \text{ m} \\ &= 1.66 \text{ A} \end{aligned}$$

in excellent agreement with the observed wavelength. The Davisson-Germer experiment thus provides direct verification of de Broglie's hypothesis of the wave nature of moving bodies.

The analysis of the Davisson-Germer experiment is actually less straightforward than indicated above, since the energy of an electron increases when it enters a crystal by an amount equal to the work function of the surface. Hence the electron speeds in the experiment were greater inside the crystal and the corresponding de Broglie wavelength shorter than the corresponding values outside. An additional complication arises from interference between waves diffracted by different families of Bragg planes, which restricts the occurrence of maxima to certain combinations of electron energy and angle of incidence rather than merely to any combination that obeys the Bragg equation.

Electrons are not the only bodies whose wave behavior can be demonstrated. The diffraction of neutrons and of whole atoms when scattered by suitable crystals has been observed, and in fact neutron diffraction, like X-ray and electron diffraction, is today a widely used tool for investigating crystal structures.

As in the case of electromagnetic waves, the wave and particle aspects of moving bodies can never be simultaneously observed, so that we cannot determine which is the "correct" description. All we can say is that in some respects a moving body exhibits wave properties and in other respects it exhibits particle properties. Which set of properties is most conspicuous depends upon how the de Broglie wavelength compares with the dimensions of the bodies involved: the 1.66 A wavelength of a 54-ev electron is of the same order of magnitude as the lattice spacing in a nickel crystal, but the wavelength of an automobile moving at 60 mi/hr is about 5×10^{-38} ft, far too small to manifest itself.

3.6 The Uncertainty Principle

The fact that a moving body must be regarded as a de Broglie wave group rather than as a localized entity suggests that there is a fundamental limit to the accuracy with which we can measure its particle properties. Figure 3-8*a* shows a de Broglie wave group: it is not easy to see how its center, which we may take as representing the position of the body, and the wavelength

FIGURE 3-8 **The position of a narrow wave group at any instant can be determined, but not its wavelength; the wavelength of a broad wave group at any instant can be determined, but not its position.**

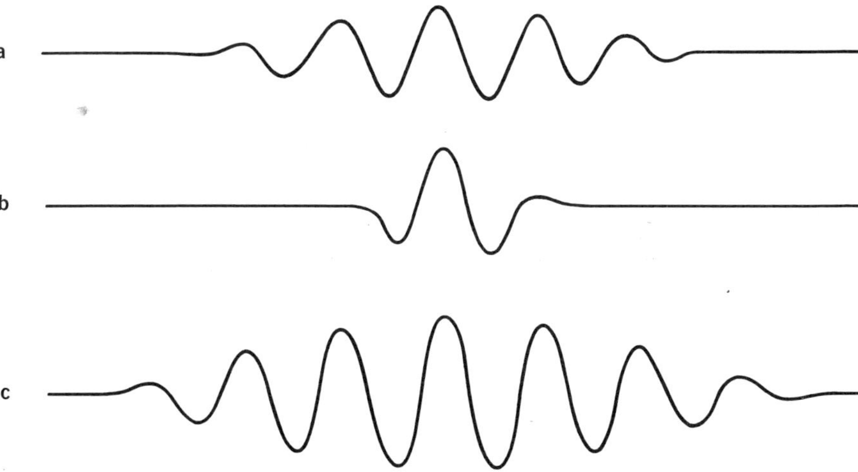

of its component waves, which is related to the body's momentum, can both be precisely determined as the group flashes by our measuring apparatus. If the group is very narrow, as in Fig. 3-8*b*, the position of its center is readily found, but the wavelength is impossible to establish. At the other extreme, a wide group, as in Fig. 3-8*c*, permits a satisfactory wavelength estimate, but where is its exact center?

A straightforward argument based upon the nature of wave groups permits us to relate the inherent uncertainty Δx in a measurement of particle position with the inherent uncertainty Δp in a simultaneous measurement of its momentum. The simplest example of the formation of wave groups is that given in Sec. 3.4, where two wave trains slightly different in angular frequency ω and propagation constant k were superimposed to yield the series of groups shown in Fig. 3-4. Here let us consider the wave groups that arise when the de Broglie waves

$$\Psi_1 = A \cos(\omega t - kx)$$
$$\Psi_2 = A[\cos(\omega + \Delta\omega)t - (k + \Delta k)x]$$

are combined. From a calculation identical with the one used in obtaining Eq. 3.11, we find that

3.17 $$\Psi = \Psi_1 + \Psi_2 \approx 2A \cos(\omega t - kx) \cos(\tfrac{1}{2}\Delta\omega\, t - \tfrac{1}{2}\Delta k\, x)$$

which is plotted in Fig. 3-9. The width of each group is evidently equal to half the wavelength λ_m of the modulation. It is reasonable to suppose that this width is of the same order of magnitude as the inherent uncertainty Δx in the position of the group, that is,

3.18 $$\Delta x \approx \tfrac{1}{2}\lambda_m$$

The modulation wavelength is related to its propagation constant k_m by

$$\lambda_m = \frac{2\pi}{k_m}$$

From Eq. 3.17 we see that the propagation constant of the modulation is

$$k_m = \tfrac{1}{2}\Delta k$$

with the result that

$$\lambda_m = \frac{2\pi}{\tfrac{1}{2}\Delta k}$$

and

3.19 $$\Delta x = \frac{2\pi}{\Delta k}$$

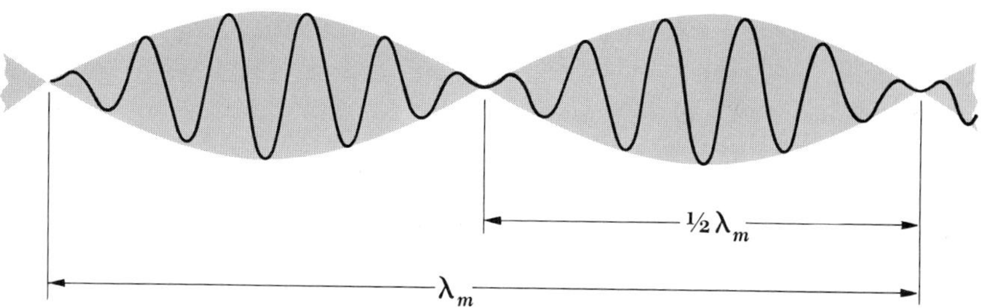

FIGURE 3-9 Wave groups that result from the interference of wave trains having the same amplitudes but different frequencies.

Because the waves that constitute the groups are a combination of waves of propagation constant k and waves of propagation constant $k + \Delta k$, the best measurement of the propagation constant we can hope to make will still have an inherent uncertainty of Δk. This uncertainty is related to the uncertainty in the position of a wave group by Eq. 3.19, namely,

3.20 $$\Delta k = \frac{2\pi}{\Delta x}$$

The de Broglie wavelength of a particle of momentum p is

$$\lambda = \frac{h}{p}$$

The propagation constant corresponding to this wavelength is

$$k = \frac{2\pi}{\lambda}$$

$$= \frac{2\pi p}{h}$$

Hence an uncertainty Δk in the propagation constant of the de Broglie waves associated with the particle results in an uncertainty Δp in the particle's momentum according to the formula

$$\Delta p = \frac{h\,\Delta k}{2\pi}$$

Substituting the value of Δk given by Eq. 3.20, we obtain

$$\Delta p = \frac{h}{\Delta x}$$

or

3.21 $$\Delta x\,\Delta p \geqslant h$$

The sign $\geqslant$ is used because the Δx and Δp of Eq. 3.21 are *irreducible minima that are consequences of the wave natures of moving bodies;* any instrumental or statistical uncertainties that arise in the actual conduct of the measurement only augment the product $\Delta x\ \Delta p$.

Equation 3.21 is one form of the *uncertainty principle* first obtained by Werner Heisenberg in 1927. It states that the product of the uncertainty Δx in the position of a body at some instant and the uncertainty Δp in its momentum at the same instant is at best equal to Planck's constant h. We cannot measure simultaneously both position and momentum with perfect accuracy. If we arrange matters so that Δx is small, corresponding to the narrow wave group of Fig. 3-8*b*, Δp will be large. If we reduce Δp in some way, corresponding to the wide wave group of Fig. 3-8*c*, Δx will be large. These uncertainties are not in our apparatus but in nature.

We have not specified precisely what we mean by an "uncertainty" in a measurable quantity because there are a number of slightly different definitions in common use. A more realistic method of approach than our own rather conservative one yields for the uncertainty principle

3.22 $$\Delta x\ \Delta p \geqslant \frac{h}{2\pi}$$ **Uncertainty principle**

an expression more widely used than Eq. 3.21.

The uncertainty principle can be derived in a variety of ways. Let us obtain it by basing our argument upon the particle nature of waves instead of upon the wave nature of particles as we did above. Suppose that we wish to measure the position and momentum of something at a certain moment. To accomplish this, we must prod it with something else that is to carry the desired information back to us; that is, we have to touch it with our finger, illuminate it with light, or interact with it in some other way. We might be examining an electron with the help of light of wavelength λ, as in Fig. 3-10. In this process photons of light strike the electron and bounce off it. Each photon possesses the momentum h/λ, and when it collides with the electron, the electron's original momentum p is changed. The precise change cannot be predicted, but it is likely to be of the same order of magnitude as the photon momentum h/λ. Hence the *act of measurement* introduces an uncertainty of

3.23 $$\Delta p = \frac{h}{\lambda}$$

in the momentum of the electron. The longer the wavelength of the light we employ in "seeing" the electron, the smaller the consequent uncertainty in its momentum.

Because light has wave properties, we cannot expect to determine the electron's position with infinite accuracy under any circumstances, but we might

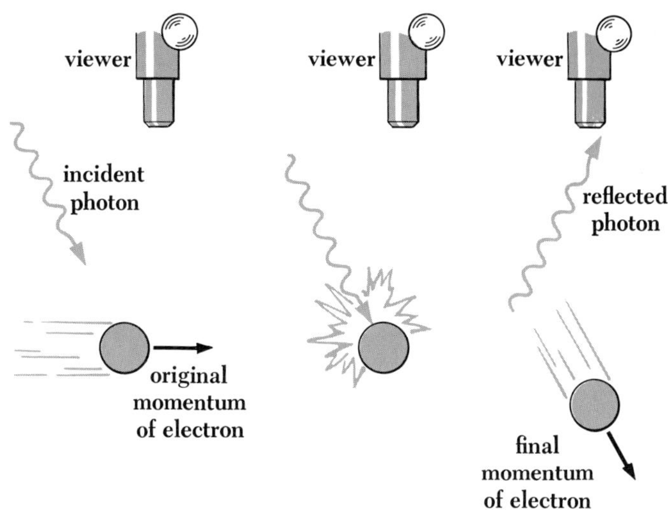

FIGURE 3-10 An electron cannot be observed without changing its momentum.

reasonably hope to keep the irreducible uncertainty Δx in its position to 1 wavelength of the light being used. That is, at best

3.24 $$\Delta x = \lambda$$

The shorter the wavelength, the smaller the uncertainty in the position of the electron.

From Eqs. 3.23 and 3.24 it is clear that, if we employ light of short wavelength to improve the accuracy of the position determination, there will be a corresponding reduction in the accuracy of the momentum determination, while light of long wavelength will yield an accurate momentum value but an inaccurate position value. Substituting $\lambda = \Delta x$ into Eq. 3.23,

$$\Delta p = \frac{h}{\Delta x}$$

from which we again obtain Eq. 3.21,

$$\Delta x \, \Delta p \geqslant h$$

Arguments like the preceding one, though superficially attractive, must as a rule be approached with caution. The above argument implies that the electron can possess a definite position and momentum at any instant, and that it is the measurement process that introduces the indeterminacy in $\Delta x \, \Delta p$. On the contrary, this indeterminacy is inherent in the nature of a moving body. The justification for the many "derivations" of this kind is, first, that they show it is impossible to imagine a way around the uncertainty principle, and second, that they present a view of the principle that can be appreciated in a more familiar context than that of wave packets.

3.7 Applications of the Uncertainty Principle

Planck's constant h is so minute—only 6.63×10^{-34} joule-sec—that the limitations imposed by the uncertainty principle are significant only in the realm of the atom. On this microscopic scale, however, there are many phenomena that can be understood in terms of this principle; we shall consider several of them here.

One interesting question is whether electrons are present in atomic nuclei. As we shall learn in the next chapter, typical nuclei are less than 10^{-14} m in radius. For an electron to be confined within such a nucleus, the uncertainty in its position may not exceed 10^{-14} m. The corresponding uncertainty in the electron's momentum is

$$\begin{aligned}\Delta p &\geqslant \frac{h}{2\pi\,\Delta x}\\ &\geqslant \frac{6.63 \times 10^{-34}\ \text{joule-sec}}{2\pi \times 10^{-14}\ \text{m}}\\ &\geqslant 1.1 \times 10^{-20}\ \text{kg-m/sec}\end{aligned}$$

If this is the uncertainty in the electron's momentum, the momentum itself must be at least comparable in magnitude. An electron whose momentum is 1.1×10^{-20} kg-m/sec has a kinetic energy T many times greater than its rest energy m_0c^2, and we may accordingly use the extreme relativistic formula

$$T = pc$$

to find T. Substituting for p and c, we obtain

$$\begin{aligned}T &= 1.1 \times 10^{-20}\ \text{kg-m/sec} \times 3 \times 10^{8}\ \text{m/sec}\\ &= 3.3 \times 10^{-12}\ \text{joule}\end{aligned}$$

Since 1 ev $= 1.6 \times 10^{-19}$ joule, the kinetic energy of the electron must be well over 20 Mev if it is to be a nuclear constituent. Experiments indicate that the electrons associated even with unstable atoms never have more than a fraction of this energy, and we conclude that electrons cannot be present within nuclei.

Let us now ask how much energy an electron needs to be confined to an atom. The hydrogen atom is about 5×10^{-11} m in radius, and therefore the uncertainty in the position of its electron may not exceed this figure.

The corresponding momentum uncertainty is

$$\Delta p = 2.2 \times 10^{-24} \text{ kg-m/sec}$$

An electron whose momentum is of this order of magnitude is nonrelativistic in behavior, and its kinetic energy is

$$\begin{aligned} T &= \frac{p^2}{2m} \\ &= \frac{(2.2 \times 10^{-24} \text{ kg-m/sec})^2}{2 \times 9.1 \times 10^{-31} \text{ kg}} \\ &= 2.7 \times 10^{-18} \text{ joule} \end{aligned}$$

or about 17 ev. This is a wholly plausible figure.

Another form of the uncertainty principle is sometimes useful. We might wish to measure the energy E emitted sometime during the time interval Δt in an atomic process. If the energy is in the form of electromagnetic waves, the limited time available restricts the accuracy with which we can determine the frequency ν of the waves. Let us assume that the irreducible uncertainty in the number of waves we count in a wave group is one wave. Since the frequency of the waves under study is equal to the number of them we count divided by the time interval, the uncertainty $\Delta\nu$ in our frequency measurement is

$$\Delta\nu = \frac{1}{\Delta t}$$

The corresponding energy uncertainty is

$$\Delta E = h\,\Delta\nu$$

and so

$$\Delta E = \frac{h}{\Delta t}$$

or

3.25 $$\Delta E\,\Delta t \geqslant h$$

A more realistic calculation changes Eq. 3.25 to

3.26 $$\Delta E\,\Delta t \geqslant \frac{h}{2\pi}$$

Equation 3.26 states that the product of the uncertainty ΔE in an energy measurement and the uncertainty Δt in the time at which the measurement was made is equal to or greater than Planck's constant divided by 2π.

As an example of the significance of Eq. 3.26 we can consider the radiation of light from an "excited" atom. Such an atom divests itself of its excess energy by emitting one or more photons of characteristic frequency. The average period that elapses between the excitation of an atom and the time it radiates is 10^{-8} sec. Thus the photon energy is uncertain by an amount

$$\begin{aligned}\Delta E &= \frac{h}{2\pi\,\Delta t} \\ &= \frac{6.63 \times 10^{-34}\text{ joule-sec}}{2\pi \times 10^{-8}\text{ sec}} \\ &= 1.1 \times 10^{-26}\text{ joule}\end{aligned}$$

and the frequency of the light is uncertain by

$$\begin{aligned}\Delta\nu &= \frac{\Delta E}{h} \\ &= 1.6 \times 10^{7}\text{ cycles/sec}\end{aligned}$$

This is the irreducible limit to the accuracy with which we can determine the frequency of the radiation emitted by an atom.

Problems

1. Show that the de Broglie wavelength of a particle approaches zero faster than $1/v$ as its speed approaches the speed of light.

2. Find the de Broglie wavelength of a 15-ev proton.

3. Find the de Broglie wavelength of a 15 kev electron.

4. What is the de Broglie wavelength of an electron whose speed is 9×10^7 m/sec?

5. Find the wavelength of a 1-kg object whose speed is 1 m/sec.

6. Neutrons in equilibrium with matter at room temperature (300°K) have average energies of about 1/25 ev. (Such neutrons are often called "thermal neutrons.") Find their de Broglie wavelength.

7. Derive a formula expressing the de Broglie wavelength (in A) of an electron in terms of the potential difference V (in volts) through which it is accelerated.

8. Assume that electromagnetic waves are a special case of de Broglie waves. Show that photons must travel with the wave velocity c and that the rest mass of the photon must be 0.

9. The velocity of ocean waves is $\sqrt{g\lambda/2\pi}$, where g is the acceleration of gravity. Find the group velocity of these waves.

10. The velocity of ripples on a liquid surface is $\sqrt{2\pi S/\lambda\rho}$, where S is the surface tension and ρ the density of the liquid. Find the group velocity of these waves.

11. The position and momentum of a 1-kev electron are simultaneously determined. If its position is located to within 1 A, what is the percentage of uncertainty in its momentum?

12. An electron microscope uses 40-kev electrons. Find its ultimate resolving power on the assumption that this is equal to the wavelength of the electrons.

13. Compare the uncertainties in the velocities of an electron and a proton confined in a 10-A box.

14. Wavelengths can be determined with accuracies of one part in 10^6. What is the uncertainty in the position of a 1-A X-ray photon when its wavelength is simultaneously measured?

2

THE THEORY OF THE ATOM

ATOMIC STRUCTURE 4

Far in the past people began to suspect that matter, despite its appearance of being continuous, possesses a definite structure on a microscopic level beyond the direct reach of our senses. This suspicion did not take on a more concrete form until a little over a century and a half ago; since then the existence of atoms and molecules, the ultimate particles of matter in its common forms, has been amply demonstrated, and their own ultimate particles, electrons, protons, and neutrons, have been identified and studied as well. In this chapter and in others to come our chief concern will be the structure of the atom, since it is this structure that is responsible for nearly all of the properties of matter that have shaped the world around us.

4.1 Atomic Models

While the scientists of the nineteenth century accepted the idea that the chemical elements consist of atoms, they knew virtually nothing about the atoms themselves. The discovery of the electron and the realization that all atoms contain electrons provided the first important insight into atomic structure. Electrons contain negative electrical charges, while atoms themselves are electrically neutral: every atom must therefore contain enough positively charged matter to balance the negative charge of its electrons. Furthermore, electrons are thousands of times lighter than whole atoms; this suggests that the positively charged constituent of atoms is what provides them with nearly all of their mass. When J. J. Thomson proposed in 1898 that atoms are uniform spheres of positively charged matter in which electrons are embedded, his hypothesis then seemed perfectly reasonable. Thomson's plum-pudding model of the atom—so called from its resemblance to that raisin-studded delicacy—is sketched in Fig. 4-1. Despite the importance of the problem, 13 years passed before a definite experimental test of the plum-pudding model was made. This experiment, as we shall see, compelled the abandon-

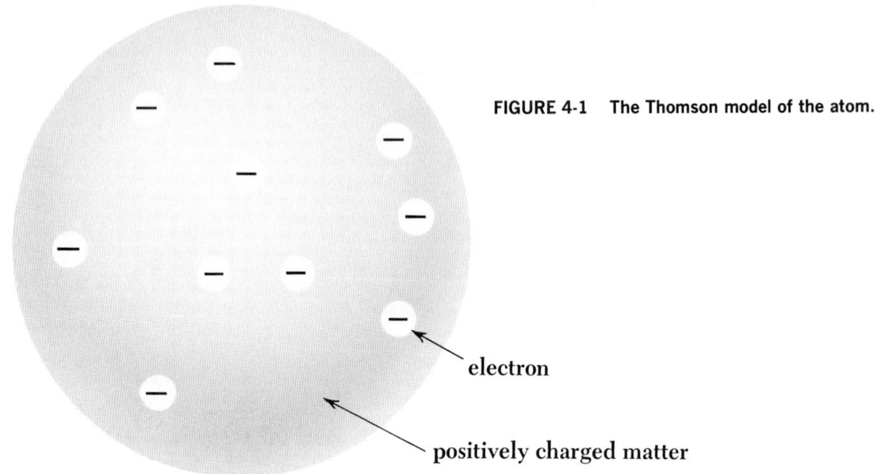

FIGURE 4-1 The Thomson model of the atom.

ment of this apparently plausible model, leaving in its place a concept of atomic structure incomprehensible in the light of classical physics.

The most direct way to find out what is inside a plum pudding is to plunge a finger into it, a technique not very different from that used by Geiger and Marsden to find out what is inside an atom. In their classic experiment, performed in 1911 at the suggestion of Ernest Rutherford, they employed as probes the fast *alpha particles* spontaneously emitted by certain radioactive elements. Alpha particles are helium atoms that have lost two electrons, leaving them with a charge of $+2e$; we shall examine their origin and properties in more detail in Chap. 12. Geiger and Marsden placed a sample of an alpha-particle-emitting substance behind a lead screen that had a small hole in it, as in Fig. 4-2, so that a narrow beam of alpha particles was produced. This beam was directed at a thin gold foil. A moveable zinc sulfide screen, which gives off a visible flash of light when struck by an alpha particle, was placed on the other side of the foil. It was anticipated that most of the alpha particles would go right through the foil, while the remainder would at most suffer only slight deflections. This behavior follows from the Thomson atomic model, in which the charges within an atom are assumed to be uniformly distributed throughout its volume. If the Thomson model is correct, only weak electric forces are exerted on alpha particles passing through a thin metal foil, and their initial momenta should be enough to carry them through with only minor departures from their original paths.

What Geiger and Marsden actually found was that, while most of the alpha particles indeed emerged without deviation, some were scattered through very large angles. A few were even scattered in the backward direction. Since alpha particles are relatively heavy (over 7,000 times more massive than electrons) and those used in this experiment traveled at high speed, it was

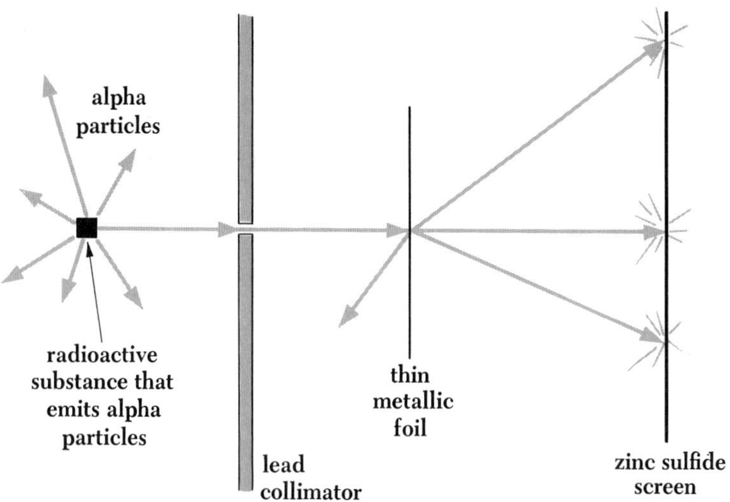

FIGURE 4-2 The Rutherford scattering experiment.

clear that strong forces had to be exerted upon them to cause such marked deflections. To explain the results, Rutherford was forced to picture an atom as being composed of a tiny *nucleus*, in which its positive charge and nearly all of its mass are concentrated, with its electrons some distance away (Fig. 4-3). Considering an atom as largely empty space, it is easy to see why most alpha particles go right through a thin foil. When an alpha particle approaches a nucleus, however, it encounters an intense electric field and is likely to be scattered through a considerable angle. The atomic electrons, being so light, do not appreciably affect the motion of incident alpha particles.

Numerical estimates of electric-field intensities within the Thomson and

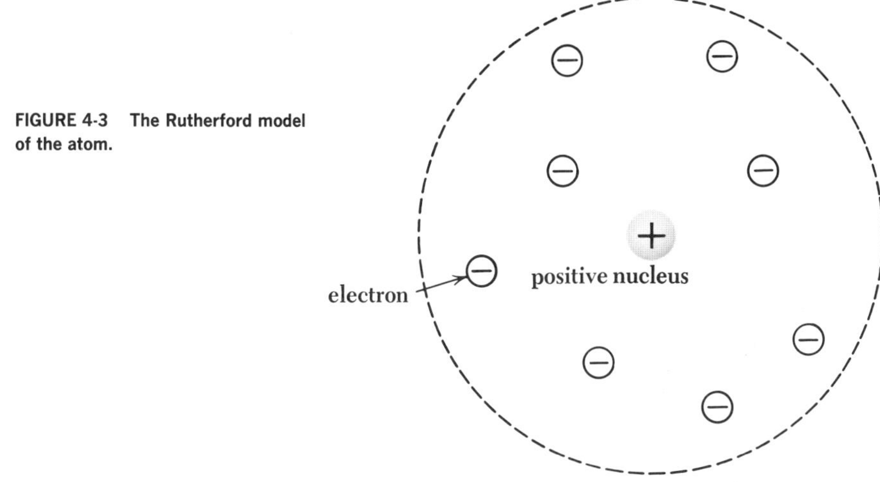

FIGURE 4-3 The Rutherford model of the atom.

Rutherford models emphasize the difference between them. If we assume with Thomson that the positive charge within a gold atom is spread evenly throughout its volume, and if we neglect the electrons completely, the electric-field intensity at the atom's surface (where it is a maximum) is about 10^{13} volts/m. On the other hand, if we assume with Rutherford that the positive charge within a gold atom is concentrated in a small nucleus at its center, the electric-field intensity at the surface of the nucleus exceeds 10^{21} volts/m—a factor of 10^8 greater. Such a strong field can deflect or even reverse the direction of an energetic alpha particle that comes near the nucleus, while the feebler field of the Thomson atom cannot.

The experiments of Geiger and Marsden and later work of a similar kind also supplied information about the nuclei of the atoms that composed the various target foils. The deflection an alpha particle experiences when it passes near a nucleus depends upon the magnitude of the nuclear charge, and so comparing the relative scattering of alpha particles by different foils provides a way of estimating the nuclear charges of the atoms involved. All of the atoms of any one element were found to have the same unique nuclear charge, and this charge increased regularly from element to element in the periodic table. The nuclear charges always turned out to be multiples of $+e$; the number of unit positive charges in the nuclei of an element is today called the *atomic number* of the element. We know now that protons, each with a charge $+e$, are responsible for the charge on a nucleus, and so the atomic number of an element is the same as the number of protons in the nuclei of its atoms.

4.2 Alpha-particle Scattering

Rutherford arrived at a formula, describing the scattering of alpha particles by thin foils on the basis of his atomic model, that agreed with the experimental results. The derivation of this formula both illustrates the application of fundamental physical laws in a novel setting and introduces certain notions, such as that of the *cross section* for an interaction, that are important in many other aspects of modern physics.

Rutherford began by assuming that the alpha particle and the nucleus it interacts with are both small enough to be considered as point masses and charges; that the electrostatic repulsive force between alpha particle and nucleus (which are both positively charged) is the only one acting; and that the nucleus is so massive compared with the alpha particle that it does not move during their interaction. Owing to the variation of the electrostatic force with $1/r^2$, where r is the instantaneous separation between alpha particle and nucleus, the alpha particle's path is a hyperbola with the nucleus at the outer focus (Fig. 4-4). The *impact parameter* b is the minimum distance to which the alpha particle would approach the nucleus if there were no force between them, and the *scattering angle* θ is the angle between the asymptotic direction of approach of the alpha particle and the asymptotic direction in which it recedes. Our first task is to find a relationship between b and θ.

As a result of the impulse $\int \mathbf{F}\, dt$ given it by the nucleus, the momentum of the alpha particle changes by $\Delta\mathbf{p}$ from the initial value $\mathbf{p}_1$ to the final value $\mathbf{p}_2$. That is,

4.1 $$\begin{aligned}\Delta\mathbf{p} &= \mathbf{p}_2 - \mathbf{p}_1 \\ &= \int \mathbf{F}\, dt\end{aligned}$$

Because the nucleus remains stationary during the passage of the alpha particle, by hypothesis, the alpha-particle kinetic energy remains constant; hence the *magnitude* of its momentum also remains constant, and

$$p_1 = p_2 = mv$$

Here v is the alpha-particle velocity far from the nucleus. From Fig. 4-5 we see that, according to the law of sines,

$$\frac{\Delta p}{\sin\theta} = \frac{mv}{\sin\dfrac{(\pi - \theta)}{2}}$$

Since

$$\sin\frac{1}{2}(\pi - \theta) = \cos\frac{\theta}{2}$$

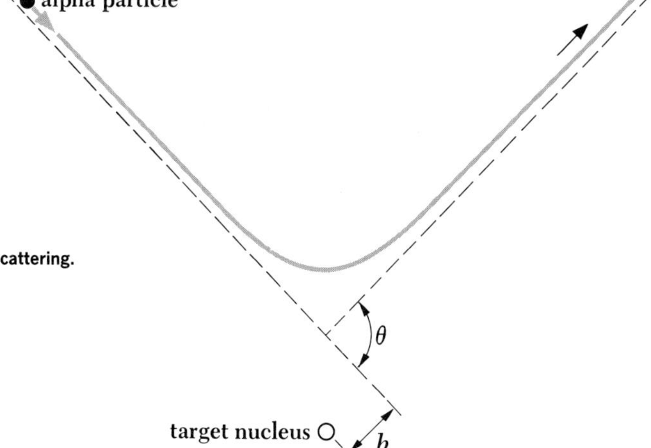

FIGURE 4-4 Rutherford scattering.

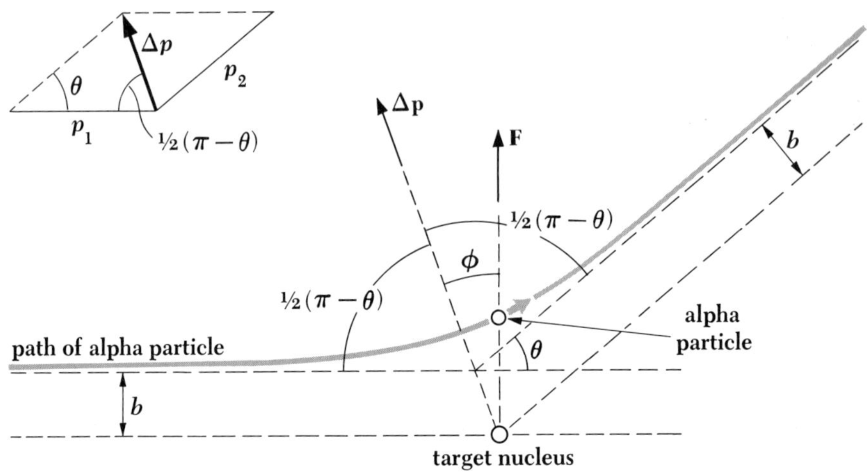

FIGURE 4-5 Geometrical relationships in Rutherford scattering.

and

$$\sin \theta = 2 \sin \frac{\theta}{2} \cos \frac{\theta}{2}$$

we have for the momentum change

$$\Delta p = 2\, mv \sin \frac{\theta}{2} \tag{4.2}$$

Because the impulse $\int \mathbf{F}\, dt$ is in the same direction as the momentum change $\Delta\mathbf{p}$, its magnitude is

$$\int F\, dt = \int F \cos \phi\, dt \tag{4.3}$$

where ϕ is the instantaneous angle between $\mathbf{F}$ and $\Delta\mathbf{p}$ along the path of the alpha particle. Inserting Eqs. 4.2 and 4.3 in Eq. 4.1,

$$2\, mv \sin \frac{\theta}{2} = \int_0^\infty F \cos \phi\, dt$$

To change the variable on the right-hand side from t to ϕ, we note that the limits of integration will change to $-\frac{1}{2}(\pi - \theta)$ and $+\frac{1}{2}(\pi - \theta)$, corresponding to ϕ at $t = 0$ and $t = \infty$ respectively, and so

$$2\, mv \sin \frac{\theta}{2} = \int_{-(\pi-\theta)/2}^{+(\pi-\theta)/2} F \cos \phi \frac{dt}{d\phi}\, d\phi \tag{4.4}$$

The quantity $d\phi/dt$ is just the angular velocity ω of the alpha particle about the nucleus (this is evident from Fig. 4-5). The electrostatic force exerted by the nucleus on the alpha particle acts along the radius vector joining them, and so there is no torque on the alpha particle and its angular momentum $m\omega r^2$ is constant. Hence

$$\begin{aligned} m\omega r^2 &= \text{constant} \\ &= mr^2 \frac{d\phi}{dt} \\ &= mvb \end{aligned}$$

from which we see that

$$\frac{dt}{d\phi} = \frac{r^2}{vb}$$

Substituting this expression for $dt/d\phi$ in Eq. 4.4,

4.5 $$2\,mv^2 b \sin\frac{\theta}{2} = \int_{-(\pi-\theta)/2}^{+(\pi-\theta)/2} Fr^2 \cos\phi \, d\phi$$

As we recall, F is the electrostatic force exerted by the nucleus on the alpha particle. The charge on the nucleus is Ze, corresponding to the atomic number Z, and that on the alpha particle is $2e$. Therefore

$$F = \frac{1}{4\pi\varepsilon_0} \frac{2Ze^2}{r^2}$$

and

$$\begin{aligned} \frac{4\pi\varepsilon_0 mv^2 b}{Ze^2} \sin\frac{\theta}{2} &= \int_{-(\pi-\theta)/2}^{+(\pi-\theta)/2} \cos\phi \, d\phi \\ &= 2\cos\frac{\theta}{2} \end{aligned}$$

The scattering angle θ is related to the impact parameter b by the equation

$$\cot\frac{\theta}{2} = \frac{2\pi\varepsilon_0 mv^2}{Ze^2} b$$

It is more convenient to specify the alpha-particle energy T instead of its mass and velocity separately; with this substitution,

4.6 $$\cot\frac{\theta}{2} = \frac{4\pi\varepsilon_0 T}{Ze^2} b$$

Figure 4-6 is a schematic representation of Eq. 4.6; the rapid decrease in θ as b increases is evident. A very near miss is required for a substantial deflection.

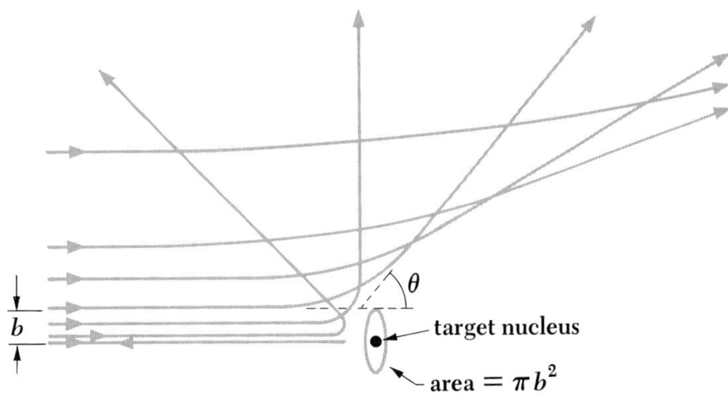

FIGURE 4-6 The scattering angle decreases with increasing impact parameter.

4.3 The Rutherford Scattering Formula

Equation 4.6 cannot be directly confronted with experiment since there is no way of measuring the impact parameter corresponding to a particular observed scattering angle. An indirect strategy is required. Our first step is to note that all alpha particles approaching a target nucleus with an impact parameter from 0 to b will be scattered through an angle of θ or more, where θ is given in terms of b by Eq. 4.6. This means that an alpha particle that is initially directed anywhere within the area πb^2 around a nucleus will be scattered through θ or more (Fig. 4-6); the area πb^2 is accordingly called the *cross section* for the interaction. The general symbol for cross section is σ, and so here

$$\sigma = \pi b^2 \tag{4.7}$$

We must keep in mind that the incident alpha particle is actually scattered before it reaches the immediate vicinity of the nucleus and hence does not necessarily pass within a distance b of it.

Now we consider a foil of thickness t that contains n atoms per unit volume. The number of target nuclei per unit area is nt, and an alpha-particle beam incident upon an area A therefore encounters ntA nuclei. The aggregate cross section for scatterings of θ or more is the number of target nuclei ntA multiplied by the cross section σ for such scattering per nucleus, or $ntA\sigma$. Hence the fraction f of incident alpha particles scattered by θ or more is the ratio between the aggregate cross section $ntA\sigma$ for such scattering and the total target area A. That is,

$$f = \frac{\text{alpha particles scattered by } \theta \text{ or more}}{\text{incident alpha particles}}$$

$$= \frac{\text{aggregate cross section}}{\text{target area}}$$

$$= \frac{ntA\sigma}{A}$$

$$= nt\pi b^2$$

Substituting for b from Eq. 4.6,

4.8 $$f = \pi nt \left(\frac{Ze^2}{4\pi\varepsilon_0 T} \right)^2 \cot^2 \frac{\theta}{2}$$

In the above calculation it was assumed that the foil is sufficiently thin so that the cross sections of adjacent nuclei do not overlap and that a scattered alpha particle receives its entire deflection from a single encounter with a nucleus.

Let us use Eq. 4.8 to determine what fraction of a beam of 7.7-Mev alpha particles is scattered through angles of more than 45° when incident upon a gold foil 3×10^{-7} m thick. (These values are typical of the alpha-particle energies and foil thicknesses used by Geiger and Marsden; for comparison, a human hair is about 10^{-4} m in diameter.) We begin by finding n, the number of gold atoms per unit volume in the foil, from the relationship

$$\frac{\text{Atoms}}{\text{Volume}} = \frac{(\text{atoms/kmole}) \times (\text{mass/volume})}{\text{mass/kmole}}$$

$$n = \frac{N_0 \rho}{w}$$

where N_0 is Avogadro's number, ρ the density of gold, and w its atomic weight. Since $N_0 = 6.03 \times 10^{26}$ atoms/kmole, $\rho = 1.93 \times 10^4$ kg/m^3, and $w = 197$, we have

$$n = \frac{6.03 \times 10^{26} \text{ atoms/kmole} \times 1.93 \times 10^4 \text{ kg/m}^3}{197 \text{ kg/kmole}}$$

$$= 5.91 \times 10^{28} \text{ atoms/m}^3$$

The atomic number Z of gold is 79, a kinetic energy of 7.7 Mev is equal to 1.23×10^{-12} joule, and $\theta = 45°$; from these figures we find that

$$f = 7 \times 10^{-5}$$

of the incident alpha particles are scattered through 45° or more—only 0.007 per cent! A foil this thin is quite transparent to alpha particles.

In an actual experiment, a detector measures alpha particles scattered between θ and $\theta + d\theta$, as in Fig. 4-7. The fraction of incident alpha particles so scattered is found by differentiating Eq. 4.8 with respect to θ, an operation that yields

4.9 $$df = -\pi nt \left(\frac{Ze^2}{4\pi\varepsilon_0 T} \right)^2 \cot \frac{\theta}{2} \csc^2 \frac{\theta}{2} \, d\theta$$

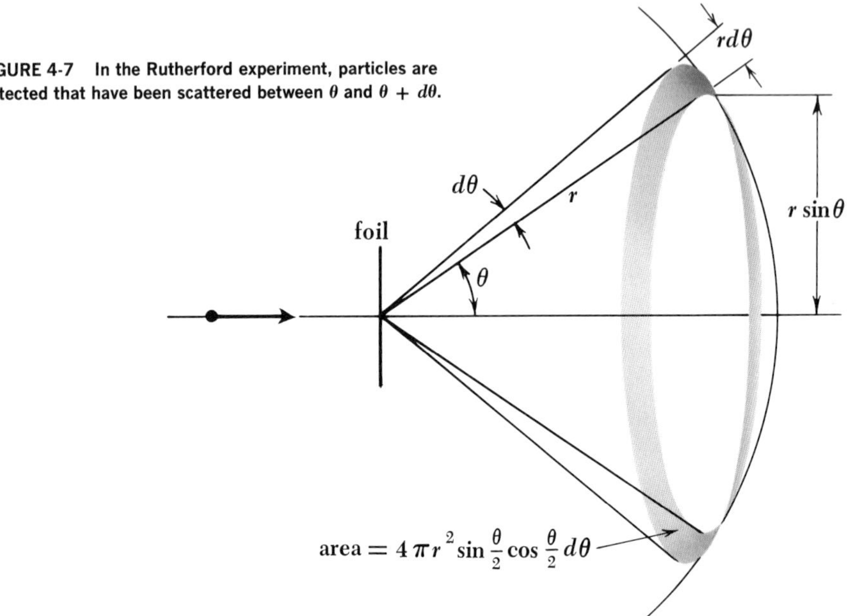

FIGURE 4-7 In the Rutherford experiment, particles are detected that have been scattered between θ and $\theta + d\theta$.

(The minus sign expresses the fact that f decreases with increasing θ.) In the experiment, a fluorescent screen was placed a distance r from the foil, and the scattered alpha particles were detected by means of the scintillations they caused. Those alpha particles scattered between θ and $\theta + d\theta$ reach a zone of a sphere of radius r whose width is $rd\theta$. The zone radius itself is $r \sin \theta$, and so the area dS of the screen struck by these particles is

$$\begin{aligned} dS &= (2\pi r \sin \theta)(rd\theta) \\ &= 2\pi r^2 \sin \theta \, d\theta \\ &= 4\pi r^2 \sin \frac{\theta}{2} \cos \frac{\theta}{2} \, d\theta \end{aligned}$$

If a total of N_i alpha particles strike the foil during the course of the experiment, the number scattered into $d\theta$ at θ is $N_i \, df$. The number $N(\theta)$ per unit area striking the screen at θ, which is the quantity actually measured, is

$$\begin{aligned} N(\theta) &= \frac{N_i \, |df|}{dS} \\ &= \frac{N_i \pi n t \left(\dfrac{Ze^2}{4\pi\varepsilon_0 T}\right)^2 \cot \dfrac{\theta}{2} \csc^2 \dfrac{\theta}{2} \, d\theta}{4\pi r^2 \sin \dfrac{\theta}{2} \cos \dfrac{\theta}{2} \, d\theta} \end{aligned}$$

4.10 $$N(\theta) = \frac{N_i n t Z^2 e^4}{(8\pi\varepsilon_0)^2 r^2 T^2 \sin^4 (\theta/2)}$$ **Rutherford scattering formula**

Equation 4.10 is the *Rutherford scattering formula.*

According to Eq. 4.10, the number of alpha particles per unit area arriving at the fluorescent screen a distance r from the scattering foil should be directly proportional to the thickness t of the foil, the number of foil atoms per unit volume n, and the square of the atomic number Z of the foil atoms, and it should be inversely proportional to the square of the kinetic energy T of the alpha particles and to $\sin^4 (\theta/2)$, where θ is the scattering angle. These predictions agreed with the measurements of Geiger and Marsden mentioned earlier, which led Rutherford to conclude that his assumptions, chief among them the hypothesis of the nuclear atom, were correct. Rutherford is therefore credited with the "discovery" of the nucleus.

4.4 Nuclear Dimensions

When we say that the experimental data on the scattering of alpha particles by thin foils verifies our assumption that atomic nuclei are point particles, what is really meant is that their dimensions are insignificant compared with the minimum distance to which the incident alpha particles approach the nuclei. Rutherford scattering therefore permits us to determine an upper limit to nuclear dimensions. Let us compute the distance of closest approach r_0 of the most energetic alpha particles employed in the early experiments. An alpha particle will have its smallest r_0 when its impact parameter is $b = 0$, corresponding to a head-on approach followed by a 180° scattering. At the instant of closest approach the initial kinetic energy T of the particle is entirely converted to electrostatic potential energy, and so at that instant

4.11 $$T = \frac{1}{4\pi\varepsilon_0}\frac{2Ze^2}{r_0}$$

since the charge of the alpha particle is $2e$ and that of the nucleus Ze. Hence

$$r_0 = \frac{2Ze^2}{4\pi\varepsilon_0 T}$$

The maximum T found in alpha particles of natural origin is 7.7 Mev, which is

$$7.7 \times 10^6 \text{ ev} \times 1.6 \times 10^{-19} \text{ joule/ev} = 1.2 \times 10^{-12} \text{ joule}$$

Since $1/4\pi\varepsilon_0 = 9 \times 10^9\ n\text{–m}^2/\text{coulomb}^2$

$$r_0 = \frac{2 \times 9 \times 10^9\ n\text{–m}^2/\text{coulomb}^2 \times (1.6 \times 10^{-19} \text{ coulomb})^2 Z}{1.2 \times 10^{-12} \text{ joule}}$$

$$= 3.8 \times 10^{-16} Z \text{ m}$$

The atomic number of gold, a typical foil material, is $Z = 79$, so that

$$r_0(\text{Au}) = 3.0 \times 10^{-14} \text{ m}$$

The radius of the gold nucleus is therefore less than 3.0×10^{-14} m, well under 1/10,000 the radius of the atom as a whole.

In more recent years particles of much higher energies than 7.7 Mev have been artificially accelerated, and it has been found that the Rutherford scattering formula does indeed eventually fail to agree with experiment. We shall discuss these experiments and the information they provide on actual nuclear dimensions in Chap. 12.

4.5 Electron Orbits

The Rutherford model of the atom, so convincingly confirmed by experiment, postulates a tiny, massive, positively charged nucleus surrounded at a relatively great distance by enough electrons to render the atom, as a whole, electrically neutral. Thomson visualized the electrons in his model atom as embedded in the positively charged matter that fills it, and thus as being unable to move. The electrons in Rutherford's model atom, however, cannot be stationary, because there is nothing that can keep them in place against the electrostatic force attracting them to the nucleus. If the electrons are in motion around the nucleus, however, dynamically stable orbits (comparable with those of the planets about the sun) are possible (Fig. 4-8).

Let us examine the classical dynamics of the hydrogen atom, whose single electron makes it the simplest of all atoms. We shall assume a circular electron orbit for convenience, though it might as reasonably be assumed elliptical in shape. The centripetal force

$$F_c = \frac{mv^2}{r}$$

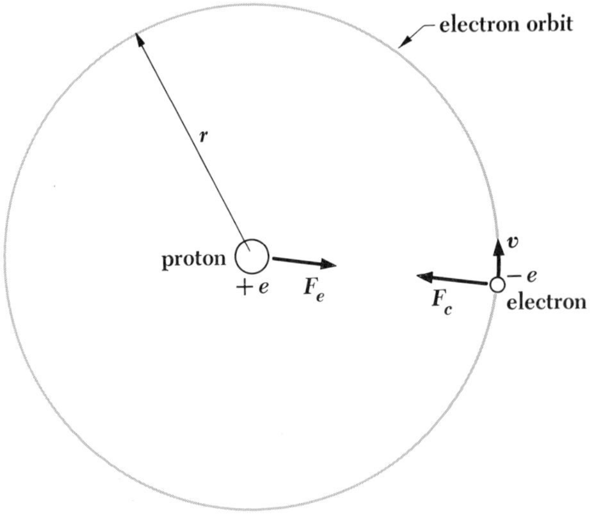

FIGURE 4-8 Force balance in the hydrogen atom.

holding the electron in an orbit r from the nucleus is provided by the electrostatic force

$$F_e = \frac{1}{4\pi\varepsilon_0}\frac{e^2}{r^2}$$

between them, and the condition for orbit stability is

$$F_c = F_e$$

4.12 $$\frac{mv^2}{r} = \frac{1}{4\pi\varepsilon_0}\frac{e^2}{r^2}$$

The electron velocity v is therefore related to its orbit radius r by the formula

4.13 $$v = \frac{e}{\sqrt{4\pi\varepsilon_0 mr}}$$

The total energy E of the electron in a hydrogen atom is the sum of its kinetic energy

$$T = \tfrac{1}{2}mv^2$$

and its potential energy

$$V = -\frac{e^2}{4\pi\varepsilon_0 r}$$

(The minus sign signifies that the force on the electron is in the $-r$ direction.) Hence

$$E = T + V$$
$$= \frac{mv^2}{2} - \frac{e^2}{4\pi\varepsilon_0 r}$$

Substituting for v from Eq. 4.12,

$$E = \frac{e^2}{8\pi\varepsilon_0 r} - \frac{e^2}{4\pi\varepsilon_0 r}$$

4.14 $$= -\frac{e^2}{8\pi\varepsilon_0 r}$$

The total energy of an atomic electron is negative; this is necessary if it is to be bound to the nucleus. If E were greater than zero, the electron would have too much energy to remain in a closed orbit about the nucleus.

Experiments indicate that 13.6 ev is required to separate a hydrogen atom into a proton and an electron; that is, its binding energy E is -13.6 ev.

Since 13.6 ev $= 2.2 \times 10^{-18}$ joule, we can find the orbital radius of the electron in a hydrogen atom from Eq. 4.14:

$$
\begin{aligned}
r &= -\frac{e^2}{8\pi\varepsilon_0 E} \\
&= -\frac{(1.6 \times 10^{-19} \text{ coulomb})^2}{8\pi \times 8.85 \times 10^{-12} \text{ farad/m} \times (-2.2 \times 10^{-18} \text{ joule})} \\
&= 5.3 \times 10^{-11} \text{ m}
\end{aligned}
$$

An atomic radius of this order of magnitude agrees with estimates made in other ways.

The above analysis is a straightforward application of Newton's laws of motion and Coulomb's law of electric force—both pillars of classical physics—and is in accord with the experimental observation that atoms are stable. However, it is *not* in accord with electromagnetic theory—another pillar of classical physics—which predicts that accelerated electric charges radiate energy in the form of electromagnetic waves. An electron pursuing a circular path is accelerated and therefore should continuously lose energy, gradually spiraling into the nucleus (Fig. 4-9). Whenever they have been directly tested, the predictions of electromagnetic theory have always agreed with experiment, yet atoms do not collapse. This contradiction can mean only one thing: The laws of physics that are valid in the macroscopic world do not hold true in the microscopic world of the atom.

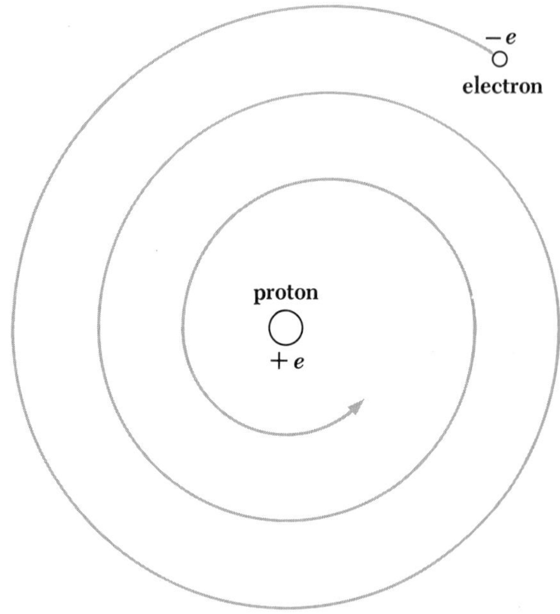

FIGURE 4-9 An atomic electron should, classically, spiral into the nucleus as it radiates energy due to its acceleration.

The reason for the failure of classical physics to yield a meaningful analysis of atomic structure is that it approaches nature exclusively in terms of the abstract concepts of "pure" particles and "pure" waves. As we learned in the two preceding chapters, particles and waves have many properties in common, though the smallness of Planck's constant renders the wave-particle duality imperceptible in the macroscopic world. The validity of classical physics decreases as the scale of the phenomena under study decreases, and full allowance must be made for the particle behavior of waves and the wave behavior of particles if the atom is to be understood. In the next chapter we shall see how the Bohr atomic model, which combines classical and modern notions, accomplishes part of the latter task. Not until we consider the atom from the point of view of quantum mechanics, which makes no compromise with intuitive notions acquired in our daily lives, will we find a really successful theory of the atom.

An interesting question arises at this point. In our derivation of the Rutherford scattering formula we made use of the same laws of physics that proved such dismal failures when applied to atomic stability. Is it not therefore possible, even likely, that the formula is not correct, and that the atom in reality does not resemble the Rutherford model of a small central nucleus surrounded by distant electrons? This question is not a trivial one, and it is, in a way, a curious coincidence that the quantum-mechanical analysis of alpha-particle scattering from thin foils results in precisely the same formula that Rutherford obtained. We are therefore correct in thinking of the atom in terms of Rutherford's model, though the dynamics of the atomic electrons requires a more sophisticated approach.

Problems

1. A 5-Mev alpha particle approaches a gold nucleus with an impact parameter of 2.6×10^{-13} m. Through what angle will it be scattered?

2. What is the impact parameter of a 5-Mev alpha particle scattered by 10° when it approaches a gold nucleus?

3. What fraction of a beam of 7.7-Mev alpha particles incident upon a gold foil 3×10^{-7} m thick is scattered by less than 1°?

4. What fraction of a beam of 7.7-Mev alpha particles incident upon a gold foil 3×10^{-7} m thick is scattered by 90° or more?

5. Show that twice as many alpha particles are scattered by a foil through angles between 60 and 90° as are scattered through angles of 90° or more.

6. A beam of 8.3-Mev alpha particles is directed at an aluminum foil. It is found that the Rutherford scattering formula ceases to be obeyed at scattering angles exceeding about 60°. If the alpha particle is assumed to have a radius of 2×10^{-15} m, find the radius of the aluminum nucleus.

7. Determine the distance of closest approach of 1-Mev protons incident upon gold nuclei.

8. Find the distance of closest approach of 8-Mev protons incident upon gold nuclei.

9. The derivation of the Rutherford scattering formula was made nonrelativistically. Justify this approximation by computing the mass ratio between an 8-Mev alpha particle and an alpha particle at rest.

10. Find the frequency of rotation of the electron in the classical model of the hydrogen atom. In what region of the spectrum are electromagnetic waves of this frequency?

11. The electric-field intensity at a distance r from the center of a uniformly charged sphere of radius R and total charge Q is $Qr/4\pi\varepsilon_0 R^3$ when $r < R$. Such a sphere corresponds to the Thomson model of the atom. Show that an electron in this sphere executes simple harmonic motion about its center and derive a formula for the frequency of this motion. Evaluate the frequency of the electron oscillations for the case of the hydrogen atom and compare it with the frequencies of the spectral lines of hydrogen (see Fig. 5-6).

THE BOHR MODEL OF THE ATOM 5

The first theory of the hydrogen atom to succeed in accounting for the more conspicuous aspects of its behavior was presented by Niels Bohr in 1913. Bohr applied quantum ideas to atomic structure to obtain a model which, despite its serious inadequacies and subsequent replacement by a quantum-mechanical description of greater accuracy and usefulness, nevertheless persists as the mental picture most scientists have of the atom. While it is not the general policy of this book to go deeply into hypotheses that have had to be discarded, we shall discuss Bohr's theory of the hydrogen atom because it provides a valuable transition to the more abstract quantum theory of the atom. For this reason our account of the Bohr theory differs somewhat from the original one given by Bohr, though all of the results are identical.

5.1 Atomic Spectra

The ability of the Bohr theory to explain the origin of spectral lines is among its most spectacular accomplishments, and so it is appropriate to preface our exposition of the theory itself with a look at atomic spectra.

We have already mentioned that heated solids emit radiation in which all wavelengths are present, though with different intensities. We shall learn in Chap. 10 that the observed features of this radiation can be explained on the basis of the quantum theory of light independently of the details of the radiation process itself or of the nature of the solid. From this fact it follows that, when a solid is heated to incandescence, we are witnessing the collective behavior of a great many interacting atoms rather than the characteristic behavior of the individual atoms of a particular element.

At the other extreme, the atoms or molecules in a rarefied gas are so far apart on the average that their only mutual interactions occur during occasional collisions. Under these circumstances we would expect any emitted radiation to be characteristic of the individual atoms or molecules present, an expectation that is realized experimentally. When an atomic gas or vapor at

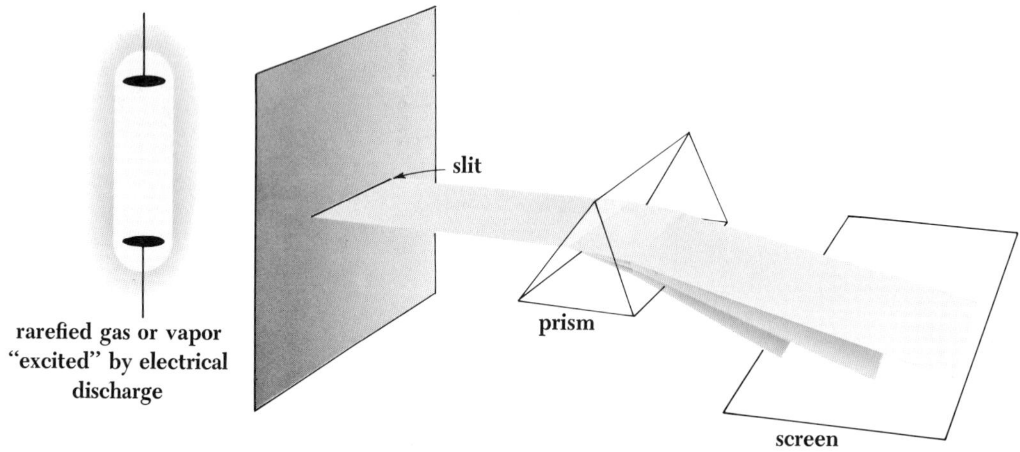

FIGURE 5-1 An idealized spectrometer.

somewhat less than atmospheric pressure is suitably "excited," usually by the passage of an electric current through it, the emitted radiation has a spectrum which contains certain discrete wavelengths only. An idealized laboratory arrangement for observing such atomic spectra is sketched in Fig. 5-1. Figure 5-2 shows the atomic spectra of several elements; they are called *emission line spectra.* Every element displays a unique line spectrum when a sample

FIGURE 5-2 Portions of the emission of hydrogen, helium, and mercury.

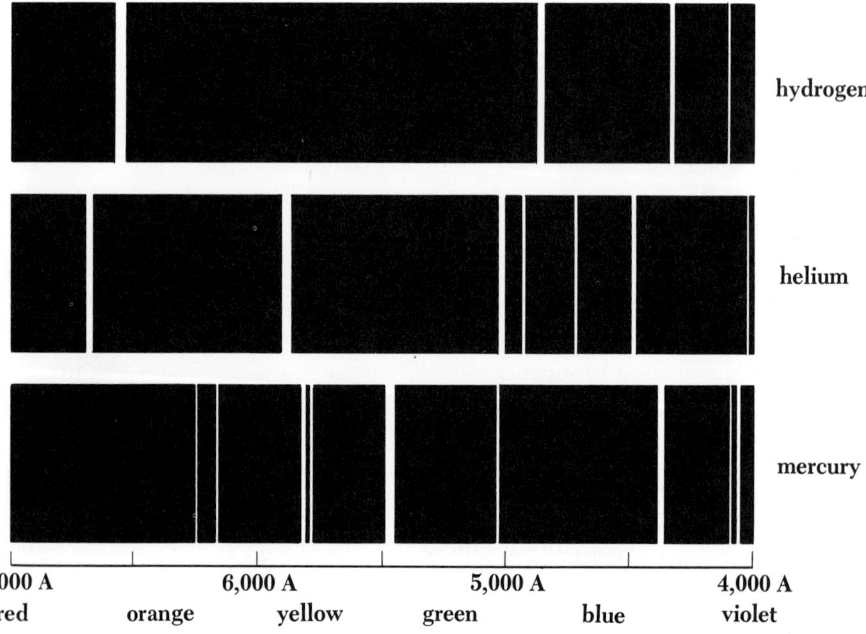

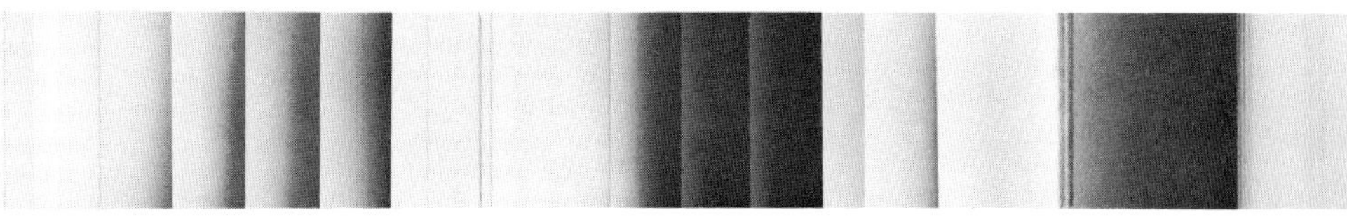

FIGURE 5-3 A portion of the band spectrum of PN.

of it in the vapor phase is excited; spectroscopy is therefore a useful tool for analyzing the composition of an unknown substance.

The spectrum of an excited molecular gas or vapor contains *bands* which consist of many separate lines very close together (Fig. 5-3). Bands owe their origin to rotations and vibrations of the atoms in an electronically excited molecule, and we shall consider their interpretation in a later chapter.

When white light is passed through a gas, it is found to absorb light of certain of the wavelengths present in its emission spectrum. The resulting *absorption line spectrum* consists of a bright background crossed by dark lines corresponding to the missing wavelengths (Fig. 5-4); emission spectra consist of bright lines on a dark background. The dark Fraunhofer lines in the solar spectrum occur because the luminous part of the sun, which radiates almost exactly according to theoretical predictions for any object heated to 5800°K, is surrounded by an envelope of cooler gas which absorbs light of certain wavelengths only.

In the latter part of the nineteenth century it was discovered that the wavelengths present in atomic spectra fall into definite sets called *spectral series.* The wavelengths in each series can be specified by a simple empirical formula, with remarkable similarity among the formulas for the various series that comprise the complete spectrum of an element. The first such spectral series was found by J. J. Balmer in 1885 in the course of a study of the visible part of the hydrogen spectrum. Figure 5-5 shows the *Balmer series.* The line with the longest wavelength, 6,563 A, is designated H_α, the next, whose wavelength is 4,863 A, is designated H_β, and so on. As the wavelength decreases, the lines are found closer together and weaker in intensity until the *series limit* at 3,646 A is reached, beyond which there are no further sep-

FIGURE 5-4 The dark lines in the absorption spectrum of an element correspond to bright lines in its emission spectrum.

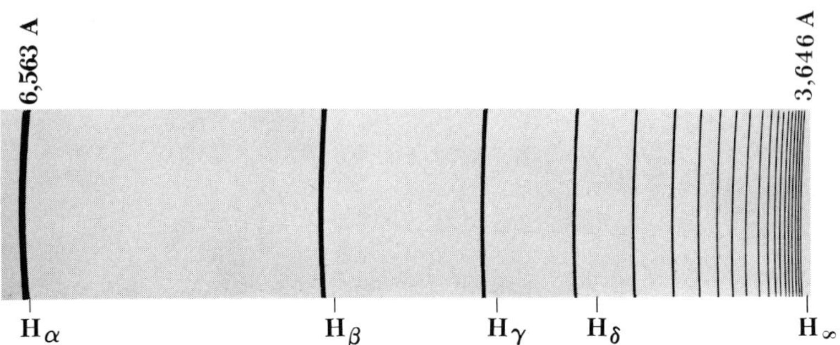

FIGURE 5-5 The Balmer series of hydrogen.

arate lines but only a faint continuous spectrum. Balmer's formula for the wavelengths of this series is

5.1 $$\frac{1}{\lambda} = R\left(\frac{1}{2^2} - \frac{1}{n^2}\right) \qquad n = 3, 4, 5, \ldots$$ **Balmer**

The quantity R, known as the *Rydberg constant,* has the value

$$\begin{aligned} R &= 1.097 \times 10^7 \text{ m}^{-1} \\ &= 1.097 \times 10^{-3} \text{ A}^{-1} \end{aligned}$$

The H_α line corresponds to $n = 3$, the H_β line to $n = 4$, and so on. The series limit corresponds to $n = \infty$, so that it occurs at a wavelength of $4/R$, in agreement with experiment.

The Balmer series contains only those wavelengths in the visible portion of the hydrogen spectrum. The spectral lines of hydrogen in the ultraviolet and infrared regions fall into several other series. In the ultraviolet the *Lyman series* contains the wavelengths specified by the formula

5.2 $$\frac{1}{\lambda} = R\left(\frac{1}{1^2} - \frac{1}{n^2}\right) \qquad n = 2, 3, 4, \ldots$$ **Lyman**

In the infrared, three spectral series have been found whose component lines have the wavelengths specified by the formulas

5.3 $$\frac{1}{\lambda} = R\left(\frac{1}{3^2} - \frac{1}{n^2}\right) \qquad n = 4, 5, 6, \ldots$$ **Paschen**

5.4 $$\frac{1}{\lambda} = R\left(\frac{1}{4^2} - \frac{1}{n^2}\right) \qquad n = 5, 6, 7, \ldots$$ **Brackett**

5.5 $$\frac{1}{\lambda} = R\left(\frac{1}{5^2} - \frac{1}{n^2}\right) \qquad n = 6, 7, 8, \ldots$$ **Pfund**

The above spectral series of hydrogen are plotted in terms of wavelength in

Fig. 5-6; the Brackett series evidently overlaps the Paschen and Pfund series. The value of R is the same in Eqs. 5.1 to 5.5.

The existence of such remarkable regularities in the hydrogen spectrum, together with similar regularities in the spectra of more complex elements, poses a definitive test for any theory of atomic structure.

5.2 The Bohr Atom

We saw in the previous chapter that the principles of classical physics are incompatible with the observed stability of the hydrogen atom. The electron in this atom is obliged to whirl around the nucleus to keep from being pulled into it and yet must radiate electromagnetic energy continuously. Because other apparently paradoxical phenomena, like the photoelectric effect and the

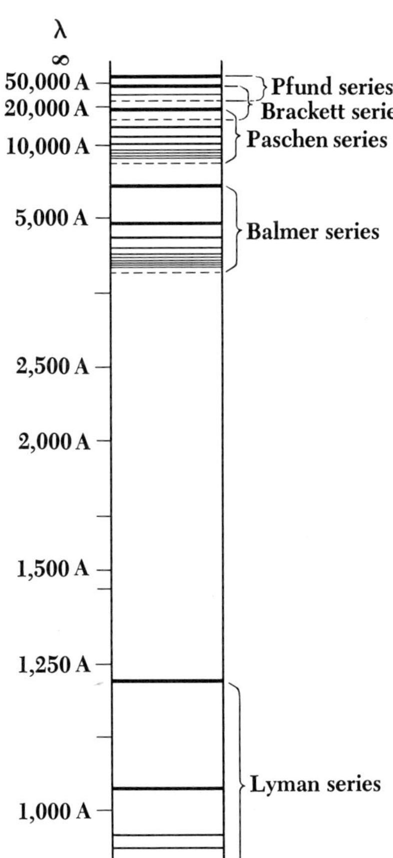

FIGURE 5-6 The spectral series of hydrogen.

diffraction of electrons, find explanation in terms of quantum concepts, it is appropriate to inquire whether this might not also be true for the atom.

Let us start by examining the wave behavior of an electron in orbit around a hydrogen nucleus. The de Broglie wavelength of this electron is

$$\lambda = \frac{h}{mv}$$

where the electron speed v is that given by Eq. 4.13:

$$v = \frac{e}{\sqrt{4\pi\varepsilon_0 mr}}$$

Hence

5.6 $$\lambda = \frac{h}{e}\sqrt{\frac{4\pi\varepsilon_0 r}{m}}$$

By substituting 5.3×10^{-11} m for the radius r of the electron orbit, we find the electron wavelength to be

$$\lambda = \frac{6.63 \times 10^{-34} \text{ joule-sec}}{1.6 \times 10^{-19} \text{ coulomb}} \sqrt{\frac{4\pi \times 8.85 \times 10^{-12} \text{ farad/m} \times 5.3 \times 10^{-11} \text{ m}}{9.1 \times 10^{-31} \text{ kg}}}$$

$$= 33 \times 10^{-11} \text{ m}$$

This wavelength is exactly the same as the circumference of the electron orbit,

$$2\pi r = 33 \times 10^{-11} \text{ m}$$

The orbit of the electron in a hydrogen atom corresponds to one complete electron wave joined on itself (Fig. 5-7).

The fact that the electron orbit in a hydrogen atom is one electron wavelength in circumference provides the clue we need to construct a theory of the atom. If we consider the vibrations of a wire loop (Fig. 5-8), we find that their wavelengths always fit an integral number of times into the loop's circumference so that each wave joins smoothly with the next. If the wire were perfectly rigid, these vibrations would continue indefinitely. Why are these the only vibrations possible in a wire loop? If a fractional number of wavelengths is placed around the loop, as in Fig. 5-9, destructive interference will occur as the waves travel around the loop, and the vibrations will die out rapidly. By considering the behavior of electron waves in the hydrogen atom as analogous to the vibrations of a wire loop, then, we may postulate that **an electron can circle a nucleus indefinitely without radiating energy provided that its orbit contains an integral number of de Broglie wavelengths.**

This postulate is the decisive one in our understanding of the atom. It combines both the particle and wave characters of the electron into a single statement, since the electron wavelength is computed from the orbital speed required to balance the electrostatic attraction of the nucleus. While we can never observe these antithetical characters simultaneously, they are inseparable in nature.

It is a simple matter to express the condition that an electron orbit contain an integral number of de Broglie wavelengths. The circumference of a circular orbit of radius r is $2\pi r$, and so we may write the condition for orbit stability as

$$n\lambda = 2\pi r_n \qquad n = 1, 2, 3, \ldots \tag{5.7}$$

FIGURE 5-7 The orbit of the electron in a hydrogen atom corresponds to a complete electron de Broglie wave joined on itself.

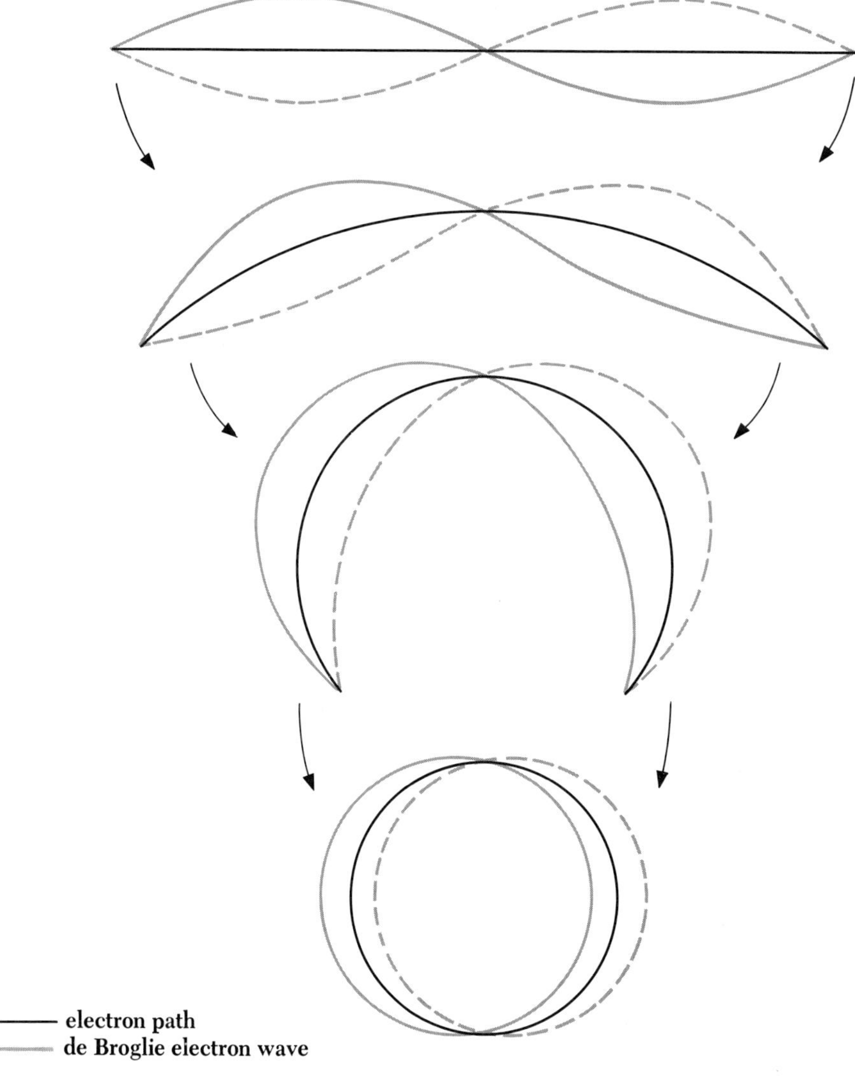

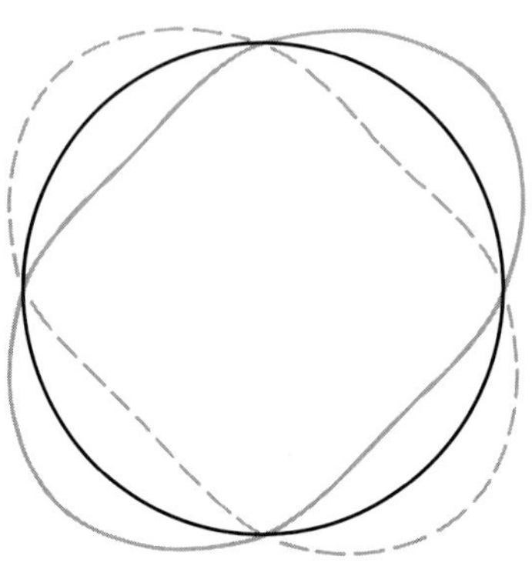
circumference = 2 wavelengths

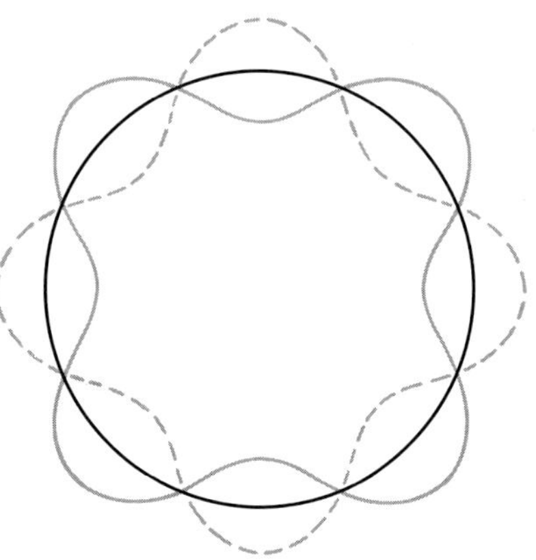
circumference = 4 wavelengths

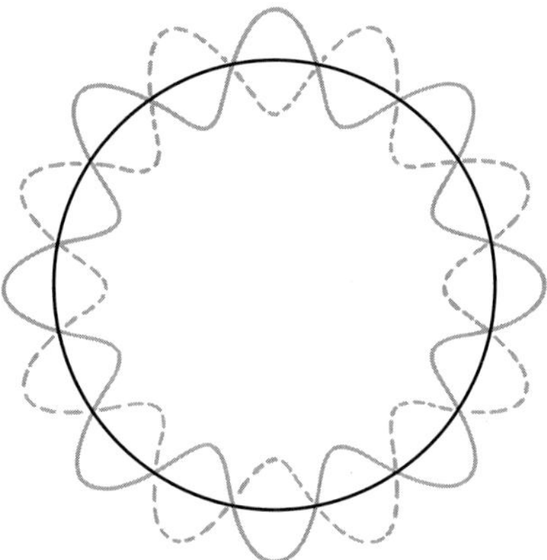
circumference = 8 wavelengths

FIGURE 5-8 The vibrations of a wire loop.

where r_n designates the radius of the orbit that contains n wavelengths. The integer n is called the *quantum number* of the orbit. Substituting for λ, the electron wavelength given by Eq. 5.6 yields

$$\frac{nh}{e}\sqrt{\frac{4\pi\varepsilon_0 r_n}{m}} = 2\pi r_n$$

and so the stable electron orbits are those whose radii are given by

5.8 $$r_n = \frac{n^2h^2\varepsilon_0}{\pi me^2} \qquad n = 1, 2, 3, \ldots$$

The innermost orbit has the radius

$$r_1 = 5.3 \times 10^{-11} \text{ m}$$

in agreement with our previous calculation. The other radii are given in terms of r_1 by the formula

$$r_n = n^2 r_1$$

so that the spacing between adjacent orbits increases progressively (Fig. 5-10).

5.3 Energy Levels and Spectra

The various permitted orbits involve different electron energies. The electron energy E_n is given in terms of the orbit radius r_n by Eq. 4.14 as

$$E_n = -\frac{e^2}{8\pi\varepsilon_0 r_n}$$

Substituting for r_n from Eq. 5.8, we see that

5.9 $$E_n = -\frac{me^4}{8\varepsilon_0{}^2h^2}\left(\frac{1}{n^2}\right) \qquad n = 1, 2, 3, \ldots$$ **Energy levels**

The energies specified by Eq. 5.9 are called the *energy levels* of the hydrogen

FIGURE 5-9 A fractional number of wavelengths cannot persist because destructive interference will occur.

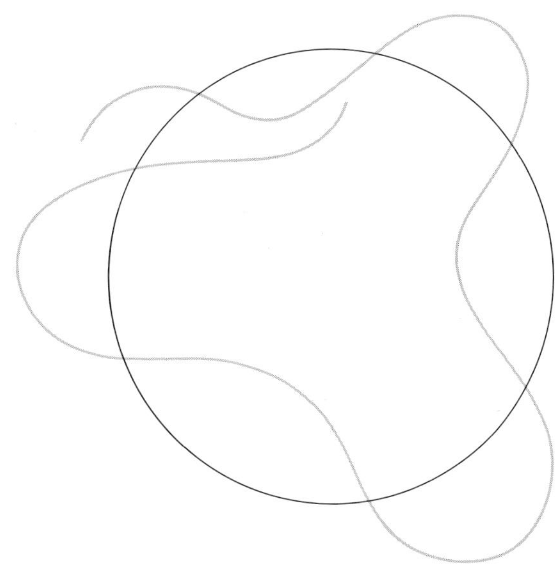

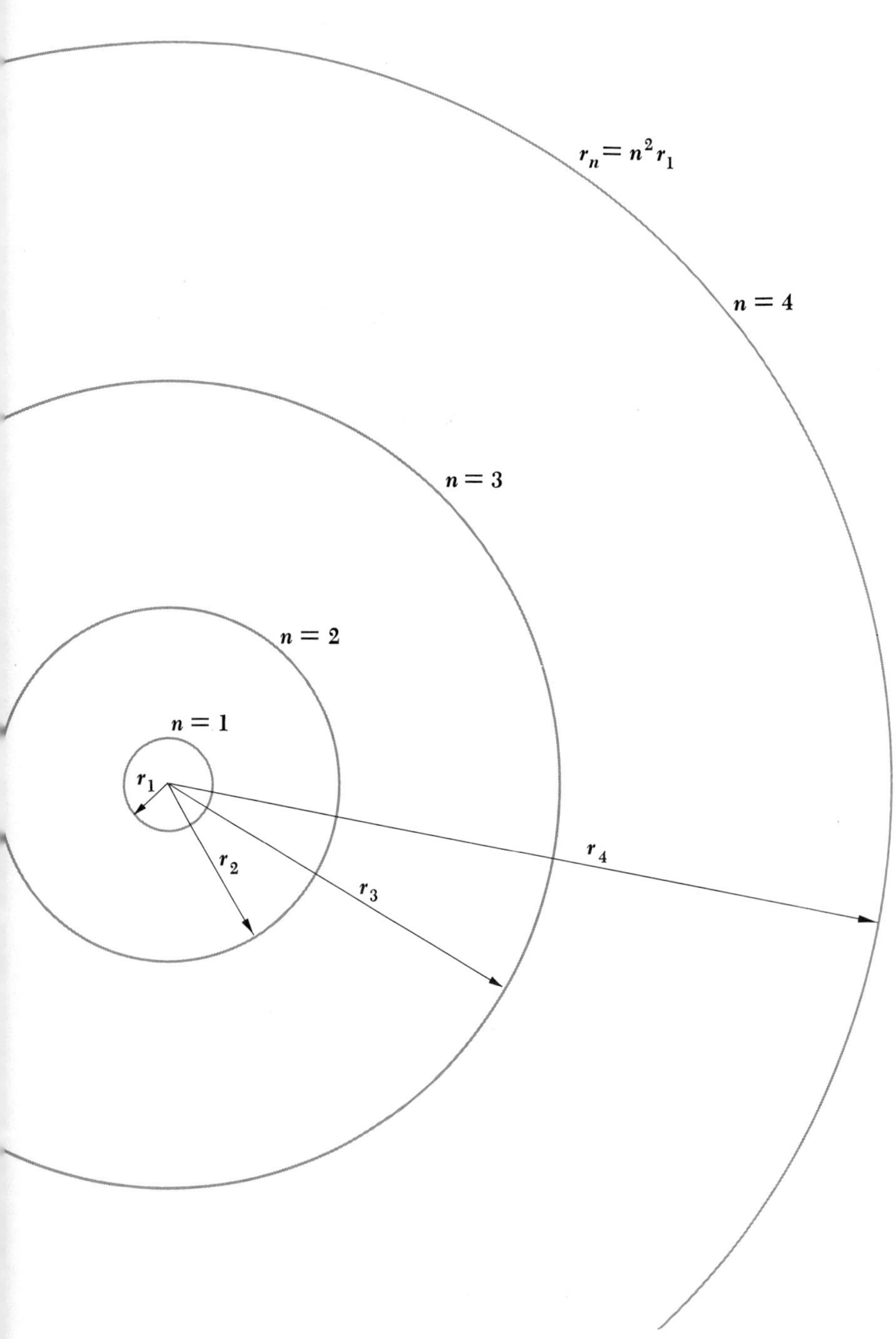

FIGURE 5-10 **Electron orbits in the Bohr model of the hydrogen atom.**

atom and are plotted in Fig. 5-11. These levels are all negative, signifying that the electron does not have enough energy to escape from the atom. The lowest energy level E_1 is called the *ground state* of the atom, and the higher levels E_2, E_3, E_4, . . . are called *excited states.* As the quantum number n increases, the corresponding energy E_n approaches closer and closer to 0; in the limit of $n = \infty$, $E_\infty = 0$ and the electron is no longer bound to the nucleus to form an atom. (A positive energy for a nucleus-electron combination means that the electron is not bound to the nucleus and has no quantum conditions to fulfill; such a combination does not constitute an atom, of course.)

It is now necessary for us to confront directly the equations we have developed with experiment. An especially striking experimental result is that atoms exhibit line spectra in both emission and absorption; do these spectra follow from our atomic model?

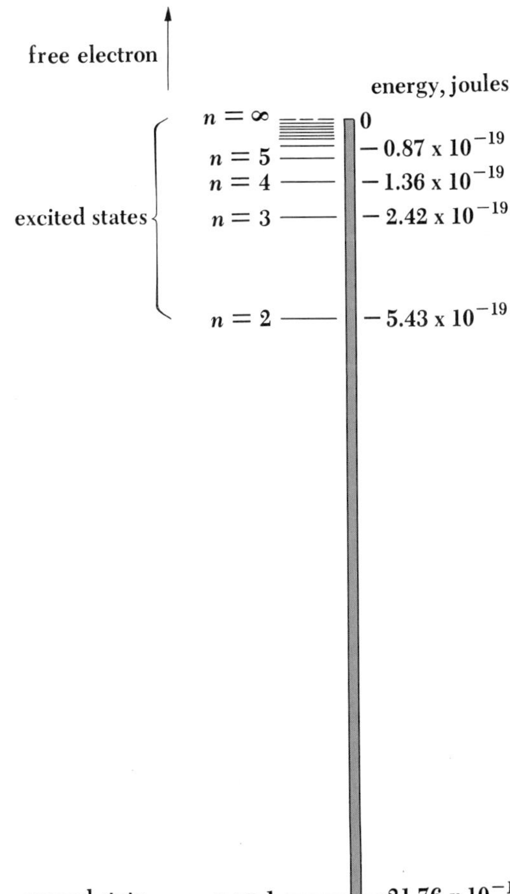

FIGURE 5-11 Energy levels of the hydrogen atom.

The presence of definite, discrete energy levels in the hydrogen atom suggests a connection with line spectra. Let us tentatively assert that, when an electron in an excited state drops to a lower state, the lost energy is emitted as a single photon of light. According to our model, electrons cannot exist in an atom except in certain specific energy levels. The jump of an electron from one level to another, with the difference in energy between the levels being given off all at once in a photon rather than in some more gradual manner, fits in well with this model. If the quantum number of the initial (higher energy) state is n_i and the quantum number of the final (lower energy) state is n_f, we are asserting that

$$\text{Initial energy} - \text{final energy} = \text{photon energy}$$

5.10 $$E_i - E_f = h\nu$$

where ν is the frequency of the emitted photon.

The initial and final states of a hydrogen atom that correspond to the quantum numbers n_i and n_f have, from Eq. 5.9, the energies

$$\text{Initial energy} = E_i = -\frac{me^4}{8\varepsilon_0{}^2h^2}\left(\frac{1}{n_i{}^2}\right)$$

$$\text{Final energy} = E_f = -\frac{me^4}{8\varepsilon_0{}^2h^2}\left(\frac{1}{n_f{}^2}\right)$$

Hence the energy difference between these states is

$$E_i - E_f = \frac{me^4}{8\varepsilon_0{}^2h^2}\left(-\frac{1}{n_i{}^2}\right) - \left(-\frac{1}{n_f{}^2}\right)$$

$$= \frac{me^4}{8\varepsilon_0{}^2h^2}\left(\frac{1}{n_f{}^2} - \frac{1}{n_i{}^2}\right)$$

The frequency ν of the photon released in this transition is

$$\nu = \frac{E_i - E_f}{h}$$

5.11 $$= \frac{me^4}{8\varepsilon_0{}^2h^3}\left(\frac{1}{n_f{}^2} - \frac{1}{n_i{}^2}\right)$$

In terms of photon wavelength λ, since

$$\lambda = \frac{c}{\nu}$$

we have

$$\frac{1}{\lambda} = \frac{\nu}{c}$$

5.12 $$= \frac{me^4}{8\varepsilon_0{}^2ch^3}\left(\frac{1}{n_f{}^2} - \frac{1}{n_i{}^2}\right)$$ **Hydrogen spectrum**

Equation 5.12 states that the radiation emitted by excited hydrogen atoms should contain certain wavelengths only. These wavelengths, furthermore, fall into definite sequences that depend upon the quantum number n_f of the final energy level of the electron. Since the initial quantum number n_i must always be greater than the final quantum number n_f in each case, in order that there be an excess of energy to be given off as a photon, the calculated formulas for the first five series are

5.13 $$n_f = 1: \quad \frac{1}{\lambda} = \frac{me^4}{8\varepsilon_0^2 ch^3}\left(\frac{1}{1^2} - \frac{1}{n^2}\right) \qquad n = 2, 3, 4, \ldots$$

5.14 $$n_f = 2: \quad \frac{1}{\lambda} = \frac{me^4}{8\varepsilon_0^2 ch^3}\left(\frac{1}{2^2} - \frac{1}{n^2}\right) \qquad n = 3, 4, 5, \ldots$$

5.15 $$n_f = 3: \quad \frac{1}{\lambda} = \frac{me^4}{8\varepsilon_0^2 ch^3}\left(\frac{1}{3^2} - \frac{1}{n^2}\right) \qquad n = 4, 5, 6, \ldots$$

5.16 $$n_f = 4: \quad \frac{1}{\lambda} = \frac{me^4}{8\varepsilon_0^2 ch^3}\left(\frac{1}{4^2} - \frac{1}{n^2}\right) \qquad n = 5, 6, 7, \ldots$$

5.17 $$n_f = 5: \quad \frac{1}{\lambda} = \frac{me^4}{8\varepsilon_0^2 ch^3}\left(\frac{1}{5^2} - \frac{1}{n^2}\right) \qquad n = 6, 7, 8, \ldots$$

These sequences are identical in form with the empirical spectral series discussed earlier. The Lyman series, Eq. 5.2, corresponds to $n_f = 1$; the Balmer series, Eq. 5.1, corresponds to $n_f = 2$; the Paschen series, Eq. 5.3, corresponds to $n_f = 3$; the Brackett series, Eq. 5.4, corresponds to $n_f = 4$; and the Pfund series, Eq. 5.5, corresponds to $n_f = 5$.

We still cannot consider our assertion that the line spectrum of hydrogen originates in electron transitions from high to low energy states as proved, however. The final step is to compare the value of the constant term in Eqs. 5.13 to 5.17 with that of the Rydberg constant R of the empirical equations 5.1 to 5.5. The value of this constant term is

$$\begin{aligned}
&\frac{me^4}{8\varepsilon_0^2 ch^3} \\
&= \frac{9.1 \times 10^{-31}\ \text{kg} \times (1.6 \times 10^{-19}\ \text{coulomb})^4}{8 \times (8.85 \times 10^{-12}\ \text{farad/m})^2 \times 3 \times 10^8\ \text{m/sec} \times (6.63 \times 10^{-34}\ \text{joule-sec})^3} \\
&= 1.097 \times 10^7\ \text{m}^{-1}
\end{aligned}$$

which is indeed the same as R! This theory of the hydrogen atom, which is essentially that developed by Bohr in 1913, therefore agrees both qualitatively and quantitatively with experiment. Figure 5-12 shows schematically how spectral lines are related to atomic energy levels.

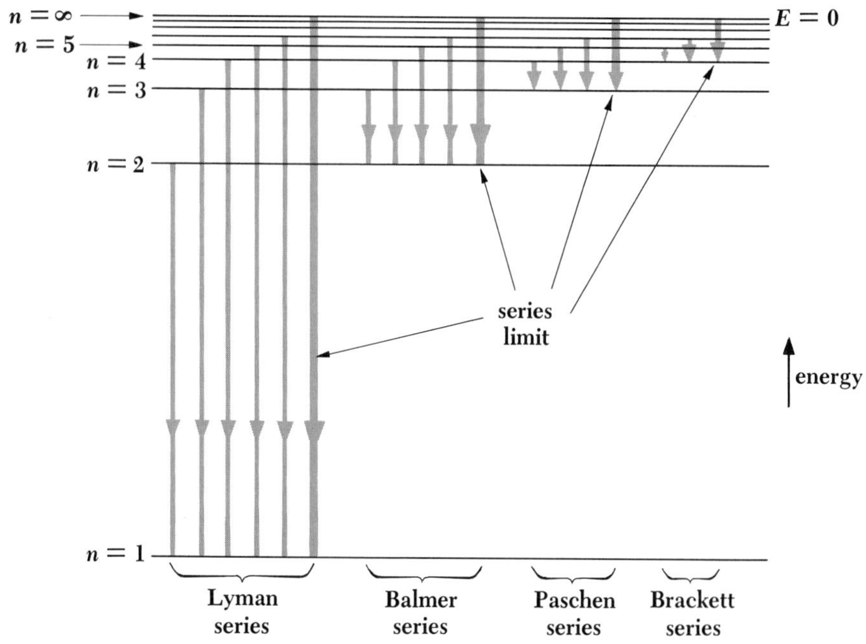

FIGURE 5-12 Spectral lines originate in transitions between energy levels.

5.4 Atomic Excitation

There are two principal mechanisms that can excite an atom to an energy level above its ground state, thereby enabling it to radiate. One mechanism is a collision with another particle during which part of their joint kinetic energy is absorbed by the atom. An atom excited in this way will return to its ground state in an average of 10^{-8} sec by emitting one or more photons. To produce an electric discharge in a rarefied gas, an electric field is established which accelerates electrons and atomic ions until their kinetic energies are sufficient to excite atoms they happen to collide with. Neon signs and mercury-vapor lamps are familiar examples of how a strong electric field applied between electrodes in a gas-filled tube leads to the emission of the characteristic spectral radiation of that gas, which happens to be reddish light in the case of neon and bluish light in the case of mercury vapor.

A different excitation mechanism is involved when an atom absorbs a photon of light whose energy is just the right amount to raise the atom to a higher energy level. For example, a photon of wavelength 6.563×10^{-7} m is emitted when a hydrogen atom in the $n = 3$ state drops to the $n = 2$ state; the absorption of a photon of wavelength 6.563×10^{-7} m by a hydrogen atom initially in the $n = 2$ state will therefore bring it up to the $n = 3$ state. This process explains the origin of absorption spectra. When white light, which contains all wavelengths, is passed through hydrogen gas, photons of those wavelengths that correspond to transitions between energy levels are absorbed. The resulting excited hydrogen atoms reradiate their excitation energy almost at once, but these photons come off in random directions with only a few in the same direction as the original beam of white light. The dark lines in an absorption spectrum are therefore never completely black, but only appear so by contrast with the bright background. We expect the absorption spectrum of any element to be identical with its emission spectrum, then, which agrees with observation.

5.5 The Franck-Hertz Experiment

Atomic spectra are not the only means of investigating the presence of discrete energy levels within atoms. A series of experiments based on the first of the excitation mechanisms of the previous section was performed by Franck and Hertz starting in 1914. These experiments provided a direct demonstration that atomic energy levels do indeed exist and, furthermore, that these levels are the same as those suggested by observations of line spectra. Franck and Hertz bombarded the vapors of various elements with electrons of known energy, using an apparatus like that shown in Fig. 5-13. A small potential difference V_0 is maintained between the grid and collecting plate, so that only electrons having energies greater than a certain minimum contribute to the current i through the galvanometer. As the accelerating potential V is increased, more and more electrons arrive at the plate and i rises (Fig. 5-14). If kinetic energy is conserved in a collision between an electron and one of the atoms in the vapor, the electron merely bounces off

FIGURE 5-13 Apparatus for the Franck-Hertz experiment.

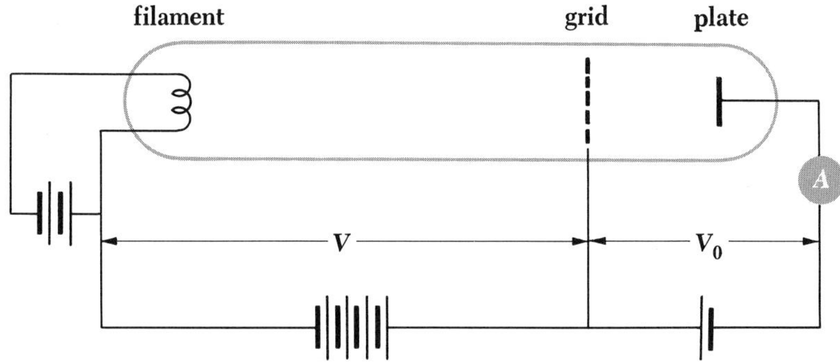

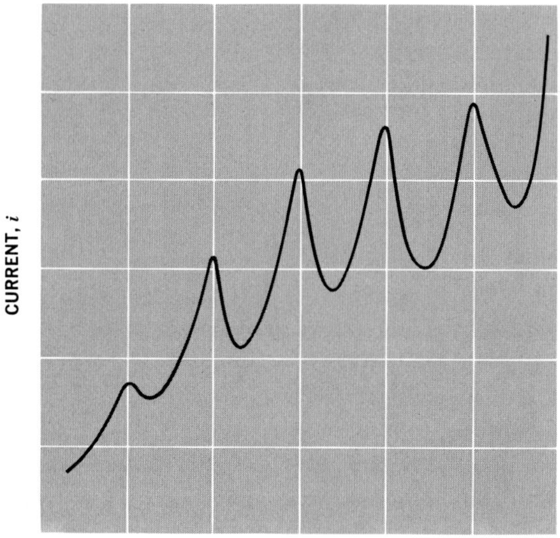

FIGURE 5-14 Results of the Franck-Hertz experiment, showing critical potentials.

CURRENT, i

ACCELERATING POTENTIAL, V

in a direction different from its original one. Because an atom is so much heavier than an electron, the latter loses almost no kinetic energy in the process. After a certain critical electron energy is reached, however, the plate current drops abruptly. The interpretation of this effect is that an electron colliding with one of the atoms gives up some or all of its kinetic energy in exciting the atom to an energy level above its ground state. Such a collision is called *inelastic*, in contrast to an *elastic* collision in which kinetic energy is conserved. The critical electron energy corresponds to the excitation energy of the atom. Then, as the accelerating potential V is raised further, the plate current again increases, since the electrons now have sufficient energy left after experiencing an inelastic collision to reach the plate. Eventually another sharp drop in plate current i occurs, which is interpreted as arising from the excitation of a higher energy level. As Fig. 5-14 indicates, a series of critical potentials for a particular atomic species is obtained in this way. (The highest potentials, of course, result from several inelastic collisions and are multiples of the lower ones.)

To check the interpretation of critical potentials as being due to discrete atomic energy levels, Franck and Hertz observed the emission spectra of vapors during electron bombardment. In the case of mercury vapor, for example, they found that a minimum electron energy of 4.9 ev was required to excite the 2,536-A spectral line of mercury—and a photon of 2,536-A light has an energy of just 4.9 ev! The Franck-Hertz experiments were performed shortly after Bohr announced his theory of the hydrogen atom, and they provided independent confirmation of his basic ideas.

5.6 The Correspondence Principle

The principles of quantum physics, so different from those of classical physics in the microscopic world that lies beyond the reach of our senses, must nevertheless yield results identical with those of classical physics in the domain where experiment indicates the latter to be valid. We have already seen that this fundamental requirement is satisfied by the theory of relativity, the quantum theory of radiation, and the wave theory of matter; we shall now show that it is satisfied also by Bohr's theory of the atom.

According to electromagnetic theory, an electron moving in a circular orbit radiates electromagnetic waves whose frequencies are equal to its frequency of revolution and to harmonics (that is, integral multiples) of that frequency. In a hydrogen atom the electron's speed is

$$v = \frac{e}{\sqrt{4\pi\varepsilon_0 m r}}$$

according to Eq. 4.13, where r is the radius of its orbit. Hence the frequency of revolution f of the electron is

$$\begin{aligned} f &= \frac{\text{electron speed}}{\text{orbit circumference}} \\ &= \frac{v}{2\pi r} \\ &= \frac{e}{2\pi\sqrt{4\pi\varepsilon_0 m r^3}} \end{aligned}$$

The radius r_n of a stable orbit is given in terms of its quantum number n by Eq. 5.8 as

$$r_n = \frac{n^2 h^2 \varepsilon_0}{\pi m e^2}$$

and so the frequency of revolution is

5.18 $$f = \frac{me^4}{8\varepsilon_0{}^2 h^3}\left(\frac{2}{n^3}\right)$$

Under what circumstances should the Bohr atom behave classically? If the electron orbit is so large that we might expect to be able to measure it directly, quantum effects should be entirely inconspicuous. An orbit 1 cm across, for example, meets this specification; its quantum number is very close to $n = 10{,}000$, and, while hydrogen atoms so grotesquely large do not actually occur because their energies would be only infinitesimally below the ionization energy, they are not prohibited in theory. What does the Bohr theory predict that such an atom will radiate? According to Eq. 5.11, a hydrogen atom dropping from the n_ith energy level to the n_fth energy level emits a photon whose frequency is

$$\nu = \frac{me^4}{8\varepsilon_0{}^2h^3}\left(\frac{1}{n_f{}^2} - \frac{1}{n_i{}^2}\right)$$

Let us write n for the initial quantum number n_i and $n - p$ (where $p = 1, 2, 3, \ldots$) for the final quantum number n_f. With this substitution,

$$\nu = \frac{me^4}{8\varepsilon_0{}^2h^3}\left(\frac{1}{(n-p)^2} - \frac{1}{n^2}\right)$$

$$= \frac{me^4}{8\varepsilon_0{}^2h^3}\left(\frac{2np - p^2}{n^2(n-p)^2}\right)$$

Now, when n_i and n_f are both very large, n is much greater than p, and

$$2np - p^2 \approx 2np$$
$$(n-p)^2 \approx n^2$$

so that

5.19 $$\nu = \frac{me^4}{8\varepsilon_0{}^2h^3}\left(\frac{2p}{n^3}\right)$$

When $p = 1$, the frequency ν of the radiation is exactly the same as the frequency of rotation f of the orbital electron given in Eq. 5.18! Harmonics of this frequency are radiated when $p = 2, 3, 4, \ldots$. Hence both quantum and classical pictures of the hydrogen atom make identical predictions in the limit of very large quantum numbers. When $n = 2$, Eq. 5.18 predicts a radiation frequency that differs from that given by Eq. 5.11 by almost 300 per cent, while when $n = 10{,}000$, the discrepancy is only about 0.01 per cent.

The requirement that quantum physics give the same results as classical physics in the limit of large quantum numbers was called by Bohr the *correspondence principle.* It has played an important role in the development of the quantum theory of matter.

5.7 Nuclear Motion and Reduced Mass

In developing the theory of the hydrogen atom, we assumed that its nucleus (a single proton) remains stationary while the orbital electron revolves around it. This is not an unreasonable assumption: the proton mass is 1,836 times greater than the electron mass, and the electrostatic force

$$F_e = \frac{1}{4\pi\varepsilon_0}\frac{e^2}{r^2}$$

that each exerts on the other has the same magnitude for both. It is nevertheless of interest to know precisely what effect nuclear motion has on the behavior of the hydrogen atom.

We begin by expressing in a different way the condition that a stable orbit in an atom consist of an integral number of electron de Broglie wavelengths. This condition states that

$$n\lambda = 2\pi r$$

and, since the de Broglie wavelength λ is given by

$$\lambda = \frac{h}{mv}$$

we may equivalently write

5.20 $$mvr = \frac{nh}{2\pi}$$

The quantity mvr we recognize as the *angular momentum* of the electron in its orbit. Hence an alternate expression of Bohr's first postulate is that the angular momentum of a hydrogen atom must be an integral multiple of $h/2\pi$. In terms of the angular velocity ω of the electron, the quantization rule of Eq. 5.20 is stated

5.21 $$m\omega r^2 = \frac{nh}{2\pi}$$

since

$$\omega = \frac{v}{r}$$

If the nucleus has a mass M that is not infinite, both it and its orbital electron revolve around a common *center of mass.* The center of mass of a body is its balance point; if we think of the nucleus and electron in a hydrogen atom as being at opposite ends of a massless rod r long, in effect constituting a lopsided dumbbell (Fig. 5-15), the center of mass, which is r_N distant from the nucleus and r_E distant from the electron, may be found from the requirement that

5.22 $$mr_E = Mr_N$$

where

5.23 $$r = r_E + r_N$$

The total angular momentum of the hydrogen atom is the sum of the angular momentum of the electron and that of the nucleus:

$$\text{Total angular momentum} = \text{electron angular momentum} + \text{nuclear angular momentum}$$
$$= m\omega r_E{}^2 + M\omega r_N{}^2$$

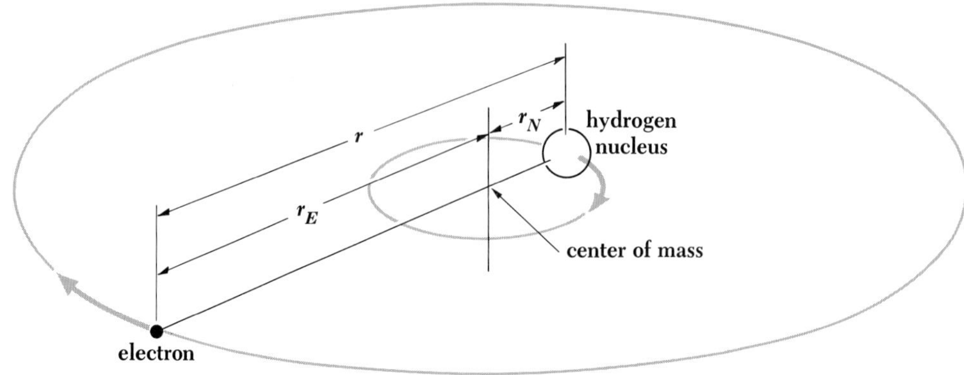

FIGURE 5-15 Both the electron and nucleus of a hydrogen atom revolve around a common center of mass.

The angular velocity ω is the same for both particles. According to Bohr's first postulate, the total angular momentum of the atom must be an integral multiple of $h/2\pi$, and so

5.24 $$m\omega {r_E}^2 + M\omega {r_N}^2 = \frac{nh}{2\pi}$$

From Eqs. 5.22 and 5.23 we find that

5.25 $$r_E = \left(\frac{M}{M+m}\right) r$$

5.26 $$r_N = \left(\frac{m}{M+m}\right) r$$

and so Eq. 5.24 becomes

5.27 $$\left(\frac{mM}{M+m}\right)\omega r^2 = \frac{nh}{2\pi}$$

The condition for force balance, Eq. 4.12, must also be generalized to take into account nuclear motion. Its proper expression is

5.28 $$m\omega^2 r_E = \frac{1}{4\pi\varepsilon_0}\frac{e^2}{r^2}$$

since the centripetal force on the electron depends upon the radius of its orbit r_E while the electrostatic force depends upon the separation r between electron and nucleus. Substituting for r_E from Eq. 5.25, we find that

5.29 $$\left(\frac{mM}{M+m}\right)\omega^2 r = \frac{1}{4\pi\varepsilon_0}\frac{e^2}{r^2}$$

If we let

5.30 $$m' = \left(\frac{mM}{M+m}\right)$$ **Reduced mass**

we see that Eqs. 5.27 and 5.29 become respectively

5.31 $$m'\omega r^2 = \frac{nh}{2\pi}$$

and

5.32 $$m'\omega^2 r = \frac{1}{4\pi\varepsilon_0}\frac{e^2}{r^2}$$

From Eqs. 5.31 and 5.32 we can proceed as we did earlier in this chapter to obtain an expression for the energy levels of the hydrogen atom. The result is

5.33 $$E_n = -\frac{m'e^4}{8\varepsilon_0{}^2h^2}\left(\frac{1}{n^2}\right)$$

which is exactly the same as Eq. 5.9 except that m' is present instead of the electron mass m. The quantity m' is known as the *reduced mass* of the electron because its value is less than m. Owing to motion of the nucleus, all of the energy levels of hydrogen are changed by the fraction

$$\begin{aligned}\frac{m'}{m} &= \frac{M}{M+m}\\ &= \frac{1{,}836}{1{,}837}\\ &= 0.99945\end{aligned}$$

an increase of 0.055 per cent since the energies E_n, being smaller in absolute value, are therefore less negative. The use of Eq. 5.33 in place of 5.9 removes a small but definite discrepancy between the predicted wavelengths of the various spectral lines of hydrogen and the actual experimentally determined wavelengths. The value of the Rydberg constant R to eight significant figures without correcting for nuclear motion is $1.0973731 \times 10^7\ \text{m}^{-1}$; the correction lowers it to $1.0967758 \times 10^7\ \text{m}^{-1}$.

The notion of reduced mass played an important part in the discovery of *deuterium*, an isotope of hydrogen whose atomic mass is almost exactly double that of ordinary hydrogen owing to the presence of a neutron as well as a proton in the nucleus. Because of the greater nuclear mass, the spectral lines of deuterium are all shifted slightly to wavelengths shorter than those of ordinary hydrogen. The H_α line of deuterium, for example, has a wavelength of 6,561 A, while that of hydrogen is 6,563 A: a small but definite difference, sufficient for the identification of deuterium.

5.8 Hydrogenic Atoms

The Bohr theory can be applied to ions containing single electrons as well as to hydrogen atoms. That is, He^+ (singly ionized helium), Li^{++} (doubly ionized lithium), and so on behave just like hydrogen except for the effects of greater nuclear charge and mass. They are accordingly called *hydrogenic atoms.* We have already seen how to take nuclear mass into account through the substitution of the reduced mass m' for the electron mass m. The nucleus of a hydrogenic atom of atomic number Z carries the charge ^+Ze, and so the electrostatic force it exerts on the orbital electron is

5.34 $$F_e = \frac{1}{4\pi\varepsilon_0}\frac{Ze^2}{r^2}$$

instead of simply $e^2/4\pi\varepsilon_0 r^2$ as in the case of hydrogen, where $Z = 1$. Consequently the formula for the energy levels of a hydrogenic atom is

5.35 $$E_n = -\frac{m'Z^2e^4}{8\varepsilon_0{}^2h^2}\left(\frac{1}{n^2}\right)$$ **Hydrogenic atoms**

These levels are different by a factor of Z^2 from those of the hydrogen atom. Figure 5-16 shows the energy levels of He^+ and H together with some possible transitions in each; evidently certain spectral lines in He^+ should correspond closely to lines found in H. Since the general formula for the spectral lines of a hydrogenic atom of atomic number Z is

5.36 $$\frac{1}{\lambda} = \frac{m'Z^2e^4}{8\varepsilon_0{}^2ch^3}\left(\frac{1}{n_f{}^2} - \frac{1}{n_i{}^2}\right)$$

a transition from the $n = 4$ state of He^+ to its $n = 2$ state involves the emission of a photon whose wavelength is nearly the same as that emitted in a transition from the $n = 2$ state of H to its $n = 1$ state. The small difference between the wavelengths arises from the difference between the nuclear masses of He^+ and H, which affects the value of the reduced mass m'.

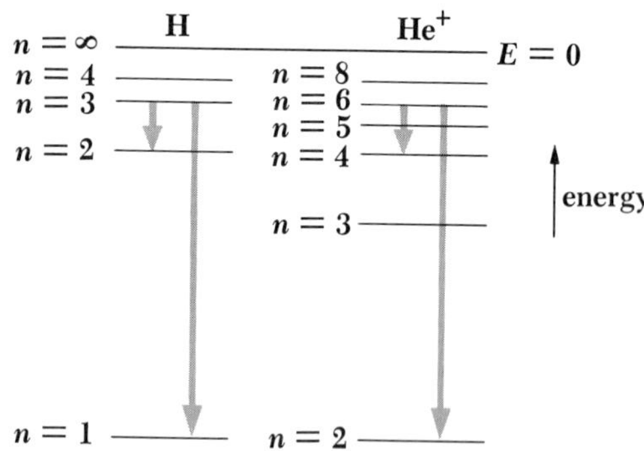

FIGURE 5-16 Energy levels in hydrogen and singly ionized helium.

Problems

1. Determine the value of the longest wavelength found in the Paschen series.

2. Find the wavelength of the spectral line corresponding to the transition in hydrogen from the $n = 6$ state to the $n = 3$ state.

3. Find the wavelength of the photon emitted when a hydrogen atom goes from the $n = 10$ state to its ground state.

4. How much energy is required to remove an electron in the $n = 2$ state from a hydrogen atom?

5. A beam of electrons bombards a sample of hydrogen. Through what potential difference must the electrons have been accelerated if the first line of the Balmer series is to be emitted?

6. A μ^- meson ($m = 207\ m_e$) can be captured by a proton to form a "mesic atom." Find the radius of the first Bohr orbit of such an atom.

7. A μ^- meson is in the $n = 2$ state of a titanium atom. Find the energy radiated when the mesic atom drops to its ground state.

8. Find the recoil speed of a hydrogen atom after it emits a photon in going from the $n = 4$ state to the $n = 1$ state.

9. How many revolutions does an electron in the $n = 2$ state of a hydrogen atom make before dropping to the $n = 1$ state? (The average lifetime of an excited state is about 10^{-8} sec.)

10. The average lifetime of an excited atomic state is 10^{-8} sec. If the wavelength of the spectral line associated with the decay of this state is 5,000 A, find the width of the line.

11. At what temperature will the average molecular kinetic energy in gaseous hydrogen equal the binding energy of a hydrogen atom?

12. An electron joins a bare helium nucleus to form a He^+ ion. Find the wavelength of the photon emitted in this process if the electron is assumed to have had no kinetic energy when it combined with the nucleus.

13. A mixture of ordinary hydrogen and tritium, a hydrogen isotope whose nucleus is approximately three times more massive than ordinary hydrogen, is excited and its spectrum observed. How far apart in wavelength will the H_α lines of the two kinds of hydrogen be?

14. A positronium atom is a system consisting of a positron (positive electron) and an electron. (*a*) Compare the wavelength of the photon emitted in the $n = 3 \rightarrow n = 2$ transition in positronium with that of the H_α line. (*b*) Compare the ionization energy in positronium with that in hydrogen.

15. Use the uncertainty principle to determine the ground-state radius r_1 of the hydrogen atom in the following way. First find a formula for the electron kinetic energy in terms of the momentum an electron must have if confined to a region of linear dimension r_1. Add this kinetic energy to the electrostatic potential energy of an electron the distance r_1 from a proton, and differentiate with respect to r_1 the resulting expression for the total electron energy E to find the value of r_1 for which E is a minimum. Compare the result with that given by Eq. 5.8 with $n = 1$.

QUANTUM MECHANICS 

The Bohr theory of the atom, discussed in the previous chapter, is able to account for certain experimental data in a convincing manner, but it has a number of severe limitations. While the Bohr theory correctly predicts the spectral series of hydrogen, hydrogen isotopes, and hydrogenic atoms, it is incapable of being extended to treat the spectra of complex atoms having two or more electrons each; it can give no explanation of why certain spectral lines are more intense than others (that is, why certain transitions between energy levels have greater probabilities of occurrence); and it cannot account for the observation that many spectral lines actually consist of several separate lines whose wavelengths differ slightly. And, perhaps most important, it does not permit us to obtain what a really successful theory of the atom should make possible: an understanding of how individual atoms interact with one another to endow macroscopic aggregates of matter with the physical and chemical properties we observe.

These objections to the Bohr theory are not put forward in an unfriendly way, for the theory was one of those seminal achievements that transform scientific thought, but rather to emphasize that an approach to atomic phenomena of greater generality is required. Such an approach was developed in 1925–1926 by Erwin Schrödinger, Werner Heisenberg, and others under the apt name of *quantum mechanics*. By the early 1930s the application of quantum mechanics to problems involving nuclei, atoms, molecules, and matter in the solid state made it possible to understand a vast body of otherwise-puzzling data and—a vital attribute of any theory—led to predictions of remarkable accuracy.

6.1 Introduction

The fundamental difference between Newtonian mechanics and quantum mechanics lies in what it is that they describe. Newtonian mechanics is concerned with the motion of a particle under the influence of applied forces, and it takes for granted that such quantities as the particle's position, mass, velocity, acceleration, etc., can be measured. This assumption is, of course, completely valid in our everyday experience, and Newtonian mechanics provides the "correct" explanation for the behavior of moving bodies in the sense that the values it predicts for observable magnitudes agree with the measured values of those magnitudes.

Quantum mechanics, too, consists of relationships between observable magnitudes, but the uncertainty principle radically alters the definition of "observable magnitude" in the atomic realm. According to the uncertainty principle, the position and momentum of a particle cannot be accurately measured at the same time, while in Newtonian mechanics both are assumed to have definite, ascertainable values at every instant. The quantities whose relationships quantum mechanics explores are *probabilities.* Instead of asserting, for example, that the radius of the electron's orbit in a ground-state hydrogen atom is always exactly 5.3×10^{-11} m, quantum mechanics states that this is the *most probable* radius; if we conduct a suitable experiment, most trials will yield a different value, either larger or smaller, but the value most likely to be found will be 5.3×10^{-11} m.

At first glance quantum mechanics seems a poor substitute for Newtonian mechanics, but closer inspection reveals a striking fact: *Newtonian mechanics is nothing but an approximate version of quantum mechanics.* The certainties proclaimed by Newtonian mechanics are illusory, and their agreement with experiment is a consequence of the fact that macroscopic bodies consist of so many individual atoms that departures from average behavior are unnoticeable. Instead of two sets of physical principles, one for the macroscopic universe and one for the microscopic universe, there is only a single set, and quantum mechanics represents our best effort to date in formulating it.

6.2 The Wave Equation

As mentioned in Chap. 3, the quantity with which quantum mechanics is concerned is the *wave function* Ψ of a body. While Ψ itself has no physical interpretation, its square Ψ^2 (or $\Psi\Psi^*$ if Ψ is complex) evaluated at a particular point at a particular time is proportional to the probability of experimentally finding the body there at that time. The problem of quantum mechanics is to determine Ψ for a body when its freedom of motion is limited by the action of external forces.

Even before we consider the actual calculation of Ψ, we can establish certain requirements it must always fulfill. For one thing, since Ψ^2 is proportional to the probability P of finding the body described by Ψ, the integral of Ψ^2 over all space must be finite—the body is *somewhere,* after all. If

6.1 $$\int_{-\infty}^{\infty} \Psi^2 \, dV$$

is 0, the particle does not exist, and if it is ∞, the particle is everywhere simultaneously; Ψ^2 cannot be negative or complex because of the way it is defined, and so the only possibility left is that its integral be a finite quantity if Ψ is to describe properly a real body. Another obvious requirement is that Ψ be single-valued, since P can have only one value at a particular place and time. A further condition that Ψ must obey is that it and its partial derivatives $\partial\Psi/\partial x$, $\partial\Psi/\partial y$, $\partial\Psi/\partial z$ be continuous everywhere.

Schrödinger's equation, which is the fundamental equation of quantum mechanics in the same sense that the second law of motion is the fundamental equation of Newtonian mechanics, is a wave equation in the variable Ψ. It is appropriate for us to review the nature and solution of a simpler wave equation, that governing wave propagation along a stretched string, before we tackle Schrödinger's equation itself.

We shall consider a stretched string lying on the x axis whose vibrations take place in the xy plane (Fig. 6-1). For simplicity, our sole concern will be with waves traveling in the $+x$ direction, though the analysis may be readily extended to waves traveling in the opposite direction as well. According to Eq. 3.5, the displacement y of the string at the position x and the time t is given by

6.2 $$y = A \cos \omega \left(t - \frac{x}{v} \right)$$

where ω is the angular frequency of the waves, v their speed, and A their amplitude. (This formula assumes that at time $t = 0$, $y = A$ at $x = 0$.)

While Eq. 6.2 completely describes the actual motion of a wave along a stretched string, it is not the most general possible solution of the differential

FIGURE 6-1 Waves in the xy plane traveling in the $+x$ direction along a stretched string lying on the x axis.

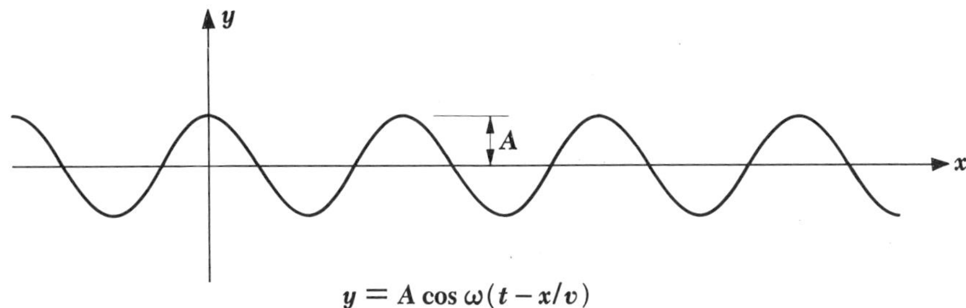

equation governing the wave. If we have forgotten this equation, a quick way to obtain it is to first differentiate y twice with respect to x, which yields

6.3 $$\frac{\partial^2 y}{\partial x^2} = -\frac{\omega^2 A}{v^2} \cos \omega \left(t - \frac{x}{v}\right)$$

Next we differentiate y twice with respect to t to obtain

6.4 $$\frac{\partial^2 y}{\partial t^2} = -\omega^2 A \cos \omega \left(t - \frac{x}{v}\right)$$

Combining Eqs. 6.3 and 6.4, we arrive at

6.5 $$\frac{\partial^2 y}{\partial x^2} = \frac{1}{v^2}\frac{\partial^2 y}{\partial t^2}$$ **Wave equation**

which is the *wave equation* for waves on a stretched string. Equation 6.5 should properly be derived directly from the physics of the situation, of course, as is done in most textbooks of mechanics, but to elaborate such a derivation here would only divert us from our immediate goal.

While Eq. 6.2 is *a* solution to Eq. 6.5, it is not the whole story. The portion of the entire solution of the wave equation for waves in the $+x$ direction is

6.6 $$y = Ae^{-i\omega(t-x/v)}$$

Hence y is a complex quantity, with a real and an imaginary part. Because

6.7 $$e^{-i\theta} = \cos \theta - i \sin \theta$$

the above formula may be written

6.8 $$y = A \cos \omega \left(t - \frac{x}{v}\right) - iA \sin \omega \left(t - \frac{x}{v}\right)$$

Since the displacement y of a stretched string when a wave passes by is a physically meaningful quantity, only the real part of Eq. 6.8 has significance, and the imaginary part is discarded as irrelevant.

6.3 Schrödinger's Equation: Time-dependent Form

In quantum mechanics the wave function Ψ corresponds to the displacement y of wave motion in a string. However, Ψ, unlike y, is not itself a measurable quantity and may therefore be complex. For this reason we shall assume that Ψ is specified in the x direction by

6.9 $$\Psi = Ae^{-i\omega(t-x/v)}$$

When we replace ω in the above formula by $2\pi\nu$ and v by $\lambda\nu$, we obtain

6.10 $$\Psi = Ae^{-2\pi i(\nu t - x/\lambda)}$$

which is convenient since we already know what ν and λ are in terms of the total energy E and momentum p of the particle being described by Ψ. Since

$$E = h\nu$$

and

$$\lambda = h/p$$

we have

6.11 $$\Psi = Ae^{-(2\pi i/h)(Et - px)}$$

Equation 6.11 is a mathematical description of the wave equivalent of an unrestricted particle of total energy E and momentum p moving in the $+x$ direction, just as Eq. 6.2 is a mathematical description of a displacement wave moving freely along a stretched string. There are circumstances in which waves may exist in a string that obey the differential equation 6.5 and yet are not described by Eq. 6.2. An important example is illustrated in Fig. 6-2, which shows the possible *standing wave* patterns that occur when a segment of a string is fastened at both ends. Here, instead of a single wave propagating indefinitely in the $+x$ direction, waves are traveling in both the $+x$ and $-x$ directions simultaneously subject to the condition that the displacement y always be zero at both ends of the string. The patterns of vibration that can exist are limited by this condition to the ones shown in Fig. 6-2. It is the *combination* of the fundamental wave equation

$$\frac{\partial^2 y}{\partial x^2} = \frac{1}{v^2}\frac{\partial^2 y}{\partial t^2}$$

and the restrictions externally imposed upon the motion that permits us to determine y as a function of x and t.

In the simplest case, where there are no restrictions on y, the waves in the $+x$ direction in the string obey the equation

$$y = A \cos \omega\left(t - \frac{x}{v}\right)$$

Despite its lack of complete generality this equation must be a valid solution of the wave equation since it is experimentally observed, and therefore it is legitimate to obtain the wave equation, Eq. 6.5, by differentiating.

The expression for the wave function Ψ given by Eq. 6.11 is correct only for freely moving particles, while we are most interested in situations where the motion of a particle is subject to various restrictions. An important con-

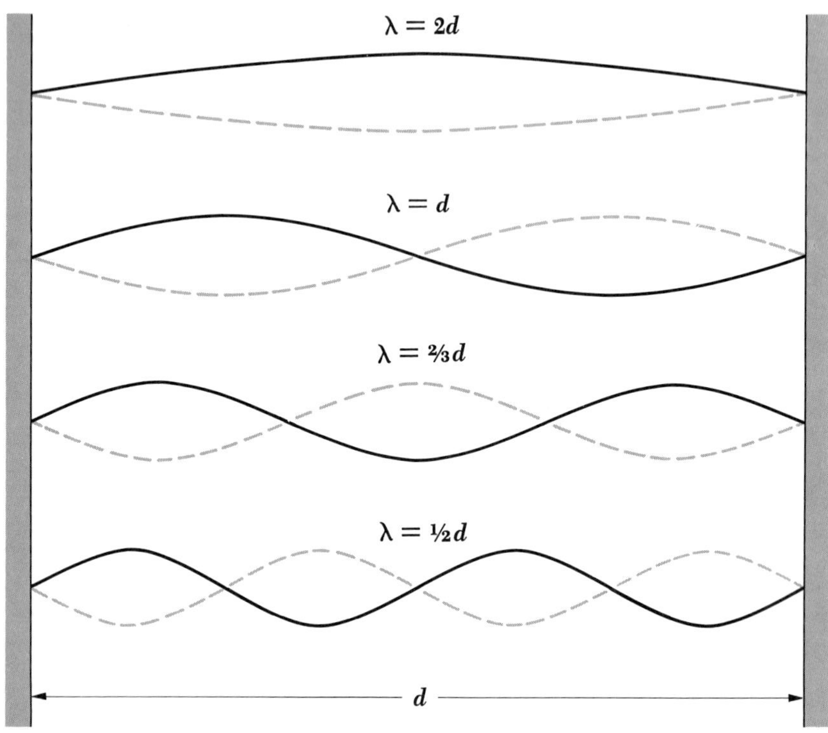

FIGURE 6-2 Standing waves in a stretched string fastened at both ends.

cern, for example, is an electron bound to an atom by the electric field of its nucleus. What we must now do is obtain the fundamental differential equation for Ψ, which we can then solve in a specific situation. We begin by differentiating Eq. 6.11 twice with respect to x, yielding

6.12 $$\frac{\partial^2 \Psi}{\partial x^2} = -\frac{4\pi^2 p^2}{h^2}\Psi$$

and once with respect to t, yielding

6.13 $$\frac{\partial \Psi}{\partial t} = -\frac{2\pi i E}{h}\Psi$$

At speeds small compared with that of light, the total energy E of a particle is the sum of its kinetic energy $p^2/2m$ and its potential energy V, where V is in general a function of position x and time t:

6.14 $$E = \frac{p^2}{2m} + V$$

Multiplying both sides of this equation by the wave function Ψ,

6.15 $$E\Psi = \frac{p^2\Psi}{2m} + V\Psi$$

From Eqs. 6.12 and 6.13 we see that

6.16 $$E\Psi = -\frac{h}{2\pi i}\frac{\partial\Psi}{\partial t}$$

and

6.17 $$p^2\Psi = -\frac{h^2}{4\pi^2}\frac{\partial^2\Psi}{\partial x^2}$$

Substituting these expressions for $E\Psi$ and $p^2\Psi$ into Eq. 6.15, we obtain

6.18 $$\frac{h}{2\pi i}\frac{\partial\Psi}{\partial t} = \frac{h^2}{8\pi^2 m}\frac{\partial^2\Psi}{\partial x^2} - V\Psi$$

Equation 6.18 is the *time-dependent form of Schrödinger's equation.* In three dimensions the time-dependent form of Schrödinger's equation is

6.19 $$\frac{h}{2\pi i}\frac{\partial\Psi}{\partial t} = \frac{h^2}{8\pi^2 m}\left(\frac{\partial^2\Psi}{\partial x^2} + \frac{\partial^2\Psi}{\partial y^2} + \frac{\partial^2\Psi}{\partial z^2}\right) - V\Psi$$ **Time-dependent Schrödinger's equation**

where the particle's potential energy V is some function of x, y, z, and t. Any restrictions that may be present on the particle's motion will affect the potential-energy function V. Once V is known, Schrödinger's equation may be solved for the wave function Ψ of the particle, from which its probability density Ψ^2 may be determined for a specified x, y, z, t.

6.4 Schrödinger's Equation: Steady-state Form

In a great many problems the potential energy V of the particle does not depend upon time t explicitly; the forces that act upon it, and hence its potential energy, vary with the position of the particle only. When this is true, Eq. 6.19 may be simplified by removing all reference to t. We note from Eq. 6.11 that the one-dimensional wave function Ψ of an unrestricted particle may be written

$$\begin{aligned}\Psi &= Ae^{-(2\pi i/h)(Et-px)}\\ &= Ae^{-(2\pi iE/h)t}e^{+(2\pi ip/h)x}\end{aligned}$$

6.20 $$= \psi e^{-(2\pi iE/h)t}$$

That is, Ψ is the product of a time-dependent function $e^{-(2\pi iE/h)t}$ and a position-dependent function ψ. As it happens, the time variations of *all* functions of particles acted upon by stationary forces have the same form as that of an unrestricted particle. Substituting the Ψ of Eq. 6.20 into the time-dependent form of Schrödinger's equation, we find that

$$-E\psi e^{-(2\pi iE/h)t} = \frac{h^2}{8\pi^2 m} e^{-(2\pi iE/h)t} \frac{\partial^2\psi}{\partial x^2} - V\psi e^{-(2\pi iE/h)t}$$

and so, dividing through by the common exponential factor,

$$\frac{\partial^2\psi}{\partial x^2} + \frac{8\pi^2 m}{h^2}(E - V)\,\psi = 0 \tag{6.21}$$

Equation 6.21 is the *steady-state form of Schrödinger's equation.* In three dimensions it is

$$\frac{\partial^2\psi}{\partial x^2} + \frac{\partial^2\psi}{\partial y^2} + \frac{\partial^2\psi}{\partial z^2} + \frac{8\pi^2 m}{h^2}(E - V)\psi = 0 \tag{6.22}$$

Steady-state Schrödinger's equation

6.5 The Particle in a Box: Energy Quantization

To solve Schrödinger's equation, even in its simpler steady-state form, usually requires sophisticated mathematical techniques. For this reason the study of quantum mechanics has traditionally been reserved for advanced students who have the required proficiency in mathematics. However, since quantum mechanics is the theoretical structure whose results are closest to experimental reality, we must explore its methods and applications if we are to achieve any understanding of modern physics. As we shall see, even a relatively limited mathematical background is sufficient for us to follow the trains of thought that have led quantum mechanics to its greatest achievements.

Our first problem using Schrödinger's equation is that of a particle bouncing back and forth between the walls of a box (Fig. 6-3). Our interest in this problem is threefold: to see how Schrödinger's equation is solved when the motion of a particle is subject to restrictions; to learn the characteristic properties of solutions of this equation, such as the limitation of particle energy to certain specific values only; and to compare the predictions of quantum mechanics with those of Newtonian mechanics.

We may specify the particle's motion by saying that it is restricted to traveling along the x axis between $x = 0$ and $x = L$ by infinitely hard walls. A particle does not lose energy when it collides with such walls, so that its total energy stays constant. From the formal point of view of quantum mechanics, the potential energy V of the particle is infinite on both sides of the box, while V is a constant—say 0 for convenience—on the inside. Since the particle cannot have an infinite amount of energy, it cannot exist outside the box, and so its wave function ψ is 0 for $x \leq 0$ and $x \geq L$. Our task is to find what ψ is within the box, namely, between $x = 0$ and $x = L$.

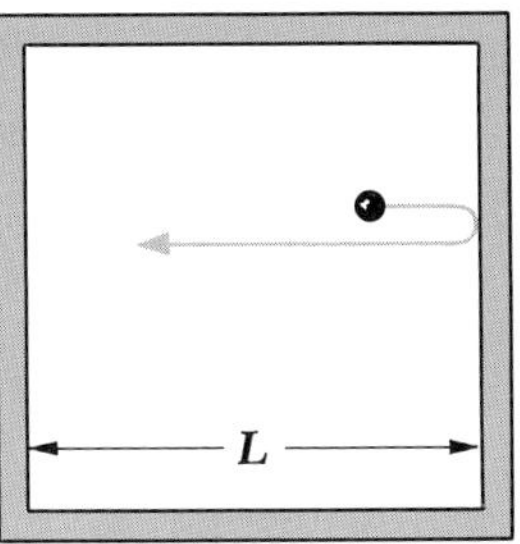

FIGURE 6-3 A particle confined to a box of width L.

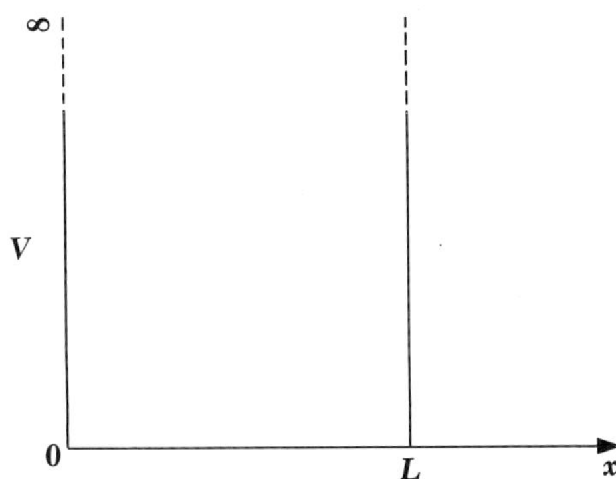

Within the box Schrödinger's equation becomes

6.23 $$\frac{d^2\psi}{dx^2} + \frac{8\pi^2 mE\psi}{h^2} = 0$$

since $V = 0$ there. This equation has the two possible solutions

6.24 $$\psi = A \cos \sqrt{\frac{8\pi^2 mE}{h^2}}\, x$$

6.25 $$\psi = B \sin \sqrt{\frac{8\pi^2 mE}{h^2}}\, x$$

which we can verify by substitution into Eq. 6.23; their sum is also a solution. A and B are constants to be evaluated. These solutions are subject to the important condition that $\psi = 0$ for $x = 0$ and $x = L$. Since $\cos 0° = 1$, the first solution cannot describe the particle because it does not permit ψ to be 0 at $x = 0$. Hence we conclude that $A = 0$. Since $\sin 0° = 0$, the second solution always yields $\psi = 0$ at $x = 0$, as required, but ψ will be 0 at $x = L$ only when

$$\sqrt{\frac{8\pi^2 mE}{h^2}}\, L = \pi, 2\pi, 3\pi, \ldots$$

6.26 $$= n\pi \qquad n = 1, 2, 3, \ldots$$

This result comes about because sines of the angles π, 2π, 3π, etc., are all 0.

From Eq. 6.26 it is clear that the energy of the particle can have only certain values. These values, constituting its *energy levels*, are

$$E_n = \frac{n^2h^2}{8mL^2} \qquad n = 1, 2, 3, \ldots \tag{6.27}$$

Particle in a box

The integer n corresponding to the energy level E_n is called its *quantum number.* A particle confined to a box cannot have an arbitrary energy: the fact of its confinement leads to restrictions on its wave function that permit it to have only those energies specified by Eq. 6.27.

It is significant that the electron cannot have zero energy; if it did, the electron wave function ψ would have to be zero everywhere in the box, and this means that the electron cannot be present there. The exclusion of $E = 0$ as a possible value for the energy of a trapped electron, like the limitation of E to a discrete set of definite values, is a quantum-mechanical result that has no counterpart in classical mechanics, where all energies, including zero, are presumed possible.

Why are we not aware of energy quantization in our own experience? Surely a marble rolling back and forth between the sides of a level box with a smooth floor can have any speed, and therefore any energy, we choose to give it, including zero. In order to assure ourselves that Eq. 6.27 does not conflict with our direct observations while providing unique insights on a microscopic scale, we shall compute the permitted energy levels of (1) an electron in a box 1 A wide and (2) a 10-g marble in a box 10 cm wide.

In case 1 we have $m = 9.1 \times 10^{-31}$ kg and $L = 1$ A $= 10^{-10}$ m, so that the permitted electron energies are

$$\begin{aligned} E_n &= \frac{n^2 \times (6.63 \times 10^{-34} \text{ joule-sec})^2}{8 \times 9.1 \times 10^{-31} \text{ kg} \times (10^{-10} \text{ m})^2} \\ &= 6 \times 10^{-18} n^2 \text{ joule} \\ &= 38n^2 \text{ ev} \end{aligned}$$

The minimum energy the electron can have is 38 ev, corresponding to $n = 1$. The sequence of energy levels continues with $E_2 = 152$ ev, $E_3 = 342$ ev, $E_4 = 608$ ev, and so on (Fig. 6-4). These energy levels are sufficiently far apart to make the quantization of electron energy in such a box conspicuous if such a box actually did exist.

In case 2 we have $m = 10$ g $= 10^{-2}$ kg and $L = 10$ cm $= 10^{-1}$ m, so that the permitted marble energies are

$$\begin{aligned} E_n &= \frac{n^2 \times (6.63 \times 10^{-34} \text{ joule-sec})^2}{8 \times 10^{-2} \text{ kg} \times (10^{-1} \text{ m})^2} \\ &= 5.5 \times 10^{-64} n^2 \text{ joule} \end{aligned}$$

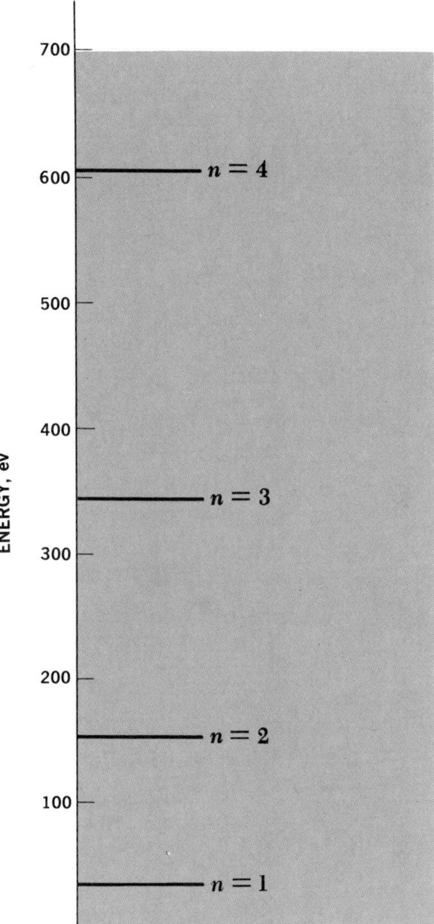

FIGURE 6-4 Energy levels of an electron confined to a box 1 A wide.

The minimum energy the marble can have is 5.5×10^{-64} joule, corresponding to $n = 1$. A marble with this kinetic energy has a speed of only 3.3×10^{-31} m/sec and is therefore experimentally indistinguishable from a stationary marble. A reasonable speed a marble might have is, say, ⅓ m/sec—which corresponds to the energy level of quantum number $n = 10^{30}$! The permissible energy levels are so very close together, then, that there is no way of determining whether the marble can take on only those energies predicted by Eq. 6.27 or any energy whatever. Hence in the domain of everyday experience quantum effects are imperceptible; this accounts for the success in this domain of Newtonian mechanics.

6.6 The Particle in a Box: Wave Functions

In the previous section we found that the wave function of a particle in a box whose energy is E is

$$\psi = B \sin \sqrt{\frac{8\pi^2 mE}{h^2}}\, x$$

Since the possible energies are

$$E_n = \frac{n^2 h^2}{8mL^2}$$

substituting E_n for E yields

6.28 $$\psi = B \sin \frac{n\pi x}{L}$$

for the wave functions corresponding to the energies E_n. It is easy to verify that these wave functions meet all the requirements we have discussed: for each quantum number n, ψ_n is a single-valued function of x, and ψ_n and $\partial\psi_n/\partial x$ are continuous. Furthermore, the integral of ψ_n^2 over all space is finite, as we can see by integrating $\psi_n^2\, dx$ from $x = 0$ to $x = L$ (since the particle, by hypothesis, is confined within these limits):

6.29 $$\begin{aligned}\int_{-\infty}^{\infty} \psi_n^2\, dx &= \int_0^L \psi_n^2\, dx \\ &= B^2 \int_0^L \sin^2\left(\frac{n\pi x}{L}\right) dx \\ &= B^2 \frac{L}{2}\end{aligned}$$

It is possible to assign a value to B such that ψ_n^2 is *equal* to the probability P of finding the particle at x, rather than merely proportional to P. If ψ_n^2 is to equal P, then it must be true that

6.30 $$\int_{-\infty}^{\infty} \psi_n^2\, dx = 1$$ **Normalization**

since

$$\int_{-\infty}^{\infty} P\, dx = 1$$

is the mathematical way of stating that the particle exists somewhere at all times. A wave function that obeys Eq. 6.30 is said to be *normalized*. Comparing Eqs. 6.29 and 6.30, we see that the wave functions of a particle in a box will be normalized if

6.31 $$B = \sqrt{\frac{2}{L}}$$

The normalized wave functions of the particle are therefore

6.32 $$\psi_n = \sqrt{\frac{2}{L}} \sin \frac{n\pi x}{L}$$

The normalized wave functions ψ_1, ψ_2, and ψ_3 together with the probability densities ψ_1^2, ψ_2^2, and ψ_3^2 are plotted in Fig. 6-5. While ψ_n may be negative as well as positive, ψ_n^2 is always positive and, since ψ_n is normalized, its value at a given x is equal to the probability P of finding the particle there. In every case $\psi_n^2 = 0$ at $x = 0$ and $x = L$, the boundaries of the box. At a particular point in the box the probability of the particle being present may be very different for different quantum numbers. For instance, ψ_1^2 has its maximum value of $\frac{2}{L}$ in the middle of the box, while $\psi_2^2 = 0$ there: a particle in the lowest energy level of $n = 1$ is most likely to

FIGURE 6-5 Wave functions and probability densities of a particle confined to a box with rigid walls.

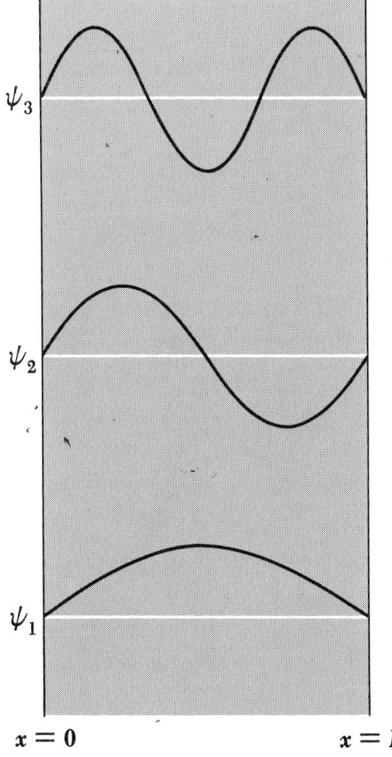

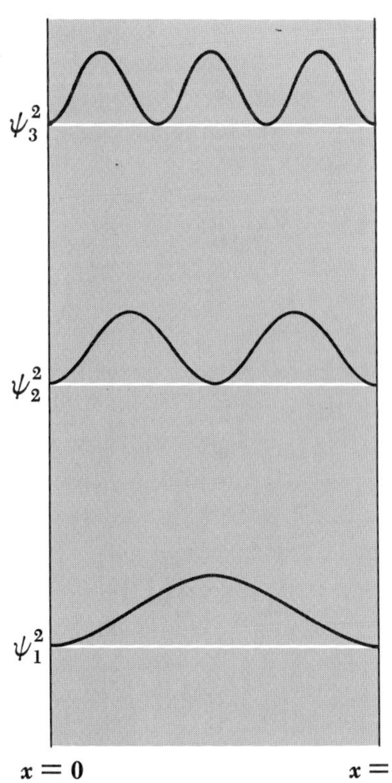

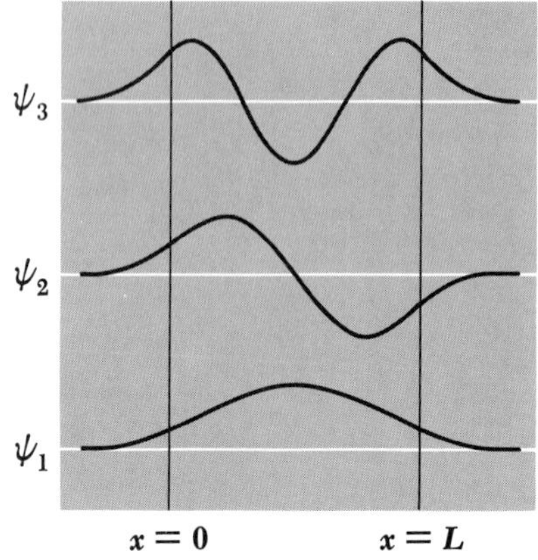

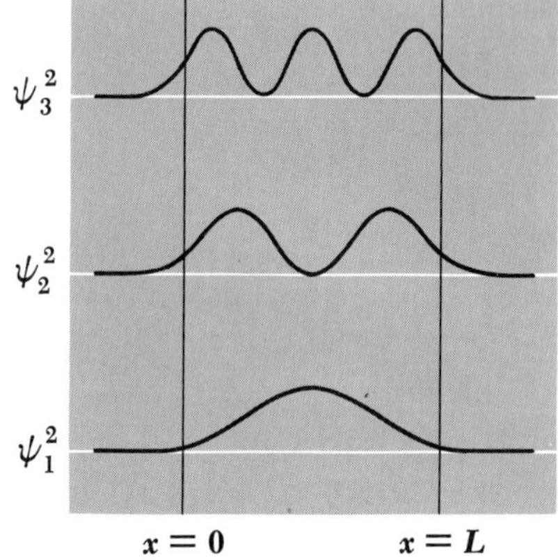

FIGURE 6-6 Wave functions and probability densities of a particle confined to a box with nonrigid walls.

be in the middle of the box, while a particle in the next higher state of $n = 2$ is *never* there! Classical physics, of course, predicts the same probability for the particle being anywhere in the box.

The wave functions shown in Fig. 6-5 resemble the possible vibrations of a string fixed at both ends, such as those of the stretched string of Fig. 6-2. This is a consequence of the fact that waves in a stretched string and the wave representing a moving particle obey wave equations of the same form, so that, when identical restrictions are placed upon each kind of wave, the solutions are identical.

6.7 The Particle in a Nonrigid Box

It is interesting to solve the problem of the particle in a box when the walls of the box are no longer assumed to be infinitely rigid. In this case the potential energy V outside the box is a finite quantity; the corresponding situation in the case of a vibrating string would involve an imperfect attachment of the string at each end, so that the ends can move slightly. This problem is more difficult to treat, and we shall simply present the result here. (We shall take another look at a particle in a nonrigid box when we examine the theory of the deuteron in Chap. 12.) The first few wave functions for a particle in such a box are shown in Fig. 6-6. The wave functions ψ_n now do *not* equal zero outside the box. Even though the particle's energy is smaller than the value of V outside the box, there is still a definite probability that it be found outside it! In other words, even though the particle does not have enough energy to break through the walls of the box according to "common sense," it may nevertheless somehow penetrate them. This peculiar situation is readily understandable in terms of the uncertainty principle. Because the uncertainty Δp in a particle's momentum is related to the uncertainty Δx in its position by the formula

$$\Delta p \, \Delta x \geqslant h/2\pi$$

an infinite uncertainty in particle momentum outside the box is the price of definitely establishing that the particle is never there. A particle requires an infinite amount of energy if its momentum is to have an infinite uncertainty, implying that $V = \infty$ outside the box. If V instead has a finite value outside the box, then, there is some probability—not necessarily great, but not zero either—that the particle will "leak" out. As we shall see in Chap. 13, the quantum-mechanical prediction that particles always have some chance of escaping from confinement (since potential energies are never infinite in the real world, our original rigid-walled box has no physical counterpart) exactly fits the observed behavior of those radioactive nuclei that emit alpha particles.

When the confining box has nonrigid walls, the particle wave function ψ_n does not equal zero at the walls. The particle wavelengths that can fit into the box are therefore somewhat longer than in the case of the box with rigid walls, corresponding to lower particle momenta and hence to lower energy levels.

The condition that the potential energy V outside the box be finite has another consequence: it is now possible for a particle to have an energy E that exceeds V. Such a particle is not trapped inside the box, since it always has enough energy to penetrate its walls, and its energy is not quantized but may have any value above V. However, the particle's kinetic energy outside the box, $E - V$, is always less than its kinetic energy inside, which is just E since $V = 0$ in the box according to our original specification. Less energy means longer wavelength, and so ψ has a longer wavelength outside the box than inside.

In the optics of light waves, it is readily observed that when a light wave reaches a region where its wavelength changes (that is, a region of different index of refraction), reflection as well as transmission occurs. This is the reason we see our reflections in shop windows. The effect is common to all types of waves, and it may be shown mathematically to follow from the requirement that the wave variable (electric-field intensity E in the case of electromagnetic waves, pressure p in the case of sound waves, wave height h in the case of water waves, etc.) and its first derivative be continuous at the boundary where the wavelength change takes place. Exactly the same considerations apply to the wave function ψ representing a moving particle. The wave function of a particle encountering a region in which it has a different potential energy, as we saw above, decreases in wavelength if V decreases and increases in wavelength if V increases. In either situation some reflection occurs at the boundaries between the regions. What does "some" reflection mean when we are discussing the motion of a single particle? Since ψ is related to the probability of finding the particle in a particular place, the partial reflection of ψ means that there is a chance that the particle will be reflected. That is, if we shoot many particles at a box with nonrigid walls, most will get through but some will be scattered.

What we have been saying, then, is that particles with enough energy to penetrate a wall nevertheless stand some chance of bouncing off instead. This prediction complements the "leaking" out of particles trapped in the box despite the fact that they have insufficient energy to penetrate its walls. Both of these predictions are unique with quantum mechanics and do not correspond to any behavior expected in classical physics. Their confirmation in numerous atomic and nuclear experiments supports the validity of the quantum-mechanical approach. An experiment relating to particle reflection at a change in V is one in which moving electrons are scattered by atoms or molecules; phenomena are found that are characteristic of wave physics rather than of particle physics.

6.8 The Harmonic Oscillator

Harmonic motion occurs when a system of some kind vibrates about an equilibrium configuration. The system may be an object supported by a spring or floating in a liquid, a diatomic molecule, an atom in a crystal lattice—there are countless examples in both the macroscopic and the microscopic realms. The condition for harmonic motion to occur is the presence of a restoring force that acts to return the system to its equilibrium configuration when it is disturbed; the inertia of the masses involved causes them to overshoot equilibrium, and the system oscillates indefinitely if no dissipative processes are also present.

In the special case of simple harmonic motion, the restoring force F on a particle of mass m is linear; that is, F is proportional to the particle's displacement from its equilibrium position, so that

6.33 $$F = -kx$$

This relationship is customarily called Hooke's law. According to the second law of motion, $\mathbf{F} = m\mathbf{a}$, and so here

$$-kx = m\frac{d^2x}{dt^2}$$

6.34 $$\frac{d^2x}{dt^2} + \frac{k}{m}x = 0$$

There are various ways to write the solution to Eq. 6.34, a convenient one being

6.35 $$x = A\cos(2\pi\nu t + \phi)$$

where

6.36 $$\nu = \frac{1}{2\pi}\sqrt{\frac{k}{m}}$$

is the frequency of the oscillations, A is their amplitude, and ϕ, the phase constant, is a constant that depends upon the value of x at the time $t = 0$.

The importance of the simple harmonic oscillator in both classical and modern physics lies not in the strict adherence of actual restoring forces to Hooke's law, which is seldom true, but in the fact that these restoring forces reduce to Hooke's law for small displacements x. To appreciate this point we note that any force which is a function of x can be expressed in a Maclaurin's series about the equilibrium position $x = 0$ as

$$F(x) = F_{x=0} + \left(\frac{dF}{dx}\right)_{x=0} x + \frac{1}{2}\left(\frac{d^2F}{dx^2}\right)_{x=0} x^2 + \frac{1}{6}\left(\frac{d^3F}{dx^3}\right)_{x=0} x^3 + \cdots$$

Since $x = 0$ is the equilibrium position, $F_{x=0} = 0$, and since for small x the values of $x^2, x^3, \ldots$ are very small compared with x, the third and higher terms of the series can be neglected. The only term of significance when x is small is therefore the second one. Hence

$$F(x) = \left(\frac{dF}{dx}\right)_{x=0} x$$

which is Hooke's law when $(dF/dx)_{x=0}$ is negative, as of course it is for any restoring force. The conclusion, then, is that *all* oscillations are simple harmonic in character when their amplitudes are sufficiently small.

The potential energy function $V(x)$ that corresponds to a Hooke's law force may be found by calculating the work needed to bring a particle from $x = 0$ to $x = x$ against such a force. The result is

6.37 $$V(x) = -\int_0^x F(x)dx = k\int_0^x x\,dx = \tfrac{1}{2}kx^2$$

and is plotted in Fig. 6-7. If the energy of the oscillator is E, the particle vibrates back and forth between $x = -A$ and $x = +A$, where E and A are related by $E = \tfrac{1}{2}kA^2$.

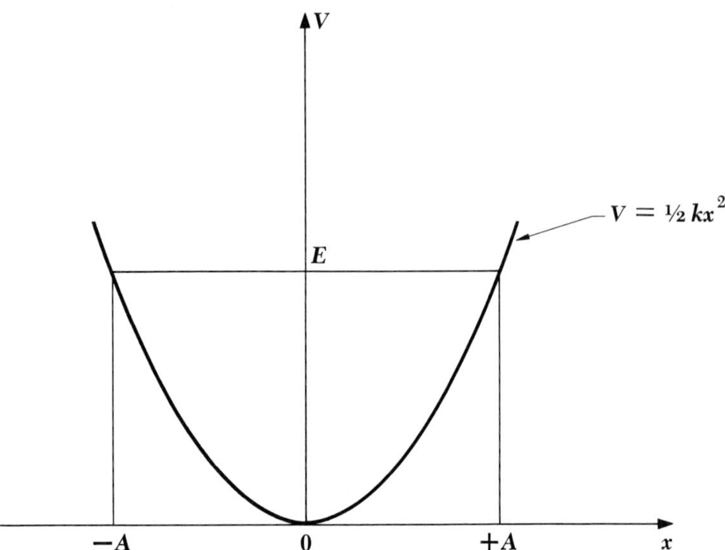

FIGURE 6-7 The potential energy of a harmonic oscillator is proportional to x^2, where x is the displacement from the equilibrium position. The amplitude A of the motion is determined by the total energy E of the oscillator, which classically can have any value.

Even before we make a detailed calculation we can anticipate three quantum-mechanical modifications to this classical picture. First, there will not be a continuous spectrum of allowed energies but a discrete spectrum consisting of certain specific values only. Second, the lowest allowed energy will not be $E = 0$ but will be some definite minimum $E = E_0$. Third, there will be a certain probability that the particle can "penetrate" the potential well it is in and go beyond the limits of $-A$ and $+A$.

Schrödinger's equation for the harmonic oscillator is, with $V = \frac{1}{2}kx^2$,

6.38
$$\frac{d^2\psi}{dx^2} + \frac{8\pi^2 m}{h^2}\left(E - \frac{1}{2}kx^2\right)\psi = 0$$

It is convenient to simplify Eq. 6.38 by introducing the dimensionless quantities

$$y = \left(\frac{2\pi}{h}\sqrt{km}\right)^{1/2} x$$

6.39
$$= 2\pi\sqrt{\frac{m\nu}{h}}\,x$$

and

$$\alpha = \frac{4\pi E}{h}\sqrt{\frac{m}{k}}$$

6.40
$$= \frac{2E}{h\nu}$$

where ν is the classical frequency of the oscillation given by Eq. 6.36. In making these substitutions, what we have essentially done is change the units in which x and E are expressed from meters and joules, respectively, to appropriate dimensionless units. In terms of y and α Schrödinger's equation becomes

6.41 $$\frac{d^2\psi}{dy^2} + (\alpha - y^2)\psi = 0$$

We begin the solution of Eq. 6.41 by finding the asymptotic form that ψ must have as $y \to \pm\infty$. If any wave function ψ is to represent an actual particle localized in space, its value must approach zero as y approaches infinity in order that $\int_{-\infty}^{\infty} \psi^2\, dy$ be a finite, nonvanishing quantity (Sec. 6.2). Let us rewrite Eq. 6.41 as follows:

$$\frac{d^2\psi}{dy^2} - (y^2 - \alpha)\psi = 0$$

$$\frac{d^2\psi}{dy^2} = (y^2 - \alpha)\psi$$

$$\frac{d^2\psi/dy^2}{(y^2 - \alpha)\psi} = 1$$

As $y \to \infty$, $y^2 \gg \alpha$ and we have

6.42 $$\lim_{y\to\infty} \frac{d^2\psi/dy^2}{y^2\psi} = 1$$

A function ψ_∞ that satisfies Eq. 6.42 is

6.43 $$\psi_\infty = e^{-y^2/2}$$

since

$$\lim_{y\to\infty} \frac{d^2\psi_\infty}{dy^2} = \lim_{y\to\infty} (y^2 - 1)e^{-y^2/2} = y^2e^{-y^2/2}$$

Equation 6.43 is the required asymptotic form of ψ.

We are now able to write

$$\psi = f(y)\psi_\infty$$

6.44 $$= f(y)e^{-y^2/2}$$

where $f(y)$ is a function of y that remains to be found. By inserting the ψ of Eq. 6.44 in Eq. 6.41 we obtain

6.45 $$\frac{d^2f}{dy^2} - 2y\frac{df}{dy} + (\alpha - 1)f = 0$$

which is the differential equation that f obeys.

The standard procedure for solving differential equations like Eq. 6.45 is to assume that $f(y)$ can be expanded in a power series in y, namely

$$f(y) = A_0 + A_1 y + A_2 y^2 + A_3 y^3 + \cdots$$

6.46
$$= \sum_{n=0}^{\infty} A_n y^n$$

and then to determine the values of the coefficients A_n. Differentiating f yields

$$\frac{df}{dy} = A_1 + 2A_2 y + 3A_3 y^2 + \cdots$$
$$= \sum_{n=1}^{\infty} nA_n y^{n-1}$$

By multiplying this equation by y we obtain

$$y\frac{df}{dy} = A_1 y + 2A_2 y^2 + 3A_3 y^3 + \cdots$$

6.47
$$= \sum_{n=0}^{\infty} nA_n y^n$$

The second derivative of f with respect to y is

$$\frac{d^2f}{dy^2} = 1 \cdot 2\ A_2 + 2 \cdot 3\ A_3 y + 3 \cdot 4\ A_4 y^2 + \cdots$$
$$= \sum_{n=2}^{\infty} n(n-1)A_n y^{n-2}$$

which is equal to

6.48
$$\frac{d^2f}{dy^2} = \sum_{n=0}^{\infty} (n+2)(n+1)A_{n+2} y^n$$

(That the latter two series are indeed equal can be verified by working out the first few terms of each.) We now substitute Eqs. 6.46, 6.47, and 6.48 in Eq. 6.45 to obtain

6.49
$$\sum_{n=0}^{\infty} [(n+2)(n+1)A_{n+2} - (2n+1-\alpha)A_n]y^n = 0$$

In order for this equation to hold for all values of y, the quantity in brackets must be zero for all values of n. Hence we have the condition that

$$(n+2)(n+1)A_{n+2} = (2n+1-\alpha)A_n$$

and so

6.50
$$A_{n+2} = \frac{2n+1-\alpha}{(n+2)(n+1)} A_n$$

This *recursion formula* enables us to find the coefficients A_2, A_3, A_4, . . . in terms of A_0 and A_1. (Since Eq. 6.45 is a second-order differential equation, its solution has two arbitrary constants, which are A_0 and A_1 here.) Starting from A_0 we obtain the sequence of coefficients A_2, A_4, A_6, . . . , and starting from A_1 we obtain the other sequence A_3, A_5, A_7, ...

6.9 The Harmonic Oscillator: Energy Levels

It is necessary for us to inquire into the behavior of

$$\psi = f(y)e^{-y^2/2}$$

as $y \to \infty$; only if $\psi \to 0$ as $y \to \infty$ can ψ be a physically acceptable wave function. Because $f(y)$ is multipled by $e^{-y^2/2}$, ψ will meet this requirement provided that

$$\lim_{y\to\infty} f(y) < e^{y^2/2}$$

(As we shall see, it is unnecessary for us to specify just how much smaller f must be in the limit than $e^{y^2/2}$.)

A suitable way to compare the asymptotic behaviors of $f(y)$ and $e^{y^2/2}$ is to express the latter in a power series (f is already in the form of a power series) and to examine the ratio between successive coefficients of each series as $n \to \infty$. From the recursion formula of Eq. 6.50 we can tell by inspection that

$$\lim_{n\to\infty} \frac{A_{n+2}}{A_n} = \frac{2}{n}$$

Since

$$e^z = 1 + z + \frac{z^2}{2!} + \frac{z^3}{3!} + \cdots$$

we can express $e^{y^2/2}$ in a power series as

$$\begin{aligned} e^{y^2/2} &= 1 + \frac{y^2}{2} + \frac{y^4}{2^2 \cdot 2!} + \frac{y^6}{2^3 \cdot 3!} + \cdots \\ &= \sum_{n=0,2,4,\ldots}^{\infty} \frac{1}{2^{n/2}\left(\frac{n}{2}\right)!} y^n \\ &= \sum_{n=0,2,4,\ldots}^{\infty} B_n y^n \end{aligned}$$

The ratio between successive coefficients of y^n here is

$$\frac{B_{n+2}}{B_n} = \frac{2^{n/2}\left(\frac{n}{2}\right)!}{2^{(n+2)/2}\left(\frac{n+2}{2}\right)!} = \frac{2^{n/2}\left(\frac{n}{2}\right)!}{2 \cdot 2^{n/2}\left(\frac{n}{2}+1\right)\left(\frac{n}{2}\right)!}$$

$$= \frac{1}{2\left(\frac{n}{2}+1\right)} = \frac{1}{n+2}$$

In the limit of $n \to \infty$ this ratio becomes

$$\lim_{n\to\infty} \frac{B_{n+2}}{B_n} = \frac{1}{n}$$

Thus successive coefficients in the power series for f decrease *less* rapidly than those in the power series for $e^{y^2/2}$ instead of *more* rapidly, which means that $f(y)e^{-y^2/2}$ does not vanish as $y \to \infty$.

There is a simple way out of this dilemma. If the series representing f terminates at a certain value of n, so that all the coefficients A_n are zero for values of n higher than this one, ψ will go to zero as $y \to \infty$ because of the $e^{-y^2/2}$ factor. In other words, if f is a polynomial with a finite number of terms instead of an infinite series, it is acceptable. From the recursion formula

$$A_{n+2} = \frac{2n+1-\alpha}{(n+2)(n+1)} A_n$$

it is clear that if

6.51 $$\alpha = 2n + 1$$

for any value of n, then $A_{n+2} = A_{n+4} = A_{n+6} = \cdots = 0$, which is what we want.

(Equation 6.51 takes care of only one sequence of coefficients, either the sequence of even n starting with A_0 or the sequence of odd n starting with A_1. If n is even, it must be true that $A_1 = 0$ and only even powers of y appear in the polynomial, while if n is odd, it must be true that $A_0 = 0$ and only odd powers of y appear. We shall see the result in the next section, where the polynomial is tabulated for various values of n.)

The condition that $\alpha = 2n + 1$ is a necessary and sufficient condition for the wave equation 6.41 to have solutions that meet the various requirements that ψ must fulfill. From Eq. 6.40, the definition of α, we have

$$\alpha_n = \frac{2E}{h\nu} = 2n + 1$$

or

6.52 $$E_n = \left(n + \frac{1}{2}\right)h\nu \qquad n = 0, 1, 2, \ldots$$ **Harmonic oscillator**

The energy of a harmonic oscillator is thus quantized in steps of $h\nu$, where ν is the classical frequency of oscillation and h is Planck's constant. The energy levels here are evenly spaced (Fig. 6-8), unlike the energy levels of a particle in a box whose spacing diverges. We note that, when $n = 0$,

6.53 $$E_0 = \tfrac{1}{2}h\nu$$ **Zero-point energy**

which is the lowest value the energy of the oscillator can have. This value is called the *zero-point energy* because a harmonic oscillator in equilibrium with its surroundings would approach an energy of $E = E_0$ and not $E = 0$ as the temperature approaches 0°K.

6.10 The Harmonic Oscillator: Wave Functions

For each choice of the parameter α_n there is a different wave function ψ_n. Each function consists of a polynomial $H_n(y)$ (called a *Hermite polynomial*) in either odd or even powers of y, the exponential factor $e^{-y^2/2}$, and a numerical coefficient which is needed for ψ_n to meet the normalization condition

$$\int_{-\infty}^{\infty} \psi_n^2 \, dy = 1 \qquad n = 0, 1, 2, \ldots$$

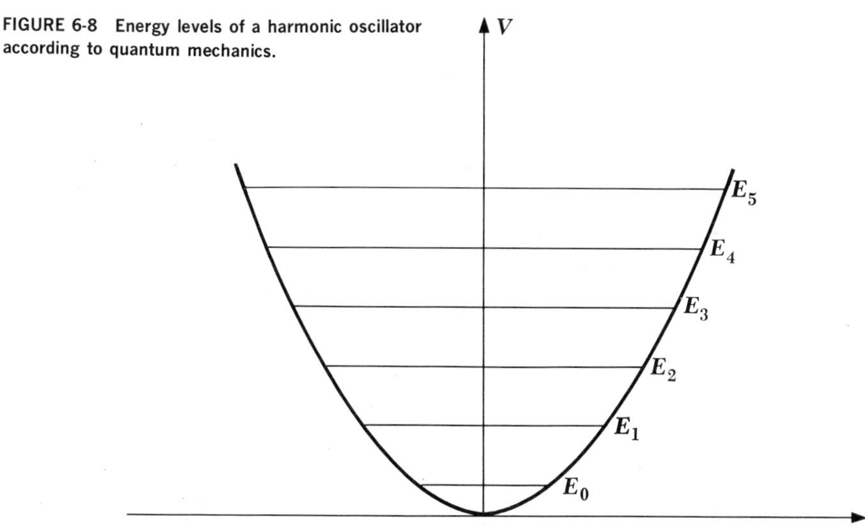

FIGURE 6-8 Energy levels of a harmonic oscillator according to quantum mechanics.

Table 6.1

SOME HERMITE POLYNOMIALS

n	$H_n(y)$	α_n	E_n
0	1	1	$\frac{1}{2}h\nu$
1	$2y$	3	$\frac{3}{2}h\nu$
2	$4y^2 - 2$	5	$\frac{5}{2}h\nu$
3	$8y^3 - 12y$	7	$\frac{7}{2}h\nu$
4	$16y^4 - 48y^2 + 12$	9	$\frac{9}{2}h\nu$
5	$32y^5 - 160y^3 + 120y$	11	$\frac{11}{2}h\nu$

The general formula for the nth wave function is

6.54 $$\psi_n = \left(\frac{4\pi m\nu}{h}\right)^{1/4} (2^n n!)^{-1/2} H_n(y) e^{-y^2/2}$$

The first six Hermite polynomials $H_n(y)$ are listed in Table 6.1, and the corresponding wave functions ψ_n are plotted in Fig. 6-9. In each case the range to which a particle oscillating classically with the same total energy E_n would be confined is indicated; evidently the particle is able to penetrate into classically forbidden regions—in other words, to exceed the amplitude A determined by the energy—with an exponentially decreasing probability, just as in the situation of a particle in a box with nonrigid walls.

It is interesting and instructive to compare the probability densities of a classical harmonic oscillator and a quantum-mechanical harmonic oscillator of the same energy. The upper graph of Fig. 6-10 shows this density for the classical oscillator: The probability P of finding the particle at a given position is greatest at the end-points of its motion, where it moves slowly, and least near the equilibrium position ($x = 0$), where it moves rapidly. Exactly the opposite behavior is manifested by a quantum-mechanical oscillator in its lowest energy state of $n = 0$. As shown, the probability density ψ_0^2 has its maximum value at $x = 0$ and drops off on either side of this position. However, this disagreement becomes less and less marked with increasing n: The lower graph of Fig. 6-10 corresponds to $n = 10$, and it is clear that ψ_{10}^2 when averaged over x has approximately the general character of the classical probability P. This is another example of the correspondence principle mentioned in Sec. 5.6: In the limit of large quantum numbers, quantum physics yields the same results as classical physics.

It might be objected that, although ψ_{10}^2 does indeed approach P when smoothed out, nevertheless ψ_{10}^2 fluctuates rapidly with x whereas P does not. However, this objection has meaning only if the fluctuations are observable, and the smaller the spacing of the peaks and hollows, the more strongly the uncertainty principle prevents their detection without altering the physical state of the oscillator. The exponential "tails" of ψ^2 beyond $x = \pm A$ also decrease in magnitude with increasing n. Thus the classical and quantum pictures begin to resemble each other more and more the larger the value of n, in agreement with the correspondence principle, although they are radically different for small n.

FIGURE 6-9 The first six harmonic-oscillator wave functions. The vertical lines show the limits $-A$ and $+A$ between which a classical oscillator with the same energy would vibrate.

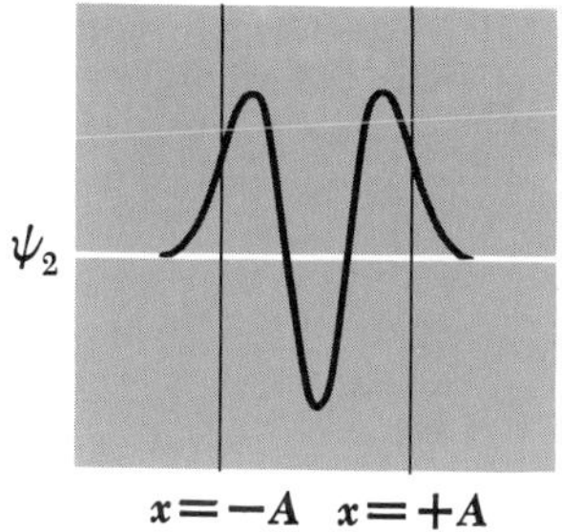

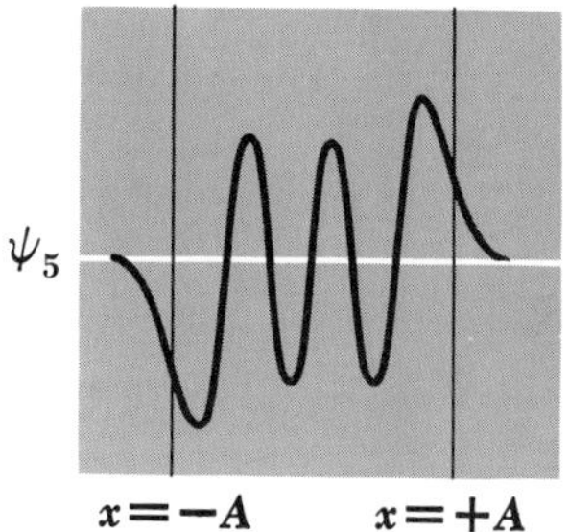

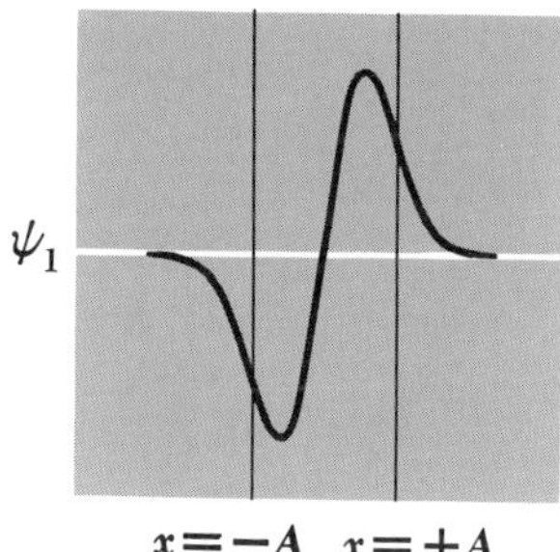

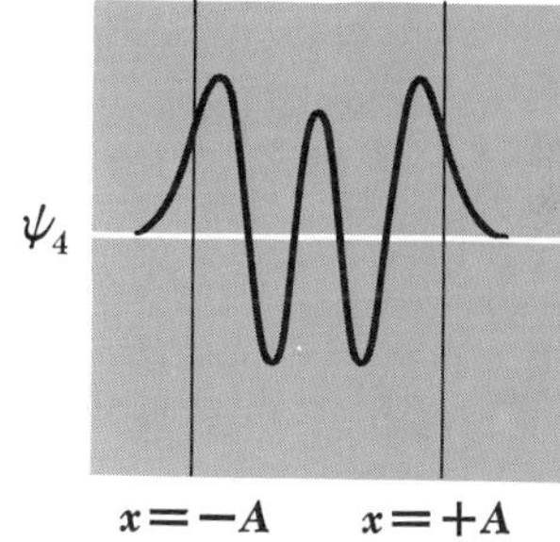

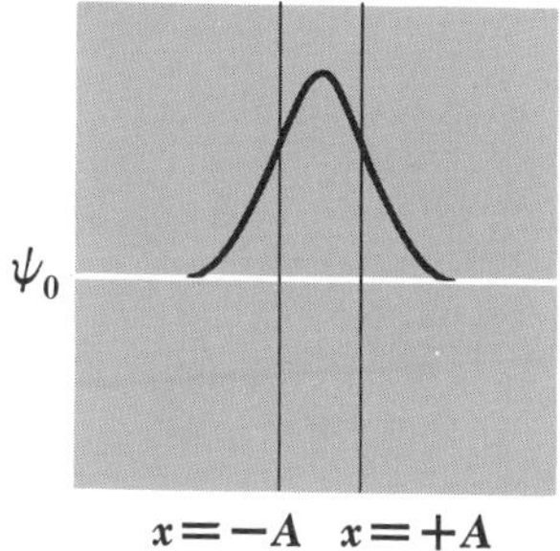

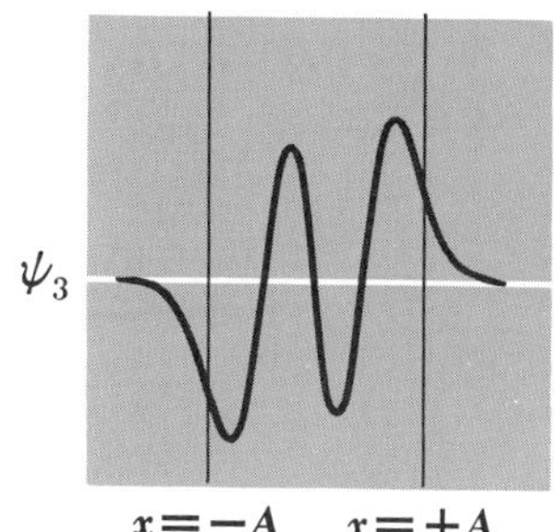

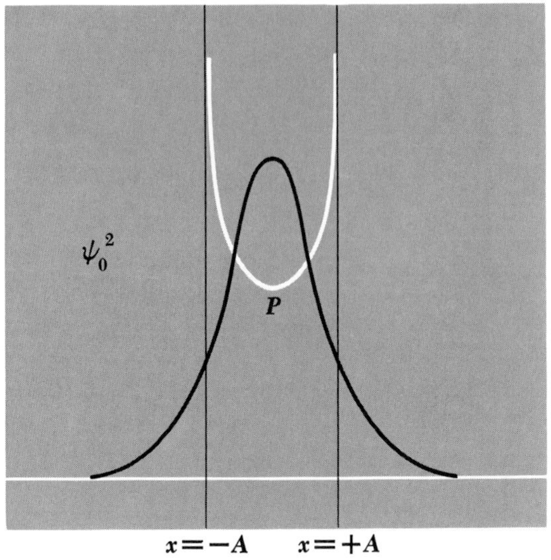

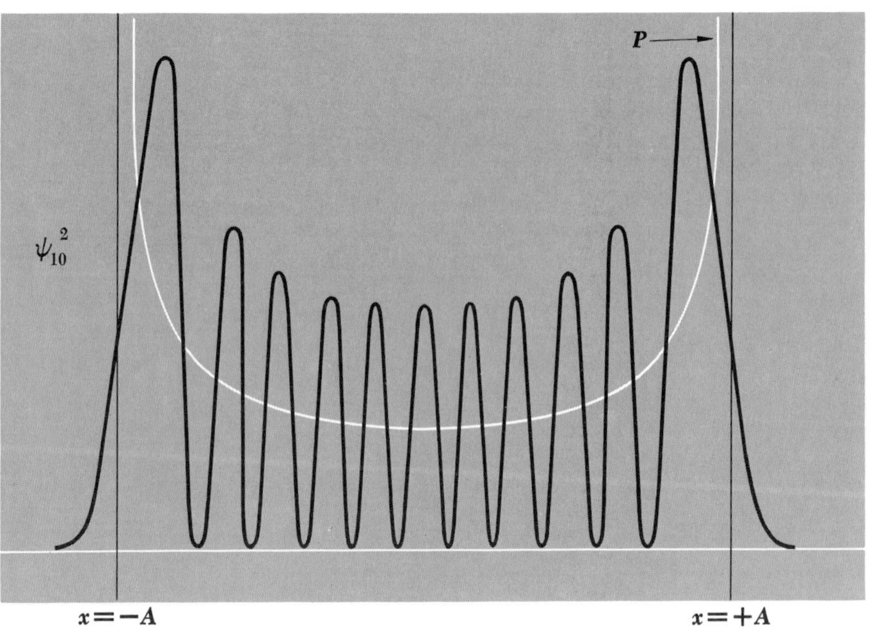

FIGURE 6-10 Probability densities for the $n = 0$ and $n = 10$ states of a quantum-mechanical harmonic oscillator. The probability densities for classical harmonic oscillators with the same energies are shown in white.

Problems

1. Find the lowest energy of a neutron confined to a box 10^{-14} m across. (The size of a nucleus is of this order of magnitude.)

2. State Schrödinger's equation for the one-dimensional motion of an electron not acted upon by any forces and show that its total energy E is not quantized.

3. Show that the uncertainty in the energy of a particle confined to a one-dimensional box is comparable with its lowest permitted energy.

4. Consider a particle of mass m trapped in a two-dimensional box L long and W wide. Starting from Schrödinger's equation, show that the permitted energies of the particle are given by

$$E = \frac{h^2}{8m}\left(\frac{a^2}{L^2} + \frac{b^2}{W^2}\right)$$

where a and b are positive integers.

5. According to the correspondence principle, quantum theory should give the same results as classical physics in the limit of large quantum numbers. Show that, as $n \to \infty$, the probability of finding a particle trapped in a box between x and $x + dx$ is independent of x, which is the classical expectation.

6. Use the fact that $\alpha \geqslant 0$ (since $E \geqslant 0$) to show that the coefficients A_n of Eq. 6.46 are all zero for negative values of n.

7. Show that the first three harmonic-oscillator wave functions are normalized solutions of Schrödinger's equation.

8. Find the zero-point energy in electron volts of a pendulum whose period is 1 sec.

7 THE QUANTUM THEORY OF THE HYDROGEN ATOM

The quantum-mechanical theory of the atom, which was developed shortly after the formulation of quantum mechanics itself, represents an epochal contribution to our knowledge of the physical universe. Besides revolutionizing our approach to atomic phenomena, this theory has made it possible for us to understand such related matters as how atoms interact with one another to form stable molecules, the origin of the periodic table of the elements, and why solids are endowed with their characteristic electrical, magnetic, and mechanical properties, all topics we shall explore in later chapters. For the moment we shall concentrate on the quantum theory of the hydrogen atom and how its formal mathematical results may be interpreted in terms of familiar concepts.

7.1 Schrödinger's Equation for the Hydrogen Atom

A hydrogen atom consists of a proton, a particle of electric charge $+e$, and an electron, a particle of charge $-e$ which is almost 2,000 times lighter than the proton. For the sake of convenience we shall consider the proton to be stationary, with the electron moving about in its vicinity but prevented from escaping because of the proton's electric field. (As in the Bohr theory, the correction for proton motion is simply a matter of replacing the electron mass m by the reduced mass m'.) Schrödinger's equation for the electron in three dimensions, which is what we must use for the hydrogen atom, is

7.1 $$\frac{\partial^2 \psi}{\partial x^2} + \frac{\partial^2 \psi}{\partial y^2} + \frac{\partial^2 \psi}{\partial z^2} + \frac{8\pi^2 m}{h^2}(E - V)\psi = 0$$

The potential energy V here is the electrostatic potential energy

7.2 $$V = -\frac{e^2}{4\pi\varepsilon_0 r}$$

of a charge $-e$ when it is the distance r from another charge $+e$. Since V is a function of r rather than of x, y, z, we cannot substitute Eq. 7.2 directly into Eq. 7.1. There are two alternatives: we can express V in terms of the cartesian coordinates x, y, z by replacing r by $\sqrt{x^2 + y^2 + z^2}$, or we can express Schrödinger's equation in terms of the spherical polar coordinates r, θ, ϕ defined in Fig. 7-1. As it happens, owing to the symmetry of the physical situation, doing the latter makes the problem considerably easier to solve.

The spherical polar coordinates r, θ, ϕ of the point P shown in Fig. 7-1 have the following interpretation:

r = length of radius vector from origin O to point P

$= \sqrt{x^2 + y^2 + z^2}$

θ = angle between radius vector and $+z$ axis

= zenith angle

$$= \cos^{-1} \frac{z}{\sqrt{x^2 + y^2 + z^2}}$$

Spherical polar coordinates

ϕ = angle between the projection of the radius vector in the xy plane and the $+x$ axis, measured in the direction shown

= azimuth angle

$$= \tan^{-1} \frac{y}{x}$$

On the surface of a sphere whose center is at O, lines of constant zenith angle θ are like parallels of latitude on a globe (but we note that the value of

FIGURE 7-1 **Spherical polar coordinates.**

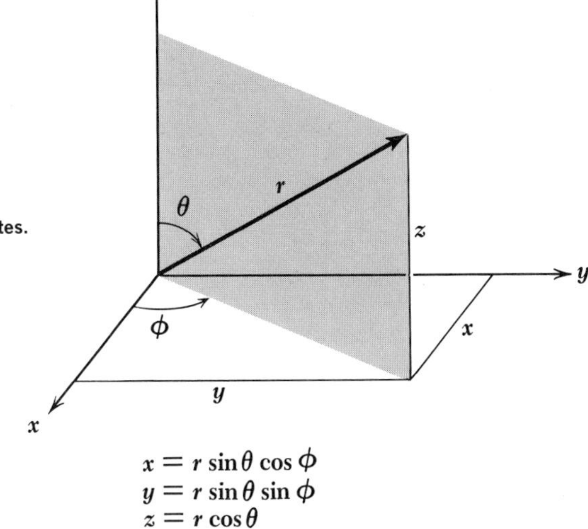

θ of a point is *not* the same as its latitude; $\theta = 90°$ at the equator, for instance, but the latitude of the equator is $0°$), and lines of constant azimuth angle ϕ are like meridians of longitude (here the definitions coincide if the axis of the globe is taken as the $+z$ axis and the $+x$ axis is at $\phi = 0°$).

In spherical polar coordinates Schrödinger's equation becomes

7.3
$$\frac{1}{r^2}\frac{\partial}{\partial r}\left(r^2\frac{\partial\psi}{\partial r}\right) + \frac{1}{r^2 \sin\theta}\frac{\partial}{\partial\theta}\left(\sin\theta\frac{\partial\psi}{\partial\theta}\right) + \frac{1}{r^2\sin^2\theta}\frac{\partial^2\psi}{\partial\phi^2} + \frac{8\pi^2 m}{h^2}(E - V)\psi = 0$$

Substituting Eq. 7.2 for the potential energy V and multiplying the entire equation by $r^2 \sin^2\theta$, we obtain

7.4
$$\sin^2\theta\frac{\partial}{\partial r}\left(r^2\frac{\partial\psi}{\partial r}\right) + \sin\theta\frac{\partial}{\partial\theta}\left(\sin\theta\frac{\partial\psi}{\partial\theta}\right) + \frac{\partial^2\psi}{\partial\phi^2} + \frac{8\pi^2 mr^2\sin^2\theta}{h^2}\left[\frac{e^2}{4\pi\varepsilon_0 r} + E\right]\psi = 0$$
Hydrogen atom

Equation 7.4 is the partial differential equation for the wave function ψ of the electron in a hydrogen atom. Together with the various conditions ψ must obey, as discussed in the previous chapter (for instance, that ψ have just one value at each point r, θ, ϕ), this equation completely specifies the behavior of the electron. In order to see just what this behavior is, we must solve Eq. 7.4 for ψ.

7.2 Separation of Variables

The virtue of writing Schrödinger's equation in spherical polar coordinates for the problem of the hydrogen atom is that in this form it may be readily separated into three independent equations, each involving only a single coordinate. The procedure is to look for solutions in which the wave function $\psi(r, \theta, \phi)$ has the form of a product of three different functions: $R(r)$, which depends upon r alone; $\Theta(\theta)$, which depends upon θ alone; and $\Phi(\phi)$, which depends upon ϕ alone. That is, we assume that

7.5
$$\psi(r, \theta, \phi) = R(r)\Theta(\theta)\Phi(\phi)$$
Electron wave function

The function $R(r)$ describes how the wave function ψ of the electron varies along a radius vector from the nucleus, with θ and ϕ constant. The function $\Theta(\theta)$ describes how ψ varies with zenith angle θ along a meridian on a sphere centered at the nucleus, with r and ϕ constant. The function $\Phi(\phi)$ describes how ψ varies with azimuth angle ϕ along a parallel on a sphere centered at the nucleus, with r and θ constant.

From Eq. 7.5, which we may write more simply as

$$\psi = R\Theta\Phi$$

we see that

$$\frac{\partial\psi}{\partial r} = \Theta\Phi\frac{\partial R}{\partial r}$$

$$\frac{\partial\psi}{\partial\theta} = R\Phi\frac{\partial\Theta}{\partial\theta}$$

$$\frac{\partial^2\psi}{\partial\phi^2} = R\Theta\frac{\partial^2\Phi}{\partial\phi^2}$$

Hence, when we substitute $R\Theta\Phi$ for ψ in Schrödinger's equation for the hydrogen atom and divide the entire equation by $R\Theta\Phi$, we find that

7.6 $$\frac{\sin^2\theta}{R}\frac{\partial}{\partial r}\left(r^2\frac{\partial R}{\partial r}\right) + \frac{\sin\theta}{\Theta}\frac{\partial}{\partial\theta}\left(\sin\theta\frac{\partial\Theta}{\partial\theta}\right) + \frac{1}{\Phi}\frac{\partial^2\Phi}{\partial\phi^2} + \frac{8\pi^2mr^2\sin^2\theta}{h^2}\left(\frac{e^2}{4\pi\varepsilon_0 r} + E\right) = 0$$

The third term of Eq. 7.6 is a function of azimuth angle ϕ only, while the other terms are functions of R and θ only. Let us rearrange Eq. 7.6 to read

7.7 $$\frac{\sin^2\theta}{R}\frac{\partial}{\partial r}\left(r^2\frac{\partial R}{\partial r}\right) + \frac{\sin\theta}{\Theta}\frac{\partial}{\partial\theta}\left(\sin\theta\frac{\partial\Theta}{\partial\theta}\right) + \frac{8\pi^2mr^2\sin^2\theta}{h^2}\left(\frac{e^2}{4\pi\varepsilon_0 r} + E\right) = -\frac{1}{\Phi}\frac{\partial^2\Phi}{\partial\phi^2}$$

This equation can be correct only if both sides of it are equal to the same constant, since they are functions of *different* variables. As we shall see, it is convenient to call this constant m_l^2. The differential equation for the function Φ is therefore

7.8 $$-\frac{1}{\Phi}\frac{d^2\Phi}{d\phi^2} = m_l^2$$

When we substitute m_l^2 for the right-hand side of Eq. 7.7, divide the entire equation by $\sin^2\theta$, and rearrange the various terms, we find that

7.9 $$\frac{1}{R}\frac{\partial}{\partial r}\left(r^2\frac{\partial R}{\partial r}\right) + \frac{8\pi^2mr^2}{h^2}\left(\frac{e^2}{4\pi\varepsilon_0 r} + E\right) = \frac{m_l^2}{\sin^2\theta} - \frac{1}{\Theta\sin\theta}\frac{\partial}{\partial\theta}\left(\sin\theta\frac{\partial\Theta}{\partial\theta}\right)$$

Again we have an equation in which different variables appear on each side, requiring that both sides be equal to the same constant. This constant we shall call $l(l+1)$, once more for reasons that will be apparent later. The equations for the functions Θ and R are therefore

$$\frac{m_l^2}{\sin^2\theta} - \frac{1}{\Theta \sin\theta}\frac{d}{d\theta}\left(\sin\theta \frac{d\Theta}{d\theta}\right) = l(l+1) \tag{7.10}$$

$$\frac{1}{R}\frac{d}{dr}\left(r^2\frac{dR}{dr}\right) + \frac{8\pi^2 m r^2}{h^2}\left(\frac{e^2}{4\pi\varepsilon_0 r} + E\right) = l(l+1) \tag{7.11}$$

Equations 7.8, 7.10, and 7.11 are usually written

$$\frac{d^2\Phi}{d\phi^2} + m_l^2\Phi = 0 \tag{7.12}$$

$$\frac{1}{\sin\theta}\frac{d}{d\theta}\left(\sin\theta\frac{d\Theta}{d\theta}\right) + \left[l(l+1) - \frac{m_l^2}{\sin^2\theta}\right]\Theta = 0 \tag{7.13}$$

$$\frac{1}{r^2}\frac{d}{dr}\left(r^2\frac{dR}{dr}\right) + \left[\frac{8\pi^2 m}{h^2}\left(\frac{e^2}{4\pi\varepsilon_0 r} + E\right) - \frac{l(l+1)}{r^2}\right]R = 0 \tag{7.14}$$

Each of these is an ordinary differential equation for a single function of a single variable. We have therefore accomplished our task of simplifying Schrödinger's equation for the hydrogen atom, which began as a partial differential equation for a function ψ of three variables.

7.3 Quantum Numbers

The first of the above equations, Eq. 7.12, is readily solved, with the result

$$\Phi(\phi) = Ae^{im_l\phi} \tag{7.15}$$

where A is the constant of integration. We have already stated that one of the conditions a wave function—and hence Φ, which is a component of the complete wave function ψ—must obey is that it have a single value at a given point in space. From Fig. 7-2 it is evident that ϕ and $\phi + 2\pi$ both identify the same meridian plane. Hence it must be true that $\Phi(\phi) = \Phi(\phi + 2\pi)$, or

$$Ae^{im_l\phi} = Ae^{im_l(\phi+2\pi)}$$

which can only happen when m_l is 0 or a positive or negative integer (±1, ±2, ±3, . . .). The constant m_l is known as the *magnetic quantum number* of the hydrogen atom.

The differential equation 7.13 for $\Theta(\theta)$ has a rather complicated solution in terms of polynomials called the *associated Legendre functions.* For our

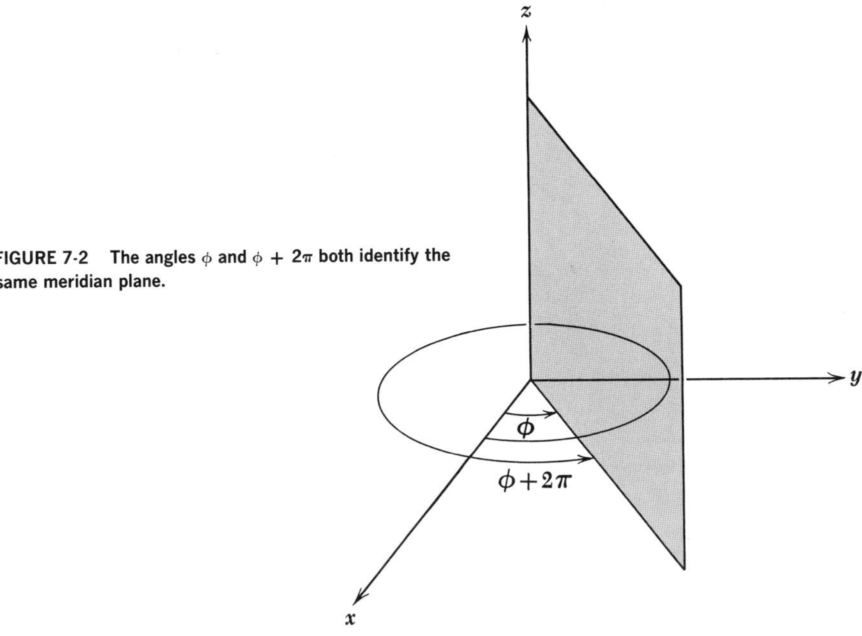

FIGURE 7-2 The angles ϕ and $\phi + 2\pi$ both identify the same meridian plane.

present purpose, the important thing about these functions is that they exist only when the constant l is an integer equal to or greater than $|m_l|$, the absolute value of m_l. This requirement can be expressed as a condition on m_l in the form

$$m_l = 0, \pm 1, \pm 2, \ldots, \pm l$$

The constant l is known as the *orbital quantum number.*

The solution of the final equation, Eq. 7.14, for the radial part $R(r)$ of the hydrogen-atom wave function ψ is also complicated, being in terms of polynomials called the *associated Laguerre functions.* Equation 7.14 can be solved only when E is positive or has one of the negative values E_n (signifying that the electron is bound to the atom) specified by

7.16 $$E_n = -\frac{me^4}{8\varepsilon_0{}^2h^2}\left(\frac{1}{n^2}\right)$$

where n is an integer. We recognize that this is precisely the same formula for the energy levels of the hydrogen atom that Bohr obtained!

Another condition that must be obeyed in order to solve Eq. 7.14 is that n, known as the *total quantum number,* must be equal to or greater than $l + 1$. This requirement may be expressed as a condition on l in the form

$$l = 0, 1, 2, \ldots, (n - 1)$$

Hence we may tabulate the three quantum numbers n, l, and m together with their permissible values as follows:

7.17
$$n = 1, 2, 3, \ldots \qquad \textbf{Total quantum number}$$
$$l = 0, 1, 2, \ldots, (n-1) \qquad \textbf{Orbital quantum number}$$
$$m_l = 0, \pm 1, \pm 2, \ldots, \pm l \qquad \textbf{Magnetic quantum number}$$

The theory of a particle in a box considered in the previous chapter showed that one quantum number arises naturally in the case of one-dimensional motion, and so it is not surprising that three quantum numbers should be needed to describe the three-dimensional motion of the electron in a hydrogen atom. It is worth noting again how inevitably quantum numbers appear in quantum-mechanical theories of particles trapped in a particular region of space.

To exhibit the dependence of R, Θ, and Φ upon the quantum numbers n, l, m, we may write for the electron wave function

7.18
$$\psi = R_{nl}\Theta_{lm_l}\Phi_{m_l}$$

7.4 Total Quantum Number

It is interesting to consider the interpretation of the hydrogen-atom quantum numbers in terms of the classical model of the atom. This model, as we saw in Chap. 4, corresponds exactly to planetary motion in the solar system except that the inverse-square force holding the electron to the nucleus is electrical rather than gravitational. Two quantities are *conserved*—that is, maintain a constant value at all times—in planetary motion, as Newton was able to show from Kepler's three empirical laws. These are the scalar *total energy* and the vector *angular momentum* of each planet. Classically the total energy can have any value whatever, but it must, of course, be negative if the planet is to be trapped permanently in the solar system. In the quantum-mechanical theory of the hydrogen atom the electron energy is also a constant, but while it may have any positive value whatever, the *only* negative values it can have are specified by the formula

(7.16)
$$E_n = -\frac{me^4}{8\varepsilon_0{}^2h^2}\left(\frac{1}{n^2}\right)$$

The theory of planetary motion can also be worked out from Schrödinger's equation, and it yields an energy restriction identical in form to Eq. 7.16. However, the total quantum number n for any of the planets turns out to be so immense that the separation of permitted energy levels is far too small to be observable. For this reason classical physics provides an adequate description of planetary motion but fails within the atom.

The quantization of electron energy in the hydrogen atom is therefore described by the total quantum number n.

7.5 Orbital Quantum Number

The interpretation of the orbital quantum number l is a bit less obvious. Let us examine the differential equation for the radial part $R(r)$ of the wave function ψ:

(7.14) $$\frac{1}{r^2}\frac{d}{dr}\left(r^2\frac{dR}{dr}\right)+\left[\frac{8\pi^2m}{h^2}\left(\frac{e^2}{4\pi\varepsilon_0 r}+E\right)-\frac{l(l+1)}{r^2}\right]R=0$$

This equation is solely concerned with the radial aspect of the electron's motion, that is, its motion toward or away from the nucleus; yet we notice the presence of E, the total electron energy, in it. The total energy E includes the electron's kinetic energy of orbital motion, which should have nothing to do with its radial motion. This contradiction may be removed by the following argument. The kinetic energy T of the electron has two parts, T_{radial} due to its motion toward or away from the nucleus, and T_{orbital} due to its motion around the nucleus. The potential energy V of the electron is the electrostatic energy

$$V=-\frac{e^2}{4\pi\varepsilon_0 r}$$

Hence the total energy of the electron is

$$\begin{aligned}E&=T_{\text{radial}}+T_{\text{orbital}}+V\\&=T_{\text{radial}}+T_{\text{orbital}}-\frac{e^2}{4\pi\varepsilon_0 r}\end{aligned}$$

Inserting this expression for E in Eq. 7.14 we obtain, after a slight rearrangement,

7.19 $$\frac{1}{r^2}\frac{d}{dr}\left(r^2\frac{dR}{dr}\right)+\frac{8\pi^2m}{h^2}\left[T_{\text{radial}}+T_{\text{orbital}}-\frac{h^2l(l+1)}{8\pi^2mr^2}\right]R=0$$

If the last two terms in the square brackets of this equation cancel each other out, we shall have what we want: a differential equation for $R(r)$ that involves functions of the radius vector r exclusively. We therefore require that

7.20 $$T_{\text{orbital}}=\frac{h^2l(l+1)}{8\pi^2mr^2}$$

The orbital kinetic energy of the electron is

$$T_{\text{orbital}} = \tfrac{1}{2}mv^2_{\text{orbital}}$$

Since the angular momentum L of the electron is

$$L = mv_{\text{orbital}}r$$

we may write for the orbital kinetic energy

$$T_{\text{orbital}} = \frac{L^2}{2mr^2}$$

Hence, from Eq. 7.20,

$$\frac{L^2}{2mr^2} = \frac{h^2l(l+1)}{8\pi^2mr^2}$$

or

7.21 $$L = \sqrt{l(l+1)}\,\frac{h}{2\pi}$$ **Electron angular momentum**

Our interpretation of this result is that, since the orbital quantum number l is restricted to the values

$$l = 0, 1, 2, \ldots, (n-1)$$

the electron can have only those particular angular momenta L specified by Eq. 7.21. Like total energy E, *angular momentum is both conserved and quantized.* The quantity

$$\frac{h}{2\pi} = 1.054 \times 10^{-34} \text{ joule-sec}$$

is the natural unit of angular momentum, and is often abbreviated $\hbar$. In the remainder of this book we shall use $\hbar$ instead of $h/2\pi$.

In macroscopic planetary motion, once again, the quantum number describing angular momentum is so large that the separation into discrete angular-momentum states cannot be experimentally observed. For example, an electron (or, for that matter, any other body) whose orbital quantum number is 2 has the angular momentum

$$\begin{aligned} L &= \sqrt{2(2+1)}\,\hbar \\ &= \sqrt{6}\,\hbar \\ &= 2.6 \times 10^{-34} \text{ joule-sec} \end{aligned}$$

By contrast the orbital angular momentum of the earth is 2.7×10^{40} joule-sec!

It is customary to specify angular-momentum states by a letter, with s corresponding to $l = 0$, p to $l = 1$, and so on according to the following scheme:

$$\begin{matrix} l = & 0 & 1 & 2 & 3 & 4 & 5 & 6 & \ldots \\ & s & p & d & f & g & h & i & \ldots \end{matrix}$$

Angular momentum states

This peculiar code originated in the empirical classification of spectra into series called sharp, principal, diffuse, and fundamental which occurred before the theory of the atom was developed. Thus an s state is one with no angular momentum, a p state has the angular momentum $\sqrt{2}\,\hbar$, etc. The combination of the total quantum number with the letter that represents orbital angular momentum provides a convenient and widely used notation for atomic states. In this notation a state in which $n = 2$, $l = 0$ is a $2s$ state, for example, and one in which $n = 4$, $l = 2$ is a $4d$ state. Table 7.1 gives the designations of atomic states in hydrogen through $n = 6$, $l = 5$.

7.6 Magnetic Quantum Number

The orbital quantum number l determines the *magnitude* of the electron's angular momentum. Angular momentum, however, like linear momentum, is a vector quantity, and so to describe it completely requires that its *direction* be specified as well as its magnitude. (The vector **L**, we recall, is perpendicular to the plane in which the rotational motion takes place, and its sense is given by the right-hand rule: when the fingers of the right hand point in the direction of the motion, the thumb is in the direction of **L**. This rule is illustrated in Fig. 7-3.)

What possible significance can a direction in space have for a hydrogen atom? The answer becomes clear when we reflect that an electron revolving about a nucleus is a minute current loop and has a magnetic field like that of a magnetic dipole. In an external magnetic field **B**, a magnetic dipole has

Table 7.1

THE SYMBOLIC DESIGNATION OF ATOMIC STATES IN HYDROGEN

	(s) $l = 0$	(p) $l = 1$	(d) $l = 2$	(f) $l = 3$	(g) $l = 4$	(h) $l = 5$
$n = 1$	$1s$					
$n = 2$	$2s$	$2p$				
$n = 3$	$3s$	$3p$	$3d$			
$n = 4$	$4s$	$4p$	$4d$	$4f$		
$n = 5$	$5s$	$5p$	$5d$	$5f$	$5g$	
$n = 6$	$6s$	$6p$	$6d$	$6f$	$6g$	$6h$

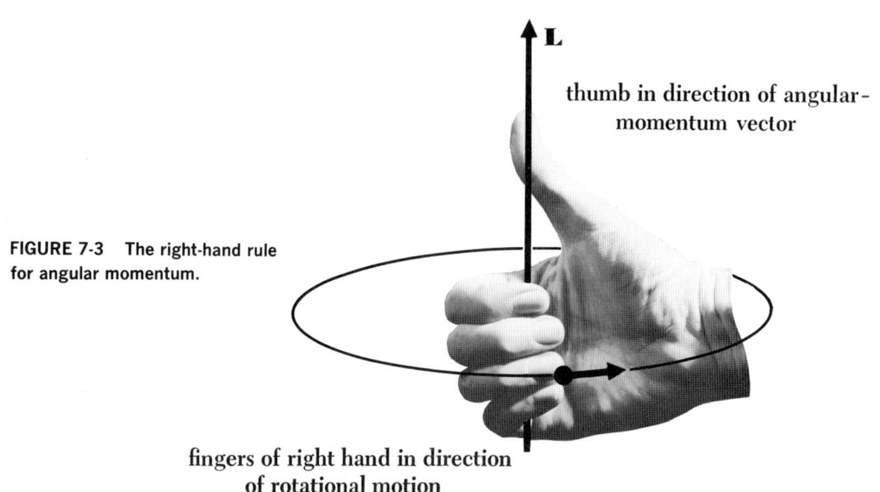

FIGURE 7-3 The right-hand rule for angular momentum.

an amount of potential energy V_m that depends upon both the magnitude μ of its magnetic moment and the orientation of this moment with respect to the field (Fig. 7-4). The torque τ on a magnetic dipole in a magnetic field of flux density **B** is

$$\tau = \mu B \sin \theta$$

where θ is the angle between $\boldsymbol{\mu}$ and **B**. The torque is a maximum when the dipole is perpendicular to the field, and zero when it is parallel or antiparallel to it. To calculate the potential energy V_m, we must first establish a reference configuration at which V_m is zero by definition. (Since only *changes* in potential energy are ever experimentally observed, the choice of a reference configuration is arbitrary.) It is convenient to set $V_m = 0$ when $\theta = 90°$, that is, when $\boldsymbol{\mu}$ is perpendicular to **B**. The potential energy at any other

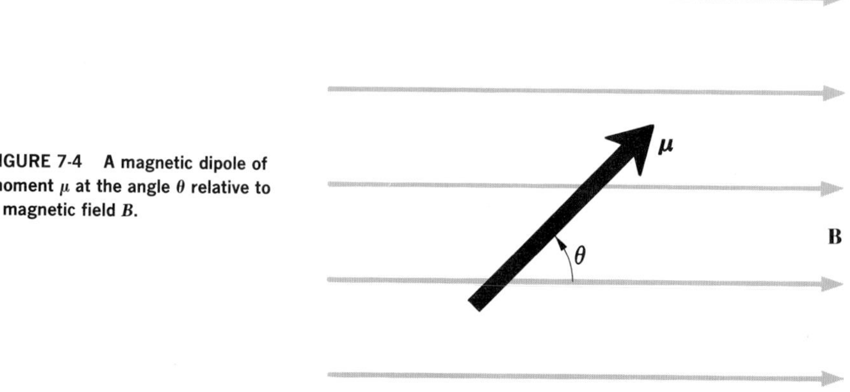

FIGURE 7-4 A magnetic dipole of moment μ at the angle θ relative to a magnetic field B.

orientation of $\boldsymbol{\mu}$ is equal to the external work that must be done to rotate the dipole from θ_0 of 90° to the angle θ characterizing that orientation. Hence

7.22
$$\begin{aligned} V_m &= \int_{90^\circ}^{\theta} \tau d\theta \\ &= \mu B \int_{90^\circ}^{\theta} \sin\theta\, d\theta \\ &= -\mu B \cos\theta \end{aligned}$$

When $\boldsymbol{\mu}$ points in the same direction as $\mathbf{B}$, then, $V_m = -\mu B$, its minimum value. This is a natural consequence of the fact that a magnetic dipole tends to align itself with an external magnetic field.

Since the magnetic moment of the orbital electron in a hydrogen atom depends upon its angular momentum $\mathbf{L}$, both the magnitude of $\mathbf{L}$ and its orientation with respect to the field determine the extent of the magnetic contribution to the total energy of the atom when it is in a magnetic field. The magnetic moment of a current loop is

$$\mu = iA$$

where i is the current and A the area it encloses. An electron which makes ν rev/sec in a circular orbit of radius r is equivalent to a current of $-e\nu$ (since the electronic charge is $-e$), and its magnetic moment is therefore

$$\mu = -e\nu\pi r^2$$

The linear speed v of the electron is $2\pi\nu r$, and so its angular momentum is

$$\begin{aligned} L &= mvr \\ &= 2\pi m\nu r^2 \end{aligned}$$

Comparing the formulas for magnetic moment μ and angular momentum L shows that

7.23
$$\boldsymbol{\mu} = -\left(\frac{e}{2m}\right)\mathbf{L}$$
Electron magnetic moment

for an orbital electron. The quantity $(-e/2m)$, which involves the charge and mass of the electron only, is called its *gyromagnetic ratio.* The minus sign means that $\boldsymbol{\mu}$ is in the opposite direction to $\mathbf{L}$. While the above expression for the magnetic moment of an orbital electron has been obtained by a classical calculation, quantum mechanics yields the same result. The magnetic potential energy of an atom in a magnetic field is therefore

7.24
$$V_m = \left(\frac{e}{2m}\right) LB \cos\theta$$

a function of both B and θ.

The magnitude L of the electron's orbital angular momentum we already know to be quantized, being given as a function of the orbital quantum number l by the formula

$$L = \sqrt{l(l+1)}\,\hbar$$

It is therefore not surprising to learn that the *direction* of **L** is also quantized with respect to an external magnetic field. This fact is often referred to as *space quantization.* The magnetic quantum number m_l specifies the direction of **L** by determining the component of **L** in the field direction. If we let the magnetic-field direction be parallel to the z axis, the component of **L** in this direction is

7.25 $$L_z = m_l\hbar$$ **Space quantization**

The possible values of m_l for a given value of l range from $+l$ through 0 to $-l$, so that the number of possible orientations of the angular-momentum vector **L** in a magnetic field is $2l + 1$. When $l = 0$, L_z can have only the single value of 0, which means that **L** is perpendicular to **B** (and $V_m = 0$); when $l = 1$, L_z may be $\hbar$, 0, or $-\hbar$; when $l = 2$, L_z may be $2\hbar$, $\hbar$, 0, $-\hbar$, or $-2\hbar$; and so on. We note that **L** can never be aligned exactly parallel or antiparallel to **B**, since L_z is always smaller than the magnitude $\sqrt{l(l+1)}\,\hbar$ of the total angular momentum. The space quantization of the orbital angular momentum of the hydrogen atom is shown in Fig. 7-5. We must regard an atom characterized by a certain value of m_l as standing ready to assume a certain orientation of its angular momentum **L** relative to an external magnetic field in the event it finds itself in such a field.

From Fig. 7-5 we see that the angle θ can have only the values specified by the relation

$$\cos\theta = \frac{m_l}{\sqrt{l(l+1)}}$$

while the permitted values of L are specified by

$$L = \sqrt{l(l+1)}\,\hbar$$

To find the magnetic energy that an atom of magnetic quantum number m_l has when it is in a magnetic field B, we insert the above expressions for $\cos\theta$ and L in Eq. 7.24, which yields

7.26 $$V_m = m_l\left(\frac{e\hbar}{2m}\right)B$$ **Magnetic energy**

The quantity $e\hbar/2m$ is called the *Bohr magneton;* its value is 9.27×10^{-24} joule/(weber/m^2). In a magnetic field, then, the energy of a particular

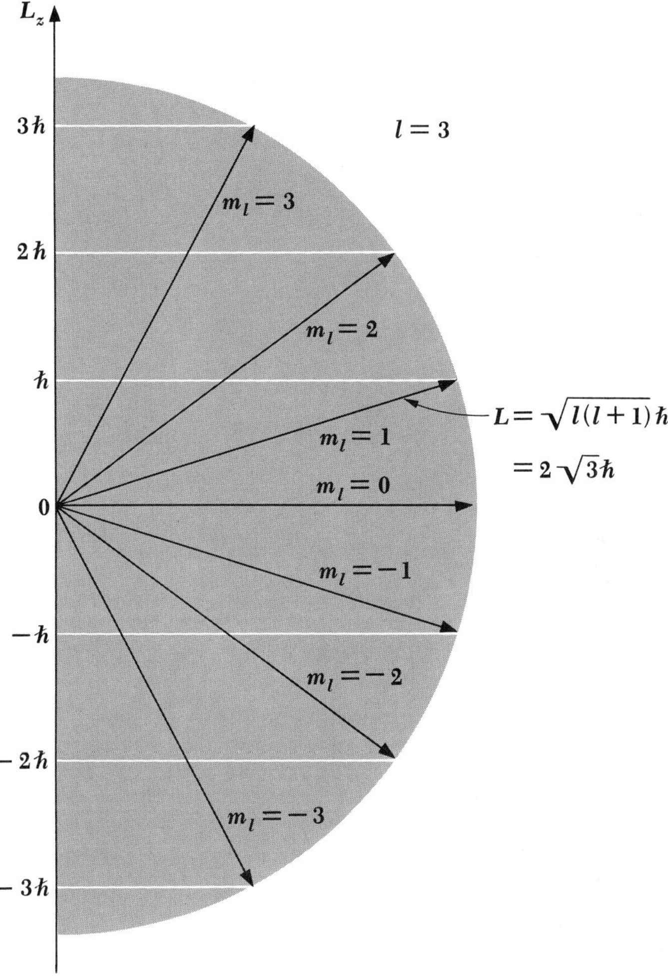

FIGURE 7-5 Space quantization of orbital angular momentum in the hydrogen atom.

atomic state depends upon the value of m_l as well as upon that of n. A state of total quantum number n breaks up into several substates when the atom is in a magnetic field, and their energies are slightly more or slightly less than the energy of the state in the absence of the field. This phenomenon leads to a "splitting" of individual spectral lines into separate lines when atoms radiate in a magnetic field, with the spacing of the lines dependent upon the magnitude of the field. The splitting of spectral lines by a magnetic field is called the *Zeeman effect* after the Dutch physicist Zeeman, who first observed it in 1896. The Zeeman effect is a vivid confirmation of space quantization.

The Stern-Gerlach Experiment

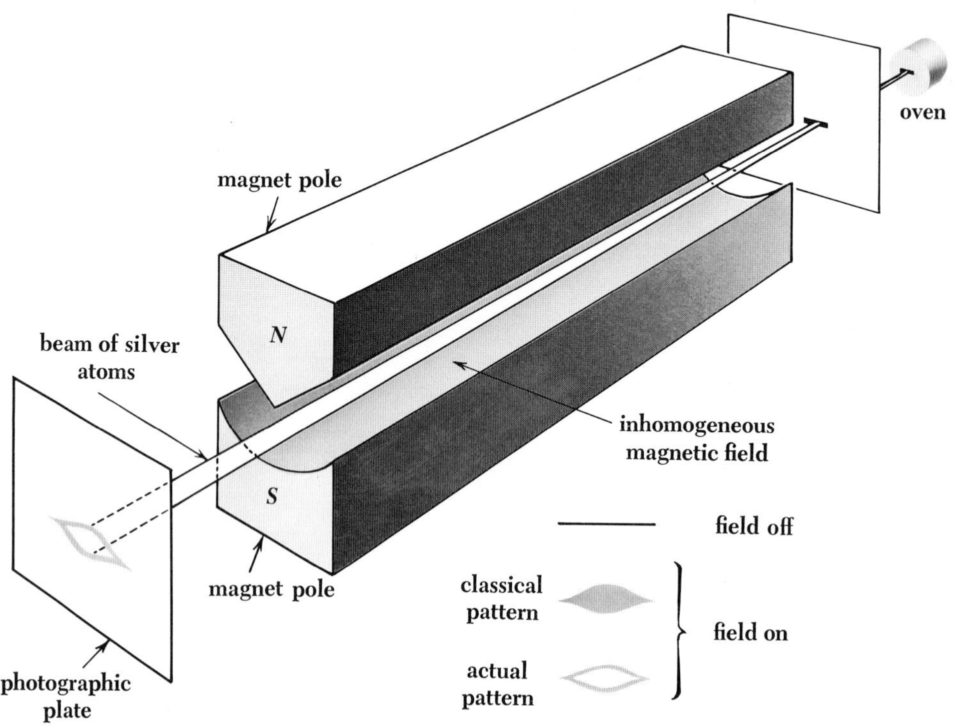

Space quantization was first explicitly demonstrated by O. Stern and W. Gerlach in 1921. They directed a beam of neutral silver atoms from an oven through a set of collimating slits into an inhomogeneous magnetic field, as shown. A photographic plate recorded the configuration of the beam after its passage through the field. In its normal state, the entire magnetic moment of a silver atom is due to the spin of one of its electrons. In a uniform magnetic field, such a dipole would merely experience a torque tending to align it with the field. In an inhomogeneous field, however, each "pole" of the dipole is subject to a force of different magnitude, and as a result there is a resultant force on the dipole that varies with its orientation relative to the field. Classically, all orientations should be present in a beam of atoms, which would result merely in a broad trace on the photographic plate instead of the thin line formed in the absence of any magnetic field. Stern and Gerlach found, however, that the initial beam split into two distinct parts, corresponding to the two opposite spin orientations in the magnetic field that are permitted by space quantization.

7.7 Electron Probability Density

In Bohr's model of the hydrogen atom the electron is visualized as revolving around the nucleus in a circular path. This model is pictured in a spherical polar coordinate system in Fig. 7-6. We see it implies that, if a suitable experiment were performed, the electron would always be found a distance of $r = n^2 r_0$ (where n is the quantum number of the orbit and $r_0 = 0.53$ A is the radius of the innermost orbit) from the nucleus and in the equatorial plane $\theta = 90°$, while its azimuth angle ϕ changes with time.

The quantum theory of the hydrogen atom modifies the straightforward prediction of the Bohr model in two ways. First, no definite values for r, θ, or ϕ can be given, but only the relative probabilities for finding the electron at various locations. This imprecision is, of course, a consequence of the wave nature of the electron. Second, we cannot even think of the electron as moving around the nucleus in any conventional sense since the probability density ψ^2 is independent of time and may vary considerably from place to place.

The electron wave function ψ in a hydrogen atom is given by

$$\psi = R\Theta\Phi$$

where

$$R = R_{nl}(r)$$

describes how ψ varies with r when the orbital and total quantum numbers have the values n and l;

$$\Theta = \Theta_{lm_l}(\theta)$$

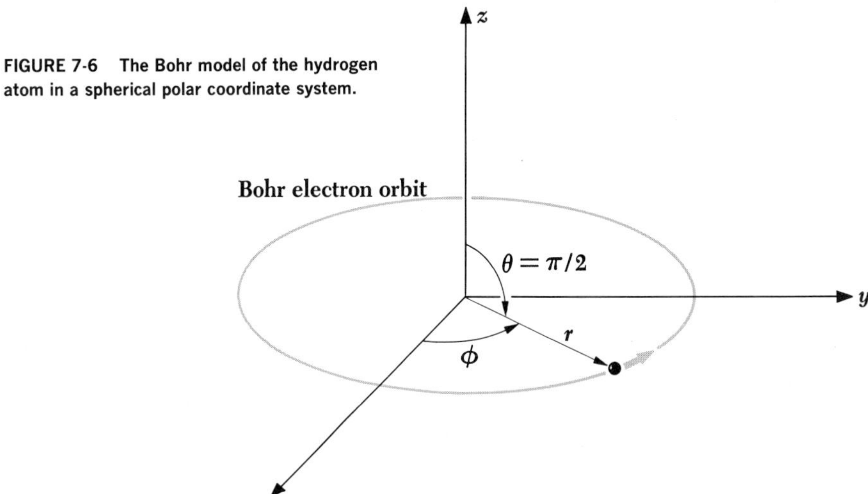

FIGURE 7-6 The Bohr model of the hydrogen atom in a spherical polar coordinate system.

describes how ψ varies with θ when the magnetic and orbital quantum numbers have the values l and m_l; and

$$\Phi = \Phi_{m_l}(\phi)$$

describes how ψ varies with ϕ when the magnetic quantum number is m_l. The probability density ψ^2 may therefore be written

$$\psi^2 = R^2\Theta^2\Phi^2$$

where it is understood that the square of any function is to be replaced by the product of it and its complex conjugate if the function is a complex quantity.

It is easy to show that the azimuthal probability density Φ^2, which is a measure of the likelihood of finding the electron at a particular azimuth angle ϕ, is a constant that does not depend upon ϕ at all. Earlier in this chapter we found that

$$\Phi = Ae^{im_l\phi} \tag{7.15}$$

where A is a constant of integration. The complex conjugate of Φ is

$$\Phi^* = Ae^{-im_l\phi} \tag{7.27}$$

and so

$$\begin{aligned}\Phi^2 &= \Phi\Phi^* \\ &= A^2e^{(im_l\phi - im_l\phi)} \\ &= A^2\end{aligned}$$

This result means that the electron's probability density is symmetrical about the z axis regardless of the quantum state it is in, so that the electron has the same chance of being found at one angle ϕ as at another.

To evaluate A, we call upon the fact that the integral of Φ^2 over all angles must equal 1, since the electron must exist somewhere. Hence

$$\begin{aligned}\int_0^{2\pi} \Phi^2 d\phi &= A^2 \int_0^{2\pi} d\phi \\ &= 2\pi A^2 \\ &= 1\end{aligned}$$

and

$$A = \frac{1}{\sqrt{2\pi}}$$

The normalized azimuthal function is therefore

$$\Phi = \frac{1}{\sqrt{2\pi}} e^{im_l\phi} \tag{7.28}$$

The radial part R of the wave function, in contrast to Φ, not only varies with r but does so in a different way for each combination of quantum numbers n and l. Figure 7-7 contains graphs of R versus r for $1s$, $2s$, $2p$, $3s$, $3p$, and $3d$ states of the hydrogen atom. Evidently R is a maximum at $r = 0$—that is, at the nucleus itself—for all s states, while it is zero at $r = 0$ for states that possess angular momentum.

The *probability density* of the electron at a distance r from the nucleus is proportional to R^2, but the *actual probability* P of finding it there is proportional to $R^2\,dV$, where dV is an infinitesimal volume element between

FIGURE 7-7 The variation with distance from the nucleus of the radial part of the electron wave function in hydrogen for various quantum states.

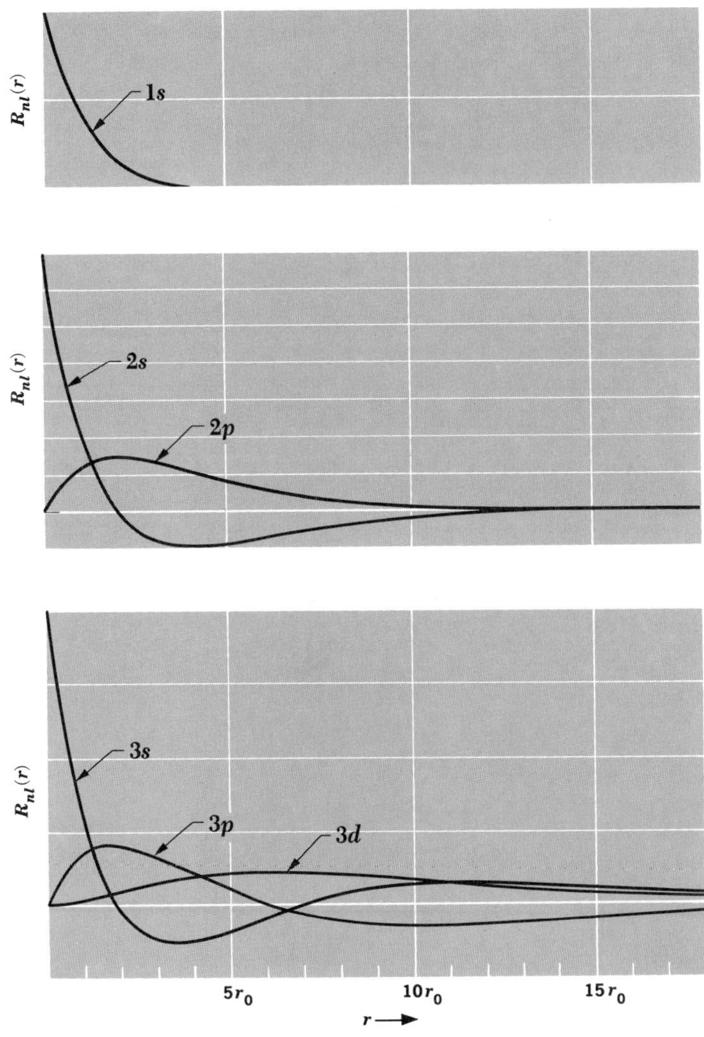

FIGURE 7-8 The volume of a spherical shell.

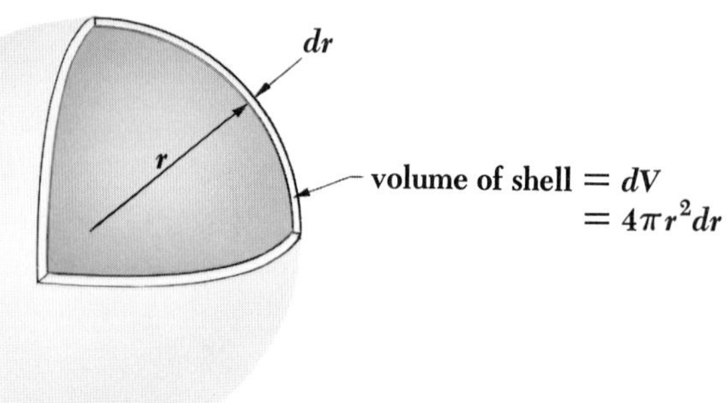

r and $r + dr$. (We recognize the parallel with mass density; water has a *mass density* of 1 g/cm³, but its *mass* depends upon the specific volume of water in question.) Now dV is the volume of a spherical shell whose inner radius is r and whose outer radius is $r + dr$, as in Fig. 7-8, so that

7.29 $$dV = 4\pi r^2\, dr$$

If R is a normalized function, then, the actual numerical probability P of finding the electron in a hydrogen atom at a distance between r and $r + dr$ from the nucleus is

7.30 $$P = 4\pi r^2 R_{nl}{}^2\, dr$$

FIGURE 7-9 The probability of finding the electron in a hydrogen atom at a distance between r and $r+dr$ from the nucleus for the quantum states of Fig. 7-7.

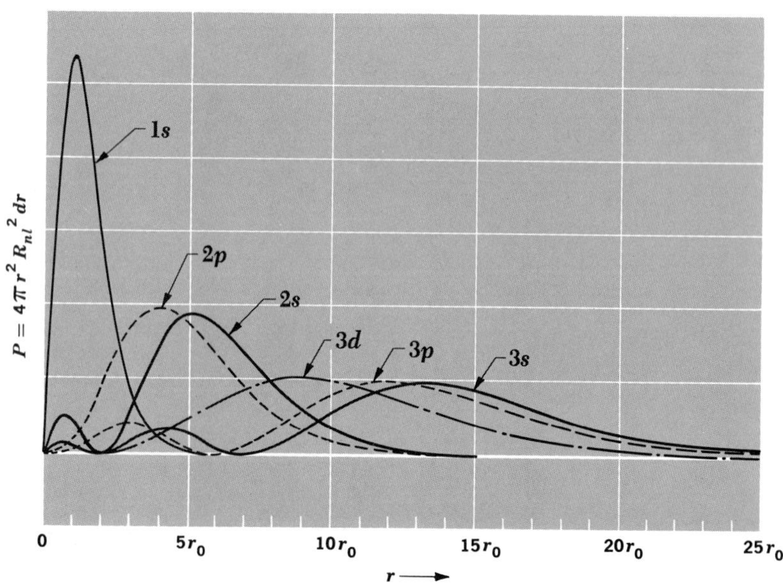

Equation 7.30 is plotted in Fig. 7-9 for the same states whose radial functions R appear in Fig. 7-7; the curves are quite different as a rule. It is interesting to note that P is not a maximum at the nucleus for s states, as R itself is, but has its maximum at a finite distance from it. Interestingly enough, the most probable value of r for a 1s electron is exactly r_0, the orbital radius of a ground-state electron in the Bohr model. However, the *average* value of r for a 1s electron is $1.5r_0$, which seems puzzling at first sight because the energy levels are the same in both the quantum-mechanical and Bohr atomic models. This apparent discrepancy is removed when we recall that the electron energy depends upon $1/r$ rather than upon r directly, and, as it happens, the average value of $1/r$ for a 1s electron is exactly $1/r_0$.

The function Θ varies with zenith angle θ for all quantum numbers l and m_l except $l = m_l = 0$, which are s states. The probability density Θ^2 for an s state is a constant (½, in fact), which means that, since Φ^2 is also a constant, the electron probability density ψ^2 has the same value at a given r in all directions. Electrons in other states, however, do have angular preferences, sometimes quite complicated ones. This can be seen in Fig. 7-10, in which electron probability densities as functions of r and θ are shown for several atomic states. (The quantity plotted is ψ^2, not $\psi^2\,dV$.) Since ψ^2 is independent of ϕ, a three-dimensional picture of ψ^2 is obtained by rotating a particular representation about a vertical axis. When this is done, the probability densities for s states are evidently spherically symmetric, while others are not. The pronounced lobe patterns characteristic of many of the states turn out to be significant in chemistry since these patterns help determine the manner in which adjacent atoms in a molecule interact; we shall refer to this notion once more in Chap. 9.

A study of Fig. 7-10 also reveals quantum-mechanical states with a remarkable resemblance to those of the Bohr model. The electron probability-density distribution for a 2p state with $m_l = \pm 1$, for instance, is like a doughnut in the equatorial plane centered at the nucleus, and calculation shows the most probable distance of the electron from the nucleus to be $4r_0$—precisely the radius of the Bohr orbit for the same total quantum number. Similar correspondences exist for 3d states with $m_l = \pm 2$, 4f states with $m_l = \pm 3$, and so on: in every case the highest angular momentum possible for that energy level, and in every case the angular momentum vector as near the z axis as possible so that the probability density be as close as possible to the equatorial plane. Thus the Bohr model predicts the most probable location of the electron in *one* of the several possible states in each energy level.

FIGURE 7-10 Photographic representation of the electron probability density distribution ψ^2 for several energy states. These may be regarded as sectional views of the distributions in a plane containing the polar axis, which is vertical and in the plane of the paper. The scale varies from figure to figure.

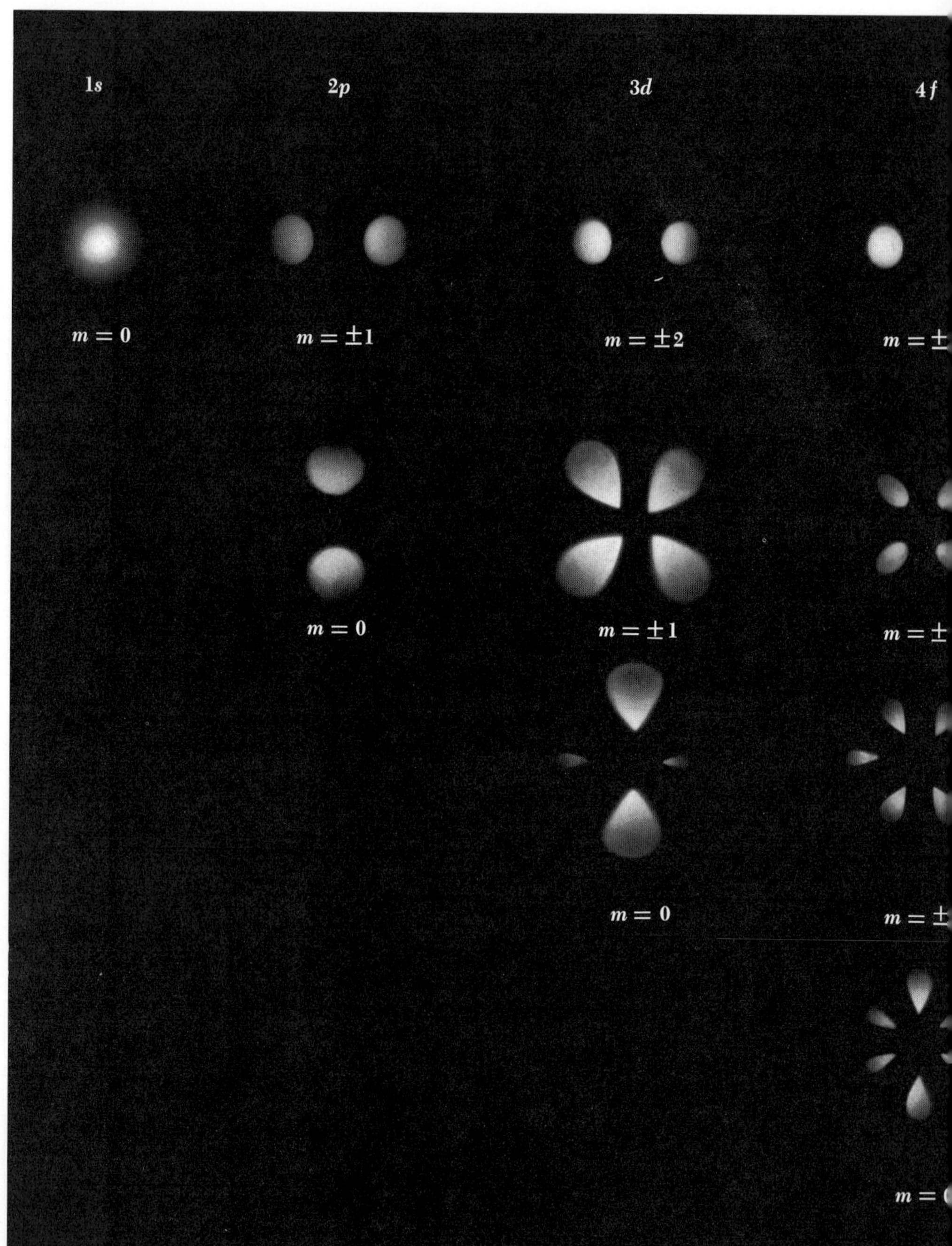

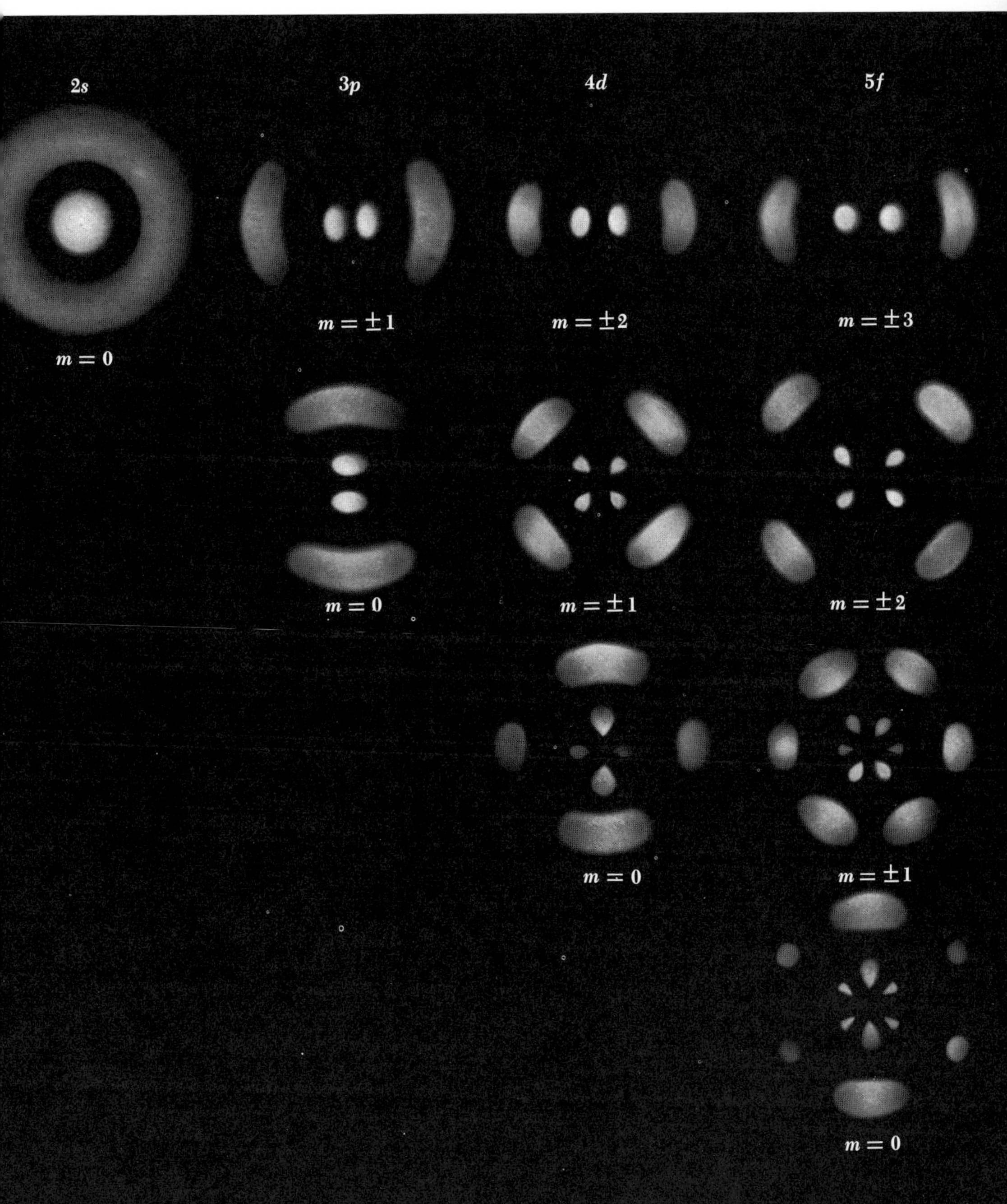
2s
3p
4d
5f
m = ±1
m = ±2
m = ±3
m = 0
m = 0
m = ±1
m = ±2
m = 0
m = ±1
m = 0

7.8 The Spectrum of Hydrogen

In formulating his theory of the hydrogen atom, Bohr was obliged to postulate that the frequency ν of the radiation emitted by an atom dropping from an energy level E_m to a lower level E_n is

$$\nu = \frac{E_m - E_n}{h}$$

It is not difficult to show that this relationship arises naturally from the quantum theory of the atom we have been discussing. Our starting point is the straightforward assertion that an atomic electron whose average position relative to the nucleus is constant in time does not radiate, while if its average position relative to the nucleus oscillates, electromagnetic waves are emitted whose frequency is the same as that of the oscillation. For simplicity we shall consider the electron to be confined to motion in the x direction only.

Our problem, then, is to determine the average position $\overline{x}$ of the electron in a hydrogen atom under various circumstances. To make the procedure clear, let us first answer a somewhat different question: what is the average position $\overline{x}$ of a number of electrons distributed along the x axis in such a way that there are N_1 electrons at x_1, N_2 electrons at x_2, and so on? The average position in this case is the same as the center of mass of the distribution, and so

$$\begin{aligned} x &= \frac{N_1x_1 + N_2x_2 + N_3x_3 + \dots}{N_1 + N_2 + N_3 + \dots} \\ &= \frac{\Sigma N_i x_i}{\Sigma N_i} \end{aligned}$$

When we are dealing with a single electron, we must replace the *number* N_i of electrons at x_i by the *probability* P_i that the electron be found in the interval dx at x_i. This probability is

$$\begin{aligned} P_i &= \Psi_i^2\, dx \\ &= \Psi_i \Psi_i^*\, dx \end{aligned}$$

where Ψ_i is the electron wave function evaluated at $x = x_i$. (We must use $\Psi\Psi^*$ here instead of Ψ^2 for probability density since all time-dependent wave functions are complex.) Making this substitution and changing the summations to integrals, we see that the average position of the single electron is

7.31 $$\overline{x} = \frac{\int_{-\infty}^{\infty} x\Psi\Psi^*\, dx}{\int_{-\infty}^{\infty} \Psi\Psi^*\, dx}$$

If Ψ is a normalized wave function, the denominator of Eq. 7.31, which is in general proportional to the probability that the electron exist somewhere between $x = -\infty$ and $x = +\infty$, is *equal* to this probability and has the value 1. Hence, with the understanding that the wave functions we shall consider are all normalized,

7.32 $$\overline{x} = \int_{-\infty}^{\infty} x\Psi\Psi^* \, dx$$

The time-dependent wave function Ψ_n of an electron in a state of quantum number n and energy E_n is the product of a time-independent wave function ψ_n and a time-varying function whose frequency is

$$\nu_n = \frac{E_n}{h}$$

Hence

7.33 $$\Psi_n = \psi_n e^{-(2\pi i E_n/h)t}$$

and

7.34 $$\Psi_n^* = \psi_n^* e^{+(2\pi i E_n/h)t}$$

The average position of such an electron is

$$\overline{x} = \int_{-\infty}^{\infty} x\Psi_n\Psi_n^* \, dx$$

$$= \int_{-\infty}^{\infty} x\psi_n\psi_n^* e^{[(2\pi i E_n/h)-(2\pi E_n/h)]t} \, dx$$

7.35 $$= \int_{-\infty}^{\infty} x\psi_n\psi_n^* \, dx$$

which is constant since ψ_n and ψ_n^* are, by definition, functions of position only. The electron does not oscillate, and no radiation occurs. Thus quantum mechanics predicts that an atom in a specific quantum state does not radiate; this agrees with observation, though not with classical physics.

We are now in a position to consider an electron changing from one energy state to another. Let us formulate a definite problem: an atom is in its ground state when, at $t = 0$, an excitation process of some kind (a beam of radiation, say, or collisions with other particles) begins to act upon it. Subsequently we find that the atom emits radiation corresponding to a transition from an excited state of energy E_m to the ground state, and we conclude that at some time during the intervening period the atom existed in the state m. The wave function Ψ of an electron capable of existing in states n or m may be written

7.36 $$\Psi = a\Psi_n + b\Psi_m$$

where aa^* is the probability that the electron is in state n and bb^* the probability that it is in state m. Of course, it must always be true that $aa^* + bb^* = 1$. At $t = 0$, $a = 1$ and $b = 0$ by hypothesis, when the electron is in the excited state $a = 0$ and $b = 1$, and ultimately $a = 1$ and $b = 0$ once more. While the electron is in either state, there is no radiation, but when it is in the midst of the transition from m to n (that is, when both a and b have nonvanishing values), electromagnetic waves are produced. Substituting the composite wave function of Eq. 7.36 into Eq. 7.32, we obtain for the average electron position

$$\bar{x} = \int_{-\infty}^{\infty} x(a\Psi_n + b\Psi_m)(a^*\Psi_n{}^* + b^*\Psi_m{}^*)\, dx$$

7.37 $$= \int_{-\infty}^{\infty} x[a^2\Psi_n\Psi_n{}^* + ab^*\Psi_n\Psi_m{}^* + a^*b\Psi_m\Psi_n{}^* + b^2\Psi_m\Psi_m{}^*]\, dx$$

(Here, as before, we let $aa^* = a^2$.) The first and last terms in the brackets are constants, according to Eq. 7.35, and so the second and third terms are the only ones capable of contributing to a time variation in $\bar{x}$.

With the help of Eqs. 7.33 to 7.35 we may expand Eq. 7.37 to obtain

7.38 $$\bar{x} = a^2\int_{-\infty}^{\infty} x\psi_n\psi_n{}^*dx + ab^*\int_{-\infty}^{\infty} x\psi_n e^{-(2\pi i E_n/h)t}\psi_m{}^*e^{+(2\pi i E_m/h)t}\, dx$$
$$+ a^*b\int_{-\infty}^{\infty} x\psi_m e^{-(2\pi i E_m/h)t}\psi_n{}^*e^{+(2\pi i E_n/h)t}\, dx + b^2\int_{-\infty}^{\infty} x\psi_m\psi_m{}^*\, dx$$

It is possible to show that

$$\psi_n\psi_m{}^* = \psi_m\psi_n{}^* \qquad \text{and} \qquad a^*b = ab^*$$

in the case of the finite bound system of two states, which permits us to combine the time-varying terms of Eq. 7.38 into the single term

7.39 $$ab^*\int_{-\infty}^{\infty} x\psi_n\psi_m{}^*[e^{(2\pi i/h)(E_m - E_n)t} + e^{-(2\pi i/h)(E_m - E_n)t}]\, dx$$

Now

$$e^{i\theta} = \cos\theta + i\sin\theta$$

and

$$e^{-i\theta} = \cos\theta - i\sin\theta$$

so that

$$e^{i\theta} + e^{-i\theta} = 2\cos\theta$$

Hence Eq. 7.39 simplifies to

$$2ab^*\cos 2\pi\left(\frac{E_m - E_n}{h}\right)t\int_{-\infty}^{\infty} x\psi_n\psi_m{}^*\, dx$$

which varies sinusoidally with time at the frequency

7.40 $$\nu = \frac{E_m - E_n}{h}$$

The full expression for $\bar{x}$, the average position of the electron, is therefore

7.41 $$\bar{x} = a^2 \int_{-\infty}^{\infty} x\psi_n\psi_n^* \, dx + b^2 \int_{-\infty}^{\infty} x\psi_m\psi_m^* \, dx + 2ab^* \cos 2\pi\nu t \int_{-\infty}^{\infty} x\psi_n\psi_m^* \, dx$$

When the electron is in state n or state m, the probabilities b^2 or a^2 respectively are zero, and the electron is, on the average, stationary. When the electron is undergoing a transition between these states, its average position oscillates with the frequency ν. This frequency is identical with that postulated by Bohr and verified by experiment, and, as we have seen, Eq. 7.40 can be derived using quantum mechanics without making any special assumptions. It is interesting to note that the frequency of the radiation is the same frequency as the beats we might imagine to be produced if the electron simultaneously existed in both the n and m states, whose characteristic frequencies are respectively E_n/h and E_m/h.

7.9 Selection Rules

It was not necessary for us to know the values of the probabilities a and b as functions of time, nor the electron wave functions ψ_n and ψ_m, in order to determine ν. We must know these quantities, however, if we wish to compute the actual rate at which radiation is emitted. The uncertainty principle prohibits us from determining a and b with any precision for a particular atom, but their average values can be ascertained for a group of many atoms. Even when both a and b are finite, though, in order for a transition between the two states to take place, the integral

$$\int_{-\infty}^{\infty} x\psi_n\psi_m^* \, dx$$

cannot be zero, since the intensity of the radiation is proportional to it. Transitions for which this integral is finite are called *allowed transitions,* while those for which it is zero are called *forbidden transitions.*

In the case of the hydrogen atom, three quantum numbers are needed to specify the initial and final states involved in a radiative transition. If the total, orbital, and magnetic quantum numbers of the initial state are n', l', m_l'

respectively and those of the final state are n, l, m_l, and the coordinate u represents either the x, y, or z coordinate, the condition for an allowed transition is

7.43 $$\int_{-\infty}^{\infty} u\psi_{n,l,m_l}\,\psi_{n',l',m_l'}{}^{*}\,du \neq 0$$

When u is taken as x, for example, the radiation referred to is that which would be produced by an ordinary dipole antenna lying along the x axis. Since the wave functions ψ_{n,l,m_l} for the hydrogen atom are known, Eq. 7.43 can be evaluated for $u = x$, $u = y$, and $u = z$ for all pairs of states differing in one or more quantum numbers. When this is done, it is found that the only transitions that can occur are those in which the orbital quantum number l changes by $+1$ or -1 and the magnetic quantum number m_l does not change or changes by $+1$ or -1; in other words, the condition for an allowed transition is that

7.44 $$\Delta l = \pm 1$$

Selection rules

7.45 $$\Delta m_l = 0, \pm 1$$

The change in total quantum number n is not restricted. Equations 7.44 and 7.45 are known as the *selection rules* for allowed transitions.

In order to get an intuitive idea of the physical basis for these selection rules, let us refer again to Fig. 7-10. There we see that, for instance, a transition from a $2p$ state to a $1s$ state involves a change from one probability-density distribution to another such that the oscillating charge during the transitions behaves like an electric dipole antenna. On the other hand, a transition from a $2s$ state to a $1s$ state involves a change from a spherically symmetric probability-density distribution to another spherically symmetric distribution, which means that the oscillations that take place are like those of a charged sphere that alternately expands and contracts. Oscillations of this kind do not lead to the radiation of electromagnetic waves.

The selection rule requiring that l change by ± 1 if an atom is to radiate means that an emitted photon carries off angular momentum equal to the difference between the angular momenta of the atom's initial and final states. The classical analog of a photon with angular momentum is a circularly polarized electromagnetic wave, so that this notion is not unique with quantum theory.

Problems

1. Prove that Eqs. 7.1 and 7.3 are equivalent.

2. Show that

$$\Theta_{20}(\theta) = \frac{\sqrt{10}}{4}(3\cos^2\theta - 1)$$

is a solution of Eq. 7.13 and that it is normalized.

3. Show that

$$R_{10}(r) = \frac{2}{{r_0}^{3/2}} e^{-r/r_0}$$

where $r_0 = 4\pi\hbar^2\varepsilon_0/me^2$, is a solution of Eq. 7.14.

4. Why is it possible for an atomic electron to have a precisely defined angular momentum even though its angular position at any time may not be at all well-defined?

5. The probability of finding an atomic electron whose radial wave function is $R(r)$ outside a sphere of radius r_0 centered on the nucleus is

$$\int_{r_0}^{\infty} R(r)^2 4\pi r^2 \, dr$$

The wave function $R_{10}(r)$ of Problem 3 corresponds to the ground state of a hydrogen atom, and r_0 there is the radius of the Bohr orbit corresponding to that state. Calculate the probability of finding a ground-state electron in a hydrogen atom at a distance greater than r_0 from the nucleus.

ELECTRON SPIN AND COMPLEX ATOMS

Despite the accuracy with which the quantum theory accounts for certain of the properties of the hydrogen atom, and despite the elegance and essential simplicity of this theory, it cannot approach a complete description of this atom or of other atoms without the further hypothesis of electron spin and the exclusion principle associated with it. In this chapter we shall be introduced to the role of electron spin in atomic phenomena and to the reason why the exclusion principle is the key to understanding the structures of complex atomic systems.

8.1 Electron Spin

Let us begin by citing two of the most conspicuous shortcomings of the theory developed in the preceding chapter. The first is the experimental fact that many spectral lines actually consist of two separate lines that are very close together. An example of this *fine structure* is the first line of the Balmer series of hydrogen, which arises from transitions between the $n = 3$ and $n = 2$ levels in hydrogen atoms. Here the theoretical prediction is for a single line of wavelength 6,563 A, while in reality there are two lines 1.4 A apart—a small effect, but a conspicuous failure for the theory.

The second major discrepancy between the simple quantum theory of the atom and the experimental data occurs in the Zeeman effect, which we briefly mentioned in Sec. 7.6. There we saw that a hydrogen atom of magnetic quantum number m_l has the magnetic energy

8.1 $$V_m = m_l \frac{e\hbar}{2m} B$$

when it is located in a magnetic field of flux density B. Now m_l can have the $2l + 1$ values of $+l$ through 0 to $-l$, so a state of given orbital quantum number l is split into $2l + 1$ substates differing in energy by $(e\hbar/2m)B$ when the atom is in a magnetic field. However, because changes in m_l are restricted to $\Delta m_l = 0, \pm 1$, a given spectral line that arises from a transition between two states of different l is split into only three components, as shown in Fig. 8-1. The *normal Zeeman effect*, then, consists of the splitting of a spectral line of frequency ν_0 into three components whose frequencies are

8.2 $$\begin{aligned} \nu_1 &= \nu_0 - \frac{e\hbar}{2m}\frac{B}{h} = \nu_0 - \frac{e}{4\pi m}B \\ \nu_2 &= \nu_0 \\ \nu_3 &= \nu_0 + \frac{e\hbar}{2m}\frac{B}{h} = \nu_0 + \frac{e}{4\pi m}B \end{aligned}$$ Normal Zeeman effect

While the normal Zeeman effect is indeed observed in the spectra of a few elements under certain circumstances, more often it is not: four, six, or even more components may appear, and even when three components are present their spacing may not agree with Eq. 8.2. Several anomalous Zeeman patterns are shown in Fig. 8-2 together with the predictions of Eq. 8.2.

FIGURE 8-1 **The normal Zeeman effect.**

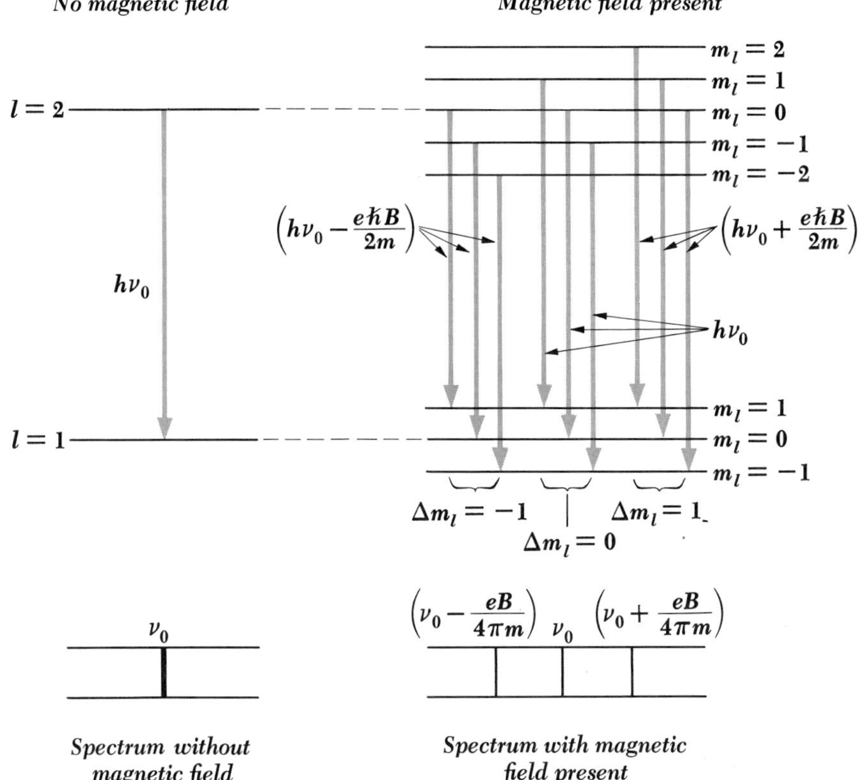

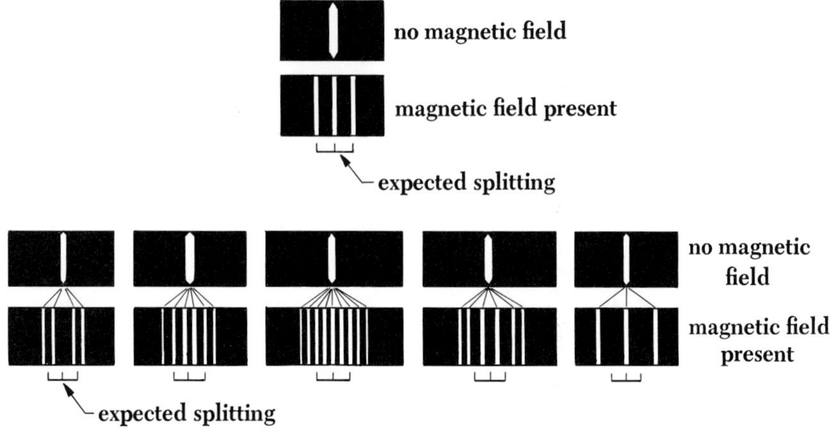

FIGURE 8-2 The normal and anomalous Zeeman effects in various spectral lines.

In an effort to account for both fine structure in spectral lines and the anomalous Zeeman effect, S. A. Goudsmit and G. E. Uhlenbeck proposed in 1925 that the electron possesses an intrinsic angular momentum independent of any orbital angular momentum it might have and, associated with this angular momentum, a certain magnetic moment. What Goudsmit and Uhlenbeck had in mind was a classical picture of an electron as a charged sphere spinning on its axis. The rotation involves angular momentum, and because the electron is negatively charged, it has a magnetic moment μ_s opposite in direction to its angular momentum vector $\mathbf{L}_s$. The notion of electron spin proved to be successful in explaining not only fine structure and the anomalous Zeeman effect but a wide variety of other atomic effects as well. Of course, the idea that electrons are spinning charged spheres is hardly in accord with quantum mechanics, but in 1928 Dirac was able to show on the basis of a relativistic quantum-theoretical treatment that particles having the charge and mass of the electron must have just the intrinsic angular momentum and magnetic moment attributed to them by Goudsmit and Uhlenbeck.

The quantum number s is used to describe the spin angular momentum of the electron. The only value s can have is $s = \frac{1}{2}$; this restriction follows from Dirac's theory and, as we shall see below, may also be obtained empirically from spectral data. The magnitude S of the angular momentum due to electron spin is given in terms of the spin quantum number s by the formula

8.3 $$S = \sqrt{s(s+1)}\,\hbar = \frac{\sqrt{3}}{2}\,\hbar$$

which is the same formula as that giving the magnitude L of the orbital angular momentum in terms of the orbital quantum number l:

$$L = \sqrt{l(l+1)}\,\hbar$$

The space quantization of electron spin is described by the spin magnetic quantum number m_s. Just as the orbital angular-momentum vector can have the $2l + 1$ orientations in a magnetic field from $+l$ to $-l$, the spin angular-momentum vector can have the $2s + 1 = 2$ orientations specified by $m_s = +\frac{1}{2}$ and $m_s = -\frac{1}{2}$ (Fig. 8-3). The component S_z of the spin angular momentum of an electron along a magnetic field in the z direction is determined by the spin magnetic quantum number, so that

8.4 $$\begin{aligned} S_z &= m_s\hbar \\ &= \pm\tfrac{1}{2}\hbar \end{aligned}$$

The gyromagnetic ratio characteristic of electron spin is almost exactly twice that characteristic of electron orbital motion. Thus, taking this ratio as equal to 2, the spin magnetic moment $\boldsymbol{\mu}_s$ of an electron is related to its spin angular momentum $\mathbf{S}$ by

8.5 $$\boldsymbol{\mu}_s = \frac{e}{m}\mathbf{S}$$

The possible components of $\boldsymbol{\mu}_s$ along any axis, say the z axis, are therefore limited to

8.6 $$\mu_{sz} = \pm\frac{e\hbar}{2m}$$

We recognize the quantity $(e\hbar/2m)$ as the Bohr magneton.

The fine-structure doubling of spectral lines may be explained on the basis of a magnetic interaction between the spin and orbital angular momenta of atomic electrons. This spin-orbit coupling can be understood in terms of a straightforward classical model. An electron revolving about a proton finds itself in a magnetic field because, in its own frame of reference, the proton is circling about *it*. This magnetic field then acts upon the electron's own spin magnetic moment to produce a kind of internal Zeeman effect. The energy V_m of a magnetic dipole of moment $\boldsymbol{\mu}$ in a magnetic field of flux density $\mathbf{B}$ is, in general,

8.7 $$V_m = -\mu B \cos\theta$$

where θ is the angle between $\boldsymbol{\mu}$ and $\mathbf{B}$. The quantity $\mu \cos\theta$ is the component of $\boldsymbol{\mu}$ parallel to $\mathbf{B}$, which in the case of the spin magnetic moment of the electron is μ_{sz}. Hence, letting

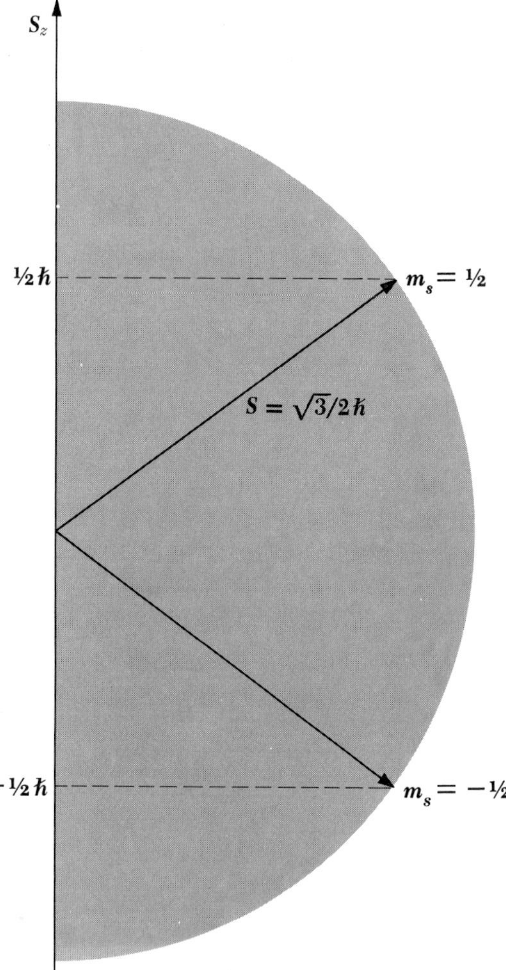

FIGURE 8-3 The two possible orientations of the spin angular-momentum vector.

$$\mu \cos \theta = \mu_{sz} = \pm \frac{e\hbar}{2m}$$

we find that

8.8 $$V_m = \pm \frac{e\hbar}{2m} B$$

Depending upon the orientation of its spin vector, the energy of an electron in a given atomic quantum state will be higher or lower by $(e\hbar/2m)B$ than its energy in the absence of spin-orbit coupling. The result is the splitting of every quantum state (except s states) into two separate substates and, consequently, the splitting of every spectral line into two component lines.

The assignment of $s = \frac{1}{2}$ is the only one that conforms to the observed fine-structure doubling. The fact that what should be single states are in fact twin states imposes the condition that the $2s + 1$ possible orientations of the spin angular-momentum vector **S** must total 2. Hence

$$2s + 1 = 2$$
$$s = \tfrac{1}{2}$$

To check whether the observed fine structure in spectral lines corresponds to the energy shifts predicted by Eq. 8.8, we must compute the magnitude B of the magnetic field experienced by an atomic electron. An estimate is easy to obtain. A circular wire loop of radius r that carries the current i has a magnetic field of flux density

$$B = \frac{\mu_0 i}{2r}$$

at its center. An orbital electron, say in a hydrogen atom, "sees" itself circled f times each second by a proton of charge $+e$, for a resulting flux density of

$$B = \frac{\mu_0 f e}{2r}$$

In the ground-state Bohr atom $f \approx 7 \times 10^{15}$ and $r \approx 5 \times 10^{-11}$ m, so that

$$B \approx 14 \text{ webers/m}^2$$

which is a very strong magnetic field. The value of the Bohr magneton is

$$\frac{e\hbar}{2m} = 9.27 \times 10^{-24} \text{ joule/(weber/m}^2)$$

Hence the magnetic energy V_m of such an electron is, from Eq. 8.8,

$$V_m = \frac{e\hbar}{2m} B$$
$$\approx 9.27 \times 10^{-24} \text{ joule/(weber/m}^2) \times 14 \text{ webers/m}^2$$
$$\approx 1.30 \times 10^{-22} \text{ joule}$$

The wavelength shift corresponding to such a change in energy is about 2 A for a spectral line of unperturbed wavelength 6,563 A, slightly more than the observed splitting of the line originating in the $n = 3 \rightarrow n = 2$ transition. However, the flux density of the magnetic field at orbits of higher order is less than for ground-state orbits, which accounts for the discrepancy.

8.2 The Exclusion Principle

In the normal configuration of a hydrogen atom, the electron is in its lowest quantum state. What are the normal configurations of more complex atoms? Are all 92 electrons of a uranium atom in the same quantum state, to be envisioned perhaps as circling the nucleus crowded together in a single Bohr orbit? Many lines of evidence make this hypothesis unlikely. One example is the great difference in chemical behavior exhibited by certain elements whose atomic structures differ by just one electron: for instance, the elements having atomic numbers 9, 10, and 11 are respectively the halogen gas fluorine, the inert gas neon, and the alkali metal sodium. Since the electronic structure of an atom controls its interactions with other atoms, it is hard to understand why the chemical properties of the elements should change so abruptly with a small change in atomic number if all of the electrons in an atom exist together in the same quantum state.

In 1925 Wolfgang Pauli discovered the fundamental principle that governs the electronic configurations of atoms having more than one electron. His *exclusion principle* states that **no two electrons in an atom can exist in the same quantum state.** Each electron in an atom must have a different set of quantum numbers n, l, m_l, m_s. Pauli was led to this conclusion from a study of atomic spectra. It is possible to determine the various states of an atom from its spectrum, and the quantum numbers of these states can be inferred. In the spectra of every element but hydrogen a number of lines are *missing* that correspond to transitions to and from states having certain combinations of quantum numbers. Thus no transitions are observed in helium to or from the ground-state configuration in which the spins of both electrons are in the same direction to give a total spin of 1, although transitions *are* observed to and from the other ground-state configuration, in which the spins are in opposite directions to give a total spin of 0. In the absent state the quantum numbers of *both* electrons would be $n = 1$, $l = 0$, $m_l = 0$, $m_s = \frac{1}{2}$, while in the state known to exist one of the electrons has $m_s = \frac{1}{2}$ and the other $m_s = -\frac{1}{2}$. Pauli showed that every unobserved atomic state involves two or more electrons with identical quantum numbers, and the exclusion principle is a statement of this empirical finding.

Before we explore the role of the exclusion principle in determining atomic structures, let us look into its quantum-mechanical implications. We saw in the previous chapter that the complete wave function ψ of the electron in a hydrogen atom can be expressed as the product of three separate wave functions, each describing that part of ψ which is a function of one of the three coordinates r, θ, ϕ. It is possible to show in an analogous way that the complete wave function $\psi(1, 2, 3, \ldots, n)$ of a system of n particles can be approximately expressed as the product of the wave functions ψ (1), ψ (2), $\psi(3)$, . . . , $\psi(n)$ of the individual particles. That is,

8.9 $$\psi(1, 2, 3, \ldots, n) = \psi(1)\, \psi(2)\, \psi(3) \ldots \psi(n)$$

We shall use this result to investigate the kinds of wave functions that can be used to describe a system of two identical particles.

Let us suppose that one of the particles is in quantum state a and the other in state b. Because the particles are identical, it should make no difference in the probability density ψ^2 of the system if the particles are exchanged, with the one in state a replacing the one in state b and vice versa. Symbolically, we require that

8.10 $$\psi^2(1,2) = \psi^2(2,1)$$

Hence the wave function $\psi(2,1)$, representing the exchanged particles, can be given by either

8.11 $$\psi(2,1) = \psi(1,2)$$ **Symmetric**

or

8.12 $$\psi(2,1) = -\psi(1,2)$$ **Antisymmetric**

and still fulfill Eq. 8.10. The wave function of the system is not itself a measurable quantity, and so it can be altered in sign by the exchange of the particles. Wave functions unaffected by an exchange of particles are said to be *symmetric*, while those reversing sign upon such an exchange are said to be *antisymmetric*.

If particle 1 is in state a and particle 2 is in state b, the wave function of the system is, according to Eq. 8.9,

8.13 $$\psi_{\text{I}} = \psi_a(1)\, \psi_b(2)$$

while if particle 2 is in state a and particle 1 is in state b, the wave function is

$$\psi_{\text{II}} = \psi_a(2)\, \psi_b(1)$$

Because the two particles are in fact indistinguishable, we have no way of knowing at any moment whether ψ_{I} or ψ_{II} describes the system. The likelihood that ψ_{I} is correct at any moment is the same as the likelihood that ψ_{II} is correct. Equivalently, we can say that the system spends half the time in the configuration whose wave function is ψ_{I} and the other half in the configuration whose wave function is ψ_{II}. Therefore a linear combination of ψ_{I} and ψ_{II} is the proper description of the system. There are two such combinations possible, the symmetric one

8.14 $$\psi_S = \frac{1}{\sqrt{2}}[\psi_a(1)\,\psi_b(2) + \psi_a(2)\,\psi_b(1)]$$

and the antisymmetric one

8.15 $$\psi_A = \frac{1}{\sqrt{2}}[\psi_a(1)\,\psi_b(2) - \psi_a(2)\,\psi_b(1)]$$

The factor $1/\sqrt{2}$ is required to normalize ψ_S and ψ_A. Exchanging particles 1 and 2 leaves ψ_S unaffected, while it reverses the sign of ψ_A. Both ψ_S and ψ_A obey Eq. 8.10.

There are a number of important distinctions between the behavior of particles in systems whose wave functions are symmetric and that of particles in systems whose wave functions are antisymmetric. The most obvious is that, in the former case, both particles 1 and 2 can simultaneously exist in the same state, with $a = b$, while in the latter case, if we set $a = b$, we find that $\psi_A = 0$: the two particles *cannot* be in the same quantum state. Comparing this conclusion with Pauli's empirical exclusion principle, which states that no two electrons in an atom can be in the same quantum state, we conclude that systems of electrons are described by wave functions that reverse sign upon the exchange of any pair of them.

The results of various experiments show that *all* particles which have a spin of ½ have wave functions that are antisymmetric to an exchange of any pair of them. Such particles, which include protons and neutrons as well as electrons, obey the exclusion principle when they are in the same system; that is, when they move in a common force field, each member of the system must be in a different quantum state. Particles of spin ½ are often referred to as *Fermi particles* or *fermions* because, as we shall learn in Chap. 10, the behavior of aggregates of them is governed by a statistical distribution law discovered by Fermi and Dirac.

Particles whose spins are 0 or an integer have wave functions that are symmetric to an exchange of any pair of them. These particles do not obey the exclusion principle. Particles of 0 or integral spin are often referred to as *Bose particles* or *bosons* because the statistical distribution law that describes aggregates of them was discovered by Bose and Einstein. Photons, alpha particles, and helium atoms are Bose particles.

There are other important consequences of the symmetry or antisymmetry of particle wave functions besides that expressed in the exclusion principle. It is these consequences that make it useful to classify particles according to the nature of their wave functions rather than simply according to whether or not they obey the exclusion principle.

8.3 Electron Configurations

A system of particles is stable when its total energy is a minimum. In the case of a complex atom, the exclusion principle prevents all of the various electrons from occupying the lowest quantum state. Before we look into how these two basic rules determine the actual electronic structures of complex atoms, let us examine the variation of electron energy with quantum state.

While the several electrons in a complex atom certainly interact directly with one another, as evidenced by the role of the exclusion principle in governing their behavior, it is not a bad approximation to consider each electron as though it exists in a constant mean force field. For a given electron this field is the electric field of the nuclear charge Ze decreased by the partial shielding of those other electrons that are closer to the nucleus. All of the electrons that have the same total quantum number n are, on the average, roughly the same distance from the nucleus. These electrons therefore interact with virtually the same electric field and have similar energies. It is conventional to speak of such electrons as occupying the same atomic *shell*. Shells are denoted by roman capital letters according to the following scheme:

$$\begin{matrix} n = 1 & 2 & 3 & 4 & 5\ldots \\ K & L & M & N & O\ldots \end{matrix}$$

Atomic shells

The energy of an electron in a particular shell also depends to a certain extent upon its orbital quantum number l, though this dependence is not so great as that upon n. In a complex atom the degree to which the full nuclear charge is shielded from a given electron by intervening shells of other electrons varies with its probability-density distribution. When l is large, the distribution has roughly circular contour lines, while when l is small, the contour lines are elliptical. An electron of small l therefore is more likely to be found near the nucleus (where it is poorly shielded by the other electrons) than one of higher l, which results in a lower total energy (that is, higher binding energy) for it. The electrons in each shell accordingly increase in energy with increasing l. This effect is illustrated in Fig. 8-4, which is a plot of the binding energies of various atomic electrons as a function of atomic number.

Electrons that share a certain value of l in a shell are said to occupy the same *subshell*. All of the electrons in a subshell have almost identical energies, since the dependence of electron energy upon m_l and m_s is comparatively minor.

The occupancy of the various subshells in an atom is usually expressed with the help of the notation introduced in the previous chapter for the various quantum states of the hydrogen atom. As indicated in Table 7.1,

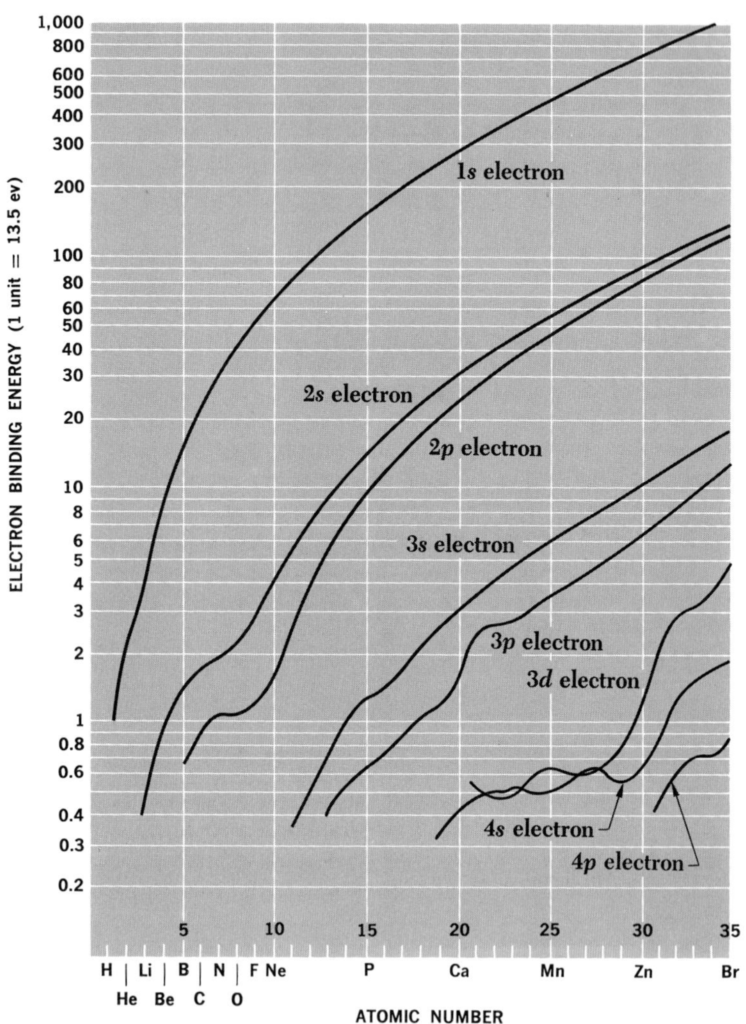

FIGURE 8-4 The binding energies of atomic electrons.

each subshell is identified by its total quantum number n followed by the letter corresponding to its orbital quantum number l. A superscript after the letter indicates the number of electrons in that subshell. For example, the electron configuration of sodium is written

$$1s^2 2s^2 2p^6 3s^1$$

which means that the 1s ($n = 1, l = 0$) and 2s ($n = 2, l = 0$) subshells contain two electrons each, the 2p ($n = 2, l = 1$) subshell contains six electrons, and the 3s ($n = 3, l = 0$) subshell contains one electron.

8.4 The Periodic Table

When the elements are listed in order of atomic number, elements with similar chemical and physical properties recur at regular intervals. This empirical observation, known as the *periodic law,* was first formulated by Dmitri Mendeleev about a century ago. A tabular arrangement of the elements exhibiting this recurrence of properties is called a *periodic table.* Table 8.1 is perhaps the simplest form of periodic table; though more elaborate periodic tables have been devised to exhibit the periodic law in finer detail, Table 8.1 is adequate for our purposes.

Elements with similar properties form the *groups* shown as vertical columns in Table 8.1. Thus group I consists of hydrogen plus the alkali metals, all of which are extremely active chemically and all of which have valences of +1. Group VII consists of the halogens, volatile, active nonmetals that have valences of −1 and form diatomic molecules in the gaseous state. Group VIII consists of the inert gases, elements so inactive that they not only almost never form compounds with other elements, but their atoms do not join together into molecules like the atoms of other gases.

The horizontal rows in Table 8.1 are called *periods.* Across each period is a more or less steady transition from an active metal through less active metals and weakly active nonmetals to highly active nonmetals and finally to an inert gas. Within each column there are also regular changes in properties, but they are far less conspicuous than those in each period. For example, increasing atomic number in the alkali metals is accompanied by greater chemical activity, while the reverse is true in the halogens.

A series of *transition elements* appears in each period after the third between the group II and group III elements. The transition elements are metals with a considerable chemical resemblance to one another but no pronounced resemblance to the elements in the major groups. Fifteen of the transition elements in period 6 are virtually indistinguishable in their properties, and are known as the *lanthanide* elements (or *rare earths*). A similar group of closely related metals, the *actinide* elements, is found in period 7.

The notion of electron shells and subshells fits perfectly into the pattern of the periodic table, which is just a mirror of the atomic structures of the elements. Let us see how this pattern arises.

The exclusion principle places definite limits on the number of electrons that can occupy a given subshell. A subshell is characterized by a certain total quantum number n and orbital quantum number l, where $l < n$. There are $2l + 1$ different values of the magnetic quantum number m_l for any l, and two possible values of the spin magnetic quantum number m_s ($+\frac{1}{2}$ and $-\frac{1}{2}$) for any m_l. Hence each subshell can contain a maximum of $2(2l + 1)$ electrons and each shell a maximum of $2n^2$ electrons. A shell or subshell con-

Table 8.1 THE PERIODIC TABLE OF THE ELEMENTS

The number above the symbol of each element is its atomic weight, and that below is its atomic number. The elements whose atomic weights are given in parentheses do not occur in nature, but have been prepared artificially in nuclear reactions. The atomic weight in such a case is the mass number of the most long-lived radioactive isotope of the element.

Period	Group I	Group II											Group III	Group IV	Group V	Group VI	Group VII	Group VIII
1	1.00 **H** 1																	4.00 **He** 2
2	6.94 **Li** 3	9.01 **Be** 4											10.81 **B** 5	12.01 **C** 6	14.01 **N** 7	16.00 **O** 8	19.00 **F** 9	20.18 **Ne** 10
3	22.99 **Na** 11	24.31 **Mg** 12											26.98 **Al** 13	28.09 **Si** 14	30.98 **P** 15	32.07 **S** 16	35.46 **Cl** 17	39.94 **Ar** 18
4	39.10 **K** 19	40.08 **Ca** 20	44.96 **Sc** 21	47.90 **Ti** 22	50.94 **V** 23	52.00 **Cr** 24	54.94 **Mn** 25	55.85 **Fe** 26	58.93 **Co** 27	58.71 **Ni** 28	63.54 **Cu** 29	65.37 **Zn** 30	69.72 **Ga** 31	72.59 **Ge** 32	74.92 **As** 33	78.96 **Se** 34	79.91 **Br** 35	83.8 **Kr** 36
5	85.47 **Rb** 37	87.66 **Sr** 38	88.91 **Y** 39	91.22 **Zr** 40	92.91 **Nb** 41	95.94 **Mo** 42	(99) **Tc** 43	101.1 **Ru** 44	102.91 **Rh** 45	106.4 **Pd** 46	107.87 **Ag** 47	112.40 **Cd** 48	114.82 **In** 49	118.69 **Sn** 50	121.75 **Sb** 51	127.60 **Te** 52	126.90 **I** 53	131.30 **Xe** 54
6	132.91 **Cs** 55	137.34 **Ba** 56	° 57–71	178.49 **Hf** 72	180.95 **Ta** 73	183.85 **W** 74	186.2 **Re** 75	190.2 **Os** 76	192.2 **Ir** 77	195.09 **Pt** 78	197.0 **Au** 79	200.59 **Hg** 80	204.37 **Tl** 81	207.19 **Pb** 82	208.98 **Bi** 83	(210) **Po** 84	(210) **At** 85	222 **Rn** 86
7	(223) **Fr** 87	226.05 **Ra** 88	°° 89–103															

	1	2	3	4	5	6	7	8	9	10	11	12	13	14	15
° Rare earths	138.91 **La** 57	140.12 **Ce** 58	140.91 **Pr** 59	144.24 **Nd** 60	(145) **Pm** 61	150.35 **Sm** 62	152.0 **Eu** 63	157.25 **Gd** 64	158.92 **Tb** 65	162.50 **Dy** 66	164.92 **Ho** 67	167.26 **Er** 68	168.93 **Tm** 69	173.04 **Yb** 70	174.97 **Lu** 71
°° Actinides	227 **Ac** 89	232.04 **Th** 90	231 **Pa** 91	238.03 **U** 92	(237) **Np** 93	(242) **Pu** 94	(243) **Am** 95	(247) **Cm** 96	(249) **Bk** 97	(251) **Cf** 98	(254) **Es** 99	(253) **Fm** 100	(256) **Md** 101	(254) **No** 102	(257) **Lw** 103

taining its full quota of electrons is said to be *closed.* A closed s subshell ($l = 0$) holds two electrons, a closed p subshell ($l = 1$) six electrons, a closed d subshell ($l = 2$) ten electrons, and so on.

The total orbital and spin angular momenta of the electrons in a closed subshell are zero, and their effective charge distributions are perfectly symmetrical. The electrons in a closed shell are all tightly bound, since the positive nuclear charge is large relative to the negative charge of the inner shielding electrons. Since an atom containing only closed shells has no dipole moment, it does not attract other electrons, and its electrons cannot be readily detached. Such atoms we expect to be passive chemically, like the inert gases—and the inert gases all turn out to have closed-shell electron configurations.

Those atoms with but a single electron in their outermost shells tend to lose this electron, which is relatively far from the nucleus and is shielded by the inner electrons from all but an effective nuclear charge of $+e$. Hydrogen and the alkali metals are in this category and accordingly have valences of $+1$. Atoms whose outer shells lack a single electron of being closed tend to acquire such an electron through the attraction of the imperfectly shielded strong nuclear charge, which accounts for the chemical behavior of the halogens. In this manner the similarities of the members of the various groups of the periodic table may be accounted for.

Table 8.2 shows the electron configurations of the elements. The origin of the transition elements evidently lies in the tighter binding of s electrons than d or f electrons in complex atoms, discussed in the previous section. The first element to exhibit this effect is potassium, whose outermost electron is in a $4s$ instead of a $3d$ substate. The difference in binding energy between $3d$ and $4s$ electrons is not very great, as can be seen in the configurations of chromium and copper. In both of these elements an additional $3d$ electron is present at the expense of a vacancy in the $4s$ subshell. In this connection another glance at Fig. 8-4 will be instructive.

The ferromagnetism of iron, cobalt, and nickel is related to the partial occupancy of their $3d$ subshells, whose electrons do not pair off to permit their spins to cancel out. In iron, for example, five of the six $3d$ electrons have parallel spins, so that each iron atom has a large resultant magnetic moment.

The order in which electron subshells are filled in atoms is

$$1s, 2s, 2p, 3s, 3p, 4s, 3d, 4p, 5s, 4d, 5p, 6s, 4f, 5d, 6p, 7s, 6d$$

as we can see from Table 8.2. The remarkable similarities in chemical behavior among the lanthanides and actinides are easy to understand on the basis of this sequence. All of the lanthanides have the same $5s^2 5p^6 6s^2$ configurations but have incomplete $4f$ subshells. The addition of $4f$ electrons

Table 8.2

ELECTRON CONFIGURATIONS OF THE ELEMENTS

	K	L		M			N				O				P			Q
	1*s*	2*s*	2*p*	3*s*	3*p*	3*d*	4*s*	4*p*	4*d*	4*f*	5*s*	5*p*	5*d*	5*f*	6*s*	6*p*	6*d*	7*s*
1 H	1																	
2 He	2																	
3 Li	2	1																
4 Be	2	2																
5 B	2	2	1															
6 C	2	2	2															
7 N	2	2	3															
8 O	2	2	4															
9 F	2	2	5															
10 Ne	2	2	6															
11 Na	2	2	6	1														
12 Mg	2	2	6	2														
13 Al	2	2	6	2	1													
14 Si	2	2	6	2	2													
15 P	2	2	6	2	3													
16 S	2	2	6	2	4													
17 Cl	2	2	6	2	5													
18 A	2	2	6	2	6													
19 K	2	2	6	2	6		1											
20 Ca	2	2	6	2	6		2											
21 Sc	2	2	6	2	6	1	2											
22 Ti	2	2	6	2	6	2	2											
23 V	2	2	6	2	6	3	2											
24 Cr	2	2	6	2	6	5	1											
25 Mn	2	2	6	2	6	5	2											
26 Fe	2	2	6	2	6	6	2											
27 Co	2	2	6	2	6	7	2											
28 Ni	2	2	6	2	6	8	2											
29 Cu	2	2	6	2	6	10	1											
30 Zn	2	2	6	2	6	10	2											
31 Ga	2	2	6	2	6	10	2	1										
32 Ge	2	2	6	2	6	10	2	2										
33 As	2	2	6	2	6	10	2	3										
34 Se	2	2	6	2	6	10	2	4										
35 Br	2	2	6	2	6	10	2	5										
36 Kr	2	2	6	2	6	10	2	6										
37 Rb	2	2	6	2	6	10	2	6			1							
38 Sr	2	2	6	2	6	10	2	6			2							
39 Y	2	2	6	2	6	10	2	6	1		2							
40 Zr	2	2	6	2	6	10	2	6	2		2							
41 Nb	2	2	6	2	6	10	2	6	4		1							
42 Mo	2	2	6	2	6	10	2	6	5		1							
43 Tc	2	2	6	2	6	10	2	6	5		2							
44 Ru	2	2	6	2	6	10	2	6	7		1							
45 Rh	2	2	6	2	6	10	2	6	8		1							
46 Pd	2	2	6	2	6	10	2	6	10									
47 Ag	2	2	6	2	6	10	2	6	10		1							
48 Cd	2	2	6	2	6	10	2	6	10		2							
49 In	2	2	6	2	6	10	2	6	10		2	1						
50 Sn	2	2	6	2	6	10	2	6	10		2	2						
51 Sb	2	2	6	2	6	10	2	6	10		2	3						

Table 8.2 (Continued)

	K	L		M			N				O				P			Q
	1s	2s	2p	3s	3p	3d	4s	4p	4d	4f	5s	5p	5d	5f	6s	6p	6d	7s
52 Te	2	2	6	2	6	10	2	6	10		2	4						
53 I	2	2	6	2	6	10	2	6	10		2	5						
54 Xe	2	2	6	2	6	10	2	6	10		2	6						
55 Cs	2	2	6	2	6	10	2	6	10		2	6			1			
56 Ba	2	2	6	2	6	10	2	6	10		2	6			2			
57 La	2	2	6	2	6	10	2	6	10		2	6	1		2			
58 Ce	2	2	6	2	6	10	2	6	10	2	2	6			2			
59 Pr	2	2	6	2	6	10	2	6	10	3	2	6			2			
60 Nd	2	2	6	2	6	10	2	6	10	4	2	6			2			
61 Pm	2	2	6	2	6	10	2	6	10	5	2	6			2			
62 Sm	2	2	6	2	6	10	2	6	10	6	2	6			2			
63 Eu	2	2	6	2	6	10	2	6	10	7	2	6			2			
64 Gd	2	2	6	2	6	10	2	6	10	7	2	6	1		2			
65 Tb	2	2	6	2	6	10	2	6	10	9	2	6			2			
66 Dy	2	2	6	2	6	10	2	6	10	10	2	6			2			
67 Ho	2	2	6	2	6	10	2	6	10	11	2	6			2			
68 Er	2	2	6	2	6	10	2	6	10	12	2	6			2			
69 Tm	2	2	6	2	6	10	2	6	10	13	2	6			2			
70 Yb	2	2	6	2	6	10	2	6	10	14	2	6			2			
71 Lu	2	2	6	2	6	10	2	6	10	14	2	6	1		2			
72 Hf	2	2	6	2	6	10	2	6	10	14	2	6	2		2			
73 Ta	2	2	6	2	6	10	2	6	10	14	2	6	3		2			
74 W	2	2	6	2	6	10	2	6	10	14	2	6	4		2			
75 Re	2	2	6	2	6	10	2	6	10	14	2	6	5		2			
76 Os	2	2	6	2	6	10	2	6	10	14	2	6	6		2			
77 Ir	2	2	6	2	6	10	2	6	10	14	2	6	7		2			
78 Pt	2	2	6	2	6	10	2	6	10	14	2	6	9		1			
79 Au	2	2	6	2	6	10	2	6	10	14	2	6	10		1			
80 Hg	2	2	6	2	6	10	2	6	10	14	2	6	10		2			
81 Tl	2	2	6	2	6	10	2	6	10	14	2	6	10		2	1		
82 Pb	2	2	6	2	6	10	2	6	10	14	2	6	10		2	2		
83 Bi	2	2	6	2	6	10	2	6	10	14	2	6	10		2	3		
84 Po	2	2	6	2	6	10	2	6	10	14	2	6	10		2	4		
85 At	2	2	6	2	6	10	2	6	10	14	2	6	10		2	5		
86 Rn	2	2	6	2	6	10	2	6	10	14	2	6	10		2	6		
87 Fr	2	2	6	2	6	10	2	6	10	14	2	6	10		2	6		1
88 Ra	2	2	6	2	6	10	2	6	10	14	2	6	10		2	6		2
89 Ac	2	2	6	2	6	10	2	6	10	14	2	6	10		2	6	1	2
90 Th	2	2	6	2	6	10	2	6	10	14	2	6	10		2	6	2	2
91 Pa	2	2	6	2	6	10	2	6	10	14	2	6	10	2	2	6	1	2
92 U	2	2	6	2	6	10	2	6	10	14	2	6	10	3	2	6	1	2
93 Np	2	2	6	2	6	10	2	6	10	14	2	6	10	4	2	6	1	2
94 Pu	2	2	6	2	6	10	2	6	10	14	2	6	10	5	2	6	1	2
95 Am	2	2	6	2	6	10	2	6	10	14	2	6	10	6	2	6	1	2
96 Cm	2	2	6	2	6	10	2	6	10	14	2	6	10	7	2	6	1	2
97 Bk	2	2	6	2	6	10	2	6	10	14	2	6	10	8	2	6	1	2
98 Cf	2	2	6	2	6	10	2	6	10	14	2	6	10	10	2	6		2
99 E	2	2	6	2	6	10	2	6	10	14	2	6	10	11	2	6		2
100 Fm	2	2	6	2	6	10	2	6	10	14	2	6	10	12	2	6		2
101 Md	2	2	6	2	6	10	2	6	10	14	2	6	10	13	2	6		2
102 No	2	2	6	2	6	10	2	6	10	14	2	6	10	14	2	6		2
103 Lw	2	2	6	2	6	10	2	6	10	14	2	6	10	14	2	6	1	2

has virtually no effect on the chemical properties of the lanthanide elements, which are determined by the outer electrons. Similarly, all of the actinides have $6s^2 6p^6 7s^2$ configurations, and differ only in the numbers of their $5f$ and $6d$ electrons.

While we have sketched the origins of only a few of the chemical and physical properties of the elements in terms of their electron configurations, many more can be quantitatively understood by similar reasoning.

8.5 Total Angular Momentum

Each electron in an atom has a certain orbital angular momentum **L** and a certain spin angular momentum **S**, both of which contribute to the total angular momentum **J** of the atom. Like all angular momenta, **J** is quantized, with a magnitude given by

8.16 $$J = \sqrt{\mathrm{J}(\mathrm{J}+1)}\,\hbar$$ **Total atomic angular momentum**

and a component J_z in the z direction given by

8.17 $$J_z = \mathrm{M_J}\hbar$$ **z component of total atomic angular momentum**

where J and $\mathrm{M_J}$ are the quantum numbers governing J and J_z. Our task in the remainder of this chapter is to look into the properties of **J** and their effect on atomic phenomena. We shall do this in terms of the semiclassical *vector model* of the atom, which provides a more intuitively accessible framework for understanding angular-momentum considerations than does a purely quantum-mechanical approach.

Let us first consider an atom whose total angular momentum is provided by a single electron. Atoms of the elements in group I of the periodic table—hydrogen, lithium, sodium, and so on—are of this kind since they have single electrons outside closed inner shells (except for hydrogen, which has no inner electrons) and the exclusion principle assures that the total angular momentum and magnetic moment of a closed shell are zero. Also in this category are the ions He^+, Be^+, Mg^+, B^{++}, Al^{++}, and so on.

The magnitude L of the orbital angular momentum **L** of an atomic electron is determined by its orbital quantum number l according to the formula

8.18 $$L = \sqrt{l(l+1)}\,\hbar$$

while the component L_z of **L** along the z axis is determined by the magnetic quantum number m_l according to the formula

8.19 $$L_z = m_l\hbar$$

Similarly the magnitude S of the spin angular momentum **S** is determined by the spin quantum number s (which has the sole value $+\frac{1}{2}$) according to the formula

8.20 $$S = \sqrt{s(s+1)}\,\hbar$$

while the component S_z of **S** along the z axis is determined by the magnetic spin quantum number m_s according to the formula

8.21 $$S_z = m_s\hbar$$

Because **L** and **S** are vectors, they must be added vectorially to yield the total angular momentum **J**:

8.22 $$\mathbf{J} = \mathbf{L} + \mathbf{S}$$

It is customary to use the symbols j and m_j for the quantum numbers that describe J and J_z for a single electron, so that

8.23 $$J = \sqrt{j(j+1)}\,\hbar$$

8.24 $$J_z = m_j\hbar$$

To obtain the relationships among the various angular-momentum quantum numbers, it is simplest to start with the z components of the vectors **J**, **L**, and **S**. Since J_z, L_z, and S_z are scalar quantities,

$$J_z = L_z \pm S_z$$
$$m_j\hbar = m_l\hbar \pm m_s\hbar$$

and

8.25 $$m_j = m_l \pm m_s$$

The possible values of m_l range from $+l$ through 0 to $-l$, and those of m_s are $\pm s$. The quantum number l is always an integer or 0 while $s = \frac{1}{2}$, and as a result m_j must be half-integral. The possible values of m_j also range from $+j$ through 0 to $-j$ in integral steps, and so, for any value of l,

8.26 $$j = l \pm s$$

Like m_j, j is always half-integral.

Because of the simultaneous quantization of **J**, **L**, and **S** they can have only certain specific relative orientations. This is a general conclusion; in the case of a one-electron atom, there are only two relative orientations possible. One of these corresponds to $j = l + s$, so that $J > L$, and the other to $j = l - s$, so that $J < L$. Figure 8-5 shows the two ways in which **L** and **S** can combine to form **J** when $l = 1$. Evidently the orbital and spin angular-momentum vectors can never be exactly parallel or antiparallel to each other or to the total angular-momentum vector.

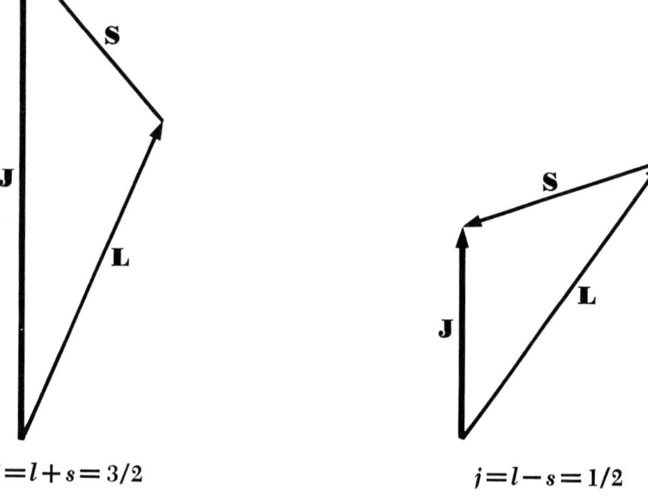

FIGURE 8-5 The two ways in which L and S can be added to form J when $l = 1$, $s = \frac{1}{2}$.

The angular momenta **L** and **S** interact magnetically, as we saw in Sec. 8.1, and as a result exert torques on each other. If there is no external magnetic field, the total angular momentum **J** is conserved in magnitude and direction, and the effect of the internal torques can only be the precession of **L** and **S** around the direction of their resultant **J** (Fig. 8-6). However, if there is an

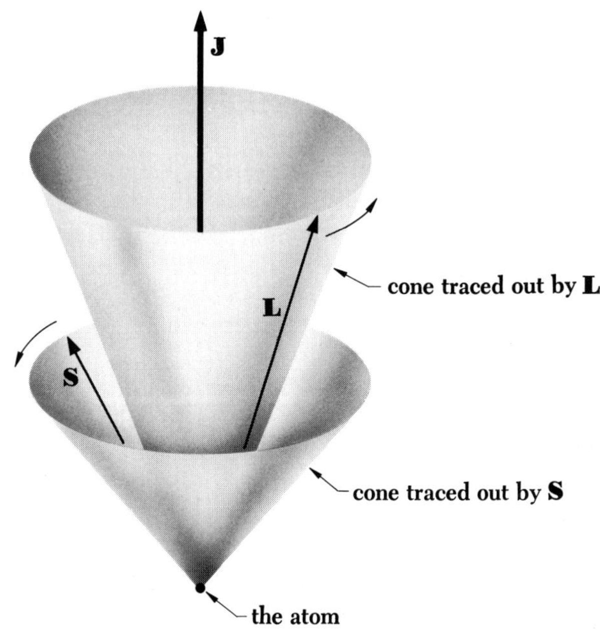

FIGURE 8-6 The orbital and spin angular-momentum vectors L and S precess about J according to the vector model of the atom.

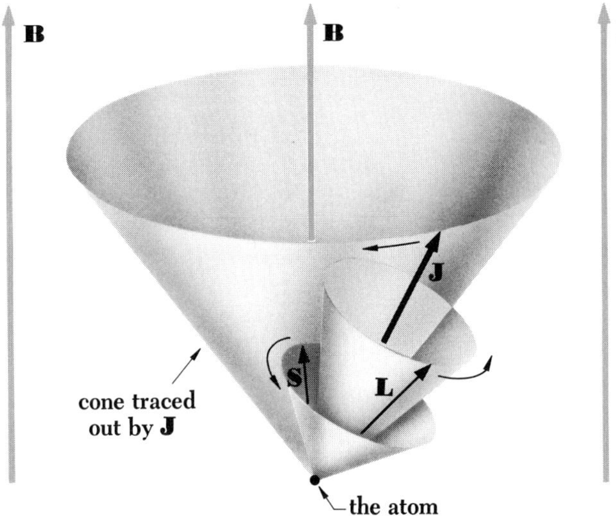

FIGURE 8-7 In the presence of an external magnetic field B, the total angular-momentum vector J precesses about B according to the vector model of the atom.

external magnetic field **B** present, then **J** precesses about the direction of **B** while **L** and **S** continue precessing about **J**, as in Fig. 8-7. The precession of **J** about **B** is what gives rise to the anomalous Zeeman effect, since different orientations of **J** involve slightly different energies in the presence of **B**.

Atomic nuclei also have intrinsic angular momenta and magnetic moments, as we shall see in Chap. 12, and these contribute to the total atomic angular momenta and magnetic moments. These contributions are small because nuclear magnetic moments are $\sim 10^{-3}$ the magnitude of electronic moments, and they lead to the *hyperfine structure* of spectral lines with typical spacings between components of $\sim 10^{-2}$ A as compared with typical fine-structure spacings of several angstroms.

When more than one electron contributes orbital and spin angular momenta to the total angular momentum **J** of an atom, **J** is still the vector sum of these individual momenta. Because the electrons involved interact with one another, the manner in which their individual momenta $\mathbf{L}_i$ and $\mathbf{S}_i$ add together to form **J** follows certain definite patterns depending upon the circumstances. The usual pattern for all but the heaviest atoms is that the orbital angular momenta $\mathbf{L}_i$ of the various electrons are coupled together electrostatically into a single resultant **L** and the spin angular momenta $\mathbf{S}_i$ are coupled together independently into another single resultant **S**; we shall examine the reasons for this behavior later in this section. The momenta **L** and **S** then interact magnetically via the spin-orbit effect to form a total angular momentum **J**. This scheme, called *LS coupling*, may be summarized as follows:

$$\mathbf{L} = \Sigma\, \mathbf{L}_i$$

8.27 $$\mathbf{S} = \Sigma\, \mathbf{S}_i$$ LS coupling

$$\mathbf{J} = \mathbf{L} + \mathbf{S}$$

As usual, L, S, J, L_z, S_z, and J_z are quantized, with the respective quantum numbers being L, S, J, $\mathsf{M_L}$, $\mathsf{M_S}$, and $\mathsf{M_J}$. Hence

8.28 $$L = \sqrt{\mathsf{L}(\mathsf{L}+1)}\,\hbar$$

8.29 $$L_z = \mathsf{M_L}\hbar$$

8.30 $$S = \sqrt{\mathsf{S}(\mathsf{S}+1)}\,\hbar$$

8.31 $$S_z = \mathsf{M_S}\hbar$$

8.32 $$J = \sqrt{\mathsf{J}(\mathsf{J}+1)}\,\hbar$$

8.33 $$J_z = \mathsf{M_J}\hbar$$

Both L and $\mathsf{M_L}$ are always integers or 0, while the other quantum numbers are half-integral if an odd number of electrons is involved and integral or 0 if an even number of electrons is involved.

As an example, let us consider two electrons, one with $l_1 = 1$ and the other with $l_2 = 2$. There are three ways in which $\mathbf{L}_1$ and $\mathbf{L}_2$ can be combined into a single vector $\mathbf{L}$ that is quantized according to Eq. 8.28, as shown in Fig. 8-8. These correspond to $\mathsf{L} = 1$, 2, and 3 since all values of L are possible from $l_1 + l_2$ to $l_1 - l_2$. The spin quantum number s is always $+\frac{1}{2}$, so there are two possibilities for the sum $\mathbf{S}_1 + \mathbf{S}_2$, as in Fig. 8-8, that correspond to $\mathsf{S} = 0$ and $\mathsf{S} = 1$. We note that $\mathbf{L}_1$ and $\mathbf{L}_2$ can never be exactly parallel to $\mathbf{L}$, nor $\mathbf{S}_1$ and $\mathbf{S}_2$ to $\mathbf{S}$, except when the vector sum is 0. The quantum number J can have all values between $\mathsf{L} + \mathsf{S}$ and $\mathsf{L} - \mathsf{S}$,

FIGURE 8-8 When $l_1 = 1$, $s_1 = \frac{1}{2}$, and $l_2 = 2$, $s_2 = \frac{1}{2}$, there are three ways in which L_1 and L_2 can combine to form L and two ways in which S_1 and S_2 can combine to form S.

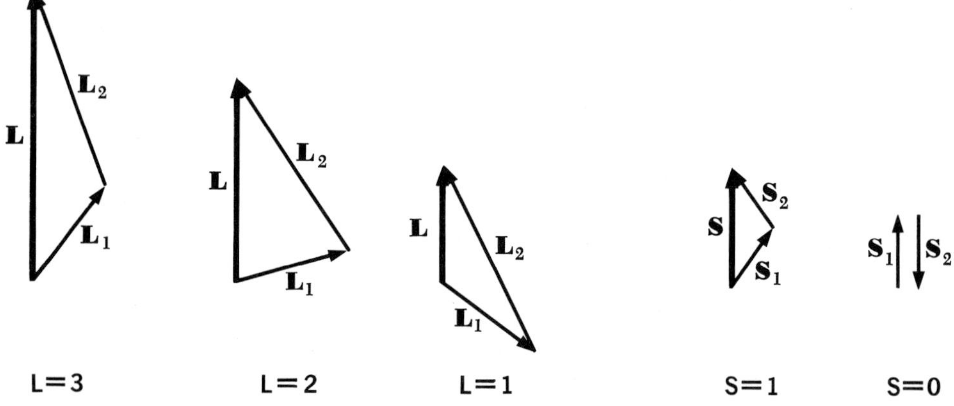

which in this case means that J can be 0, 1, 2, 3, or 4. In working out the possible values of L, S, and J for a many-electron atom, it is necessary to keep in mind that the exclusion principle limits the possible sets of electron quantum numbers.

The *LS* scheme owes its existence to the relative strengths of the electrostatic forces that couple the individual orbital angular momenta into a resultant **L** and the individual spin angular momenta into a resultant **S**. The origins of these forces are interesting. The coupling between orbital angular momenta can be understood by reference to Fig. 7-10, which shows how the electron probability density ψ^2 varies in space for different quantum states in hydrogen. The corresponding patterns for electrons in more complex atoms will be somewhat different, of course, but it remains true in general that ψ^2 is not spherically symmetric except for *s* states. (In the latter case $l = 0$ and the electron has no orbital angular momentum to contribute anyway.) Because of the asymmetrical distributions of their charge densities, the electrostatic forces between the electrons in an atom vary with the relative orientations of their angular-momentum vectors, and only certain relative orientations are stable. These stable configurations correspond to a total orbital angular momentum that is quantized according to the formula $L = \sqrt{\mathsf{L}(\mathsf{L} + 1)}\,\hbar$.

The origin of the strong coupling between electron spins is harder to visualize because it is a purely quantum-mechanical effect with no classical analog. (The direct interaction between the intrinsic electron magnetic moments, it should be noted, is insignificant and not responsible for the coupling between electron spin angular momenta.) The basic idea is that the complete wave function $\psi(1, 2, \ldots, n)$ of a system of n electrons is the product of a wave function $u(1, 2, \ldots, n)$ that describes the coordinates of the electrons and a spin function $s(1, 2, \ldots, n)$ that describes the orientations of their spins. As we saw in Sec. 8.2, the complete function $\psi(1, 2, \ldots, n)$ must be antisymmetric, which means that $u(1, 2, \ldots, n)$ is not independent of $s(1, 2, \ldots, n)$. A change in the relative orientations of the electron spin angular-momentum vectors must therefore be accompanied by a change in the electronic configuration of the atom, which means a change in the atom's electrostatic potential energy. To go from one total spin angular momentum **S** to a different one involves altering the structure of the atom, and therefore a strong electrostatic force, besides altering the directions of the spin angular momenta $\mathbf{S}_1, \mathbf{S}_2, \ldots, \mathbf{S}_n$, which requires only a weak magnetic force. This situation is what is described when it is said that the spin momenta $\mathbf{S}_i$ are strongly coupled together electrostatically.

The electrostatic forces that couple the $\mathbf{L}_i$ into a single vector **L** and the $\mathbf{S}_i$ into another vector **S** are stronger than the magnetic spin-orbit forces that couple **L** and **S** to form **J** in light atoms, and dominate the situation even when a moderate external magnetic field is applied. (In the latter case the precession of **J** around **B** is accordingly slower than the precession of **L** and **S** around **J**.) However, in heavy atoms the nuclear charge becomes great enough to produce spin-orbit interactions comparable in magnitude to the electrostatic ones between the $\mathbf{L}_i$ and between the $\mathbf{S}_i$, and the *LS* coupling scheme begins to break down. A similar breakdown occurs in strong external magnetic fields (typically ~10 webers/m^2), which produces the Paschen-Back effect in atomic spectra. In the limit of the failure of *LS* coupling, the total angular momenta $\mathbf{J}_i$ of the individual electrons add together directly to form the angular momentum **J** of the entire atom, a situation referred to as *jj coupling* since each $\mathbf{J}_i$ is described by a quantum number j in the manner described in the previous section.

In Sec. 7-5 we saw that individual orbital angular-momentum states are customarily described by a lower-case letter, with s corresponding to $l = 0$, p to $l = 1$, d to $l = 2$, and so on. A similar scheme using capital letters is used to designate the entire electronic state of an atom according to its total orbital angular-momentum quantum number L, as follows:

$$\begin{array}{cccccccc} \mathsf{L} = 0 & 1 & 2 & 3 & 4 & 5 & 6 \ldots \\ S & P & D & F & G & H & I \ldots \end{array}$$

A superscript number before the letter (^2P for instance) is used to indicate the *multiplicity* of the state, which is the number of different possible orientations of **L** and **S** and hence the number of different possible values of J. The multiplicity is equal to 2S + 1 in the usual situation where L > S, since J ranges from L + S through 0 to L − S. Thus when S = 0, the multiplicity is 1 (a *singlet* state) and J = L; when S = ½, the multiplicity is 2 (a *doublet* state) and J = L ± ½; when S = 1, the multiplicity is 3 (a *triplet* state) and J = L + 1, L, or L − 1; and so on. (In a configuration in which S > L, the multiplicity is given by 2L + 1.) The total angular-momentum quantum number J is used as a subscript after the letter, so that a $^2P_{3/2}$ state (read as "doublet P three-halves") refers to an electronic configuration in which S = ½, L = 1, and J = 3/2. In the event that the angular momentum of the atom arises from a single outer electron, the total quantum number n of this electron can be used as a prefix: thus the ground state of the sodium atom is described by $3^2S_{1/2}$, since its electronic configuration has an electron with $n = 3$, $l = 0$, and $s = ½$ (and hence $j = ½$) outside closed $n = 1$ and $n = 2$ shells. For consistency it is conventional to denote the above state by $3^2S_{1/2}$ with the superscript 2 indicating a doublet, even though there is only a single possibility for J since L = 0.

8.6 Atomic Spectra

We are now in a position to understand the chief features of the spectra of the various elements. Before we examine some representative examples, it should be mentioned that further complications exist which have not been considered here, for instance those that originate in relativistic effects and in the coupling between electrons and vacuum fluctuations in the electromagnetic field (see Sec. 10.8). These additional factors split certain energy states into closely spaced substates and therefore represent other sources of fine structure in spectral lines.

Figure 8-9 shows the various states of the hydrogen atom classified by their total quantum number n and orbital angular-momentum quantum number l. The selection rule for allowed transitions here is $\Delta l = \pm 1$, which is illustrated by the transitions shown. To indicate some of the detail that is omitted in a

FIGURE 8-9 Energy-level diagram for hydrogen showing the origins of some of the more prominent spectral lines. The detailed structures of the $n = 2$ and $n = 3$ levels and the transitions that lead to the various components of the H_α line are pictured in the inset.

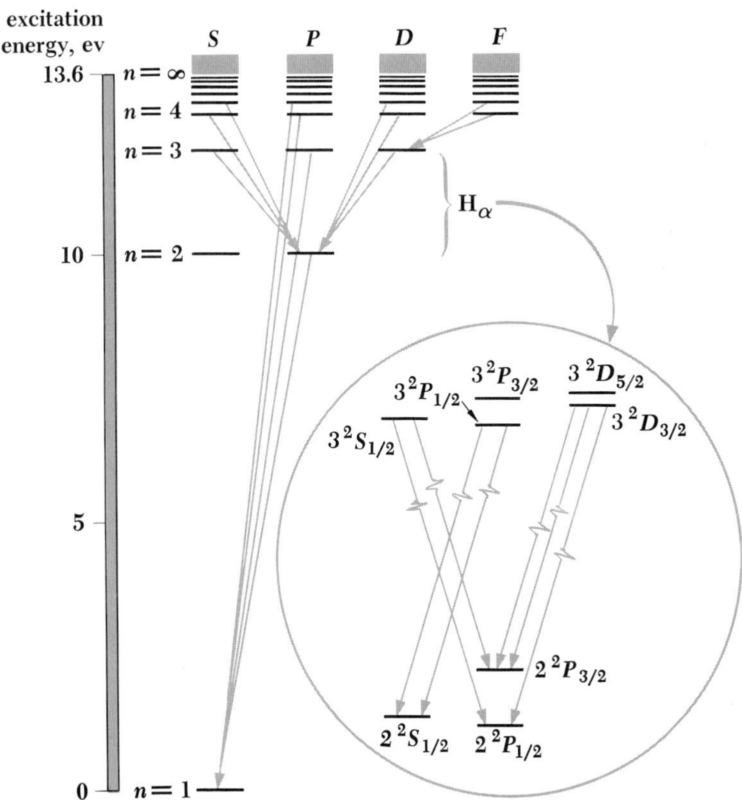

simple diagram of this kind, the detailed structures of the $n = 2$ and $n = 3$ levels are pictured; not only are all substates of the same n and different j separated in energy, but the same is true of states of the same n and j but with different l. The latter effect is most marked for states of small n and l, and was first established in the "Lamb shift" of the $2^2S_{1/2}$ state relative to the $2^2P_{1/2}$ state. The various separations conspire to split the H_α spectral line (Fig. 5-5) into seven closely spaced components.

The sodium atom has a single 3s electron outside closed inner shells, and so, if we assume that the 10 electrons in its inner core completely shield $+10e$ of nuclear charge, the outer electron is acted upon by an effective nuclear charge of $+e$ just as in the hydrogen atom. Hence we expect, as a first approximation, that the energy levels of sodium will be the same as those of hydrogen except that the lowest one will correspond to $n = 3$ instead of $n = 1$ because of the exclusion principle. Figure 8-10 is the energy-level diagram for sodium and, by comparison with the hydrogen levels also shown, there is indeed agreement for the states of highest l, that is, for the states of highest angular momentum. To understand the reason for the discrepancies at lower values of l, we need only refer to Fig. 7-9 to see how the probability for finding the electron in a hydrogen atom varies with distance from the nucleus. The smaller the value of l for a given n, the closer the electron gets to the nucleus on occasion; this is perhaps even more obvious in Fig. 7-10. Although the sodium wave functions are not identical with those of hydrogen, their general behavior is similar, and accordingly we expect the outer electron in a sodium atom to penetrate the core of inner electrons most often when it is in an s state, less often when it is in a p state, still less often when it is in a d state, and so on. The less shielded an outer electron is from the full nuclear charge, the greater the average force acting on it, and the smaller (that is, the more negative) its total energy. For this reason the states of small l in sodium are displaced downward from their equivalents in hydrogen, as in Fig. 8-10, and there are pronounced differences in energy between states of the same n but different l.

A single electron is responsible for the energy levels of both hydrogen and sodium. However, there are two 1s electrons in the ground state of helium, and it is interesting to consider the effect of LS coupling on the properties and behavior of the helium atom. To do this we first note the selection rules for allowed transitions under LS coupling:

8.34 $$\Delta L = 0, \pm 1$$

8.35 $$\Delta J = 0, \pm 1$$

8.36 $$\Delta S = 0$$

When only a single electron is involved, $\Delta L = 0$ is prohibited and $\Delta L =$

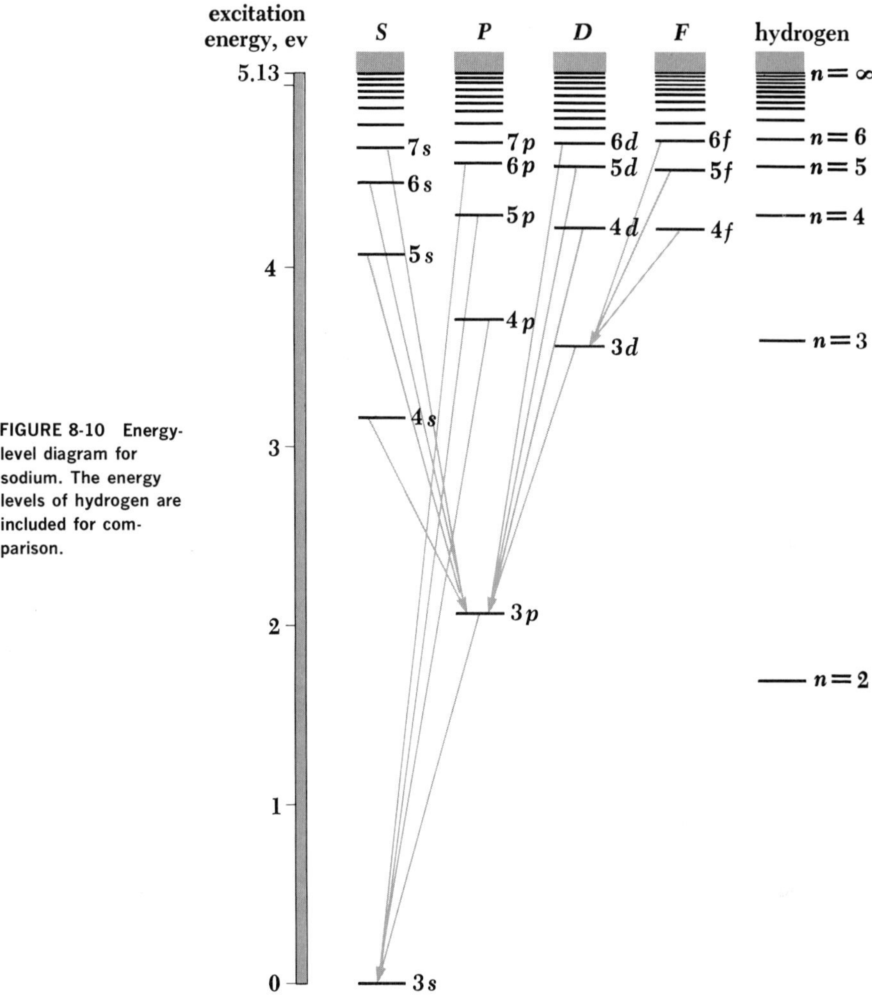

FIGURE 8-10 Energy-level diagram for sodium. The energy levels of hydrogen are included for comparison.

$\Delta l = \pm 1$ is the only possibility. Furthermore, J must change when the initial state has J = 0, so that J = 0 → J = 0 is prohibited.

The helium energy-level diagram is shown in Fig. 8-11. The various levels represent configurations in which one electron is in its ground state and the other is in an excited state, but because the angular momenta of the two electrons are coupled, it is proper to consider the levels as characteristic of the entire atom. Three differences between this diagram and the corresponding ones for hydrogen and sodium are conspicuous. First, there is the division into singlet and triplet states, which are, respectively, states in which the spins of the two electrons are antiparallel (to give S = 0) and parallel (to give S = 1). Because of the selection rule ΔS = 0, no allowed transitions

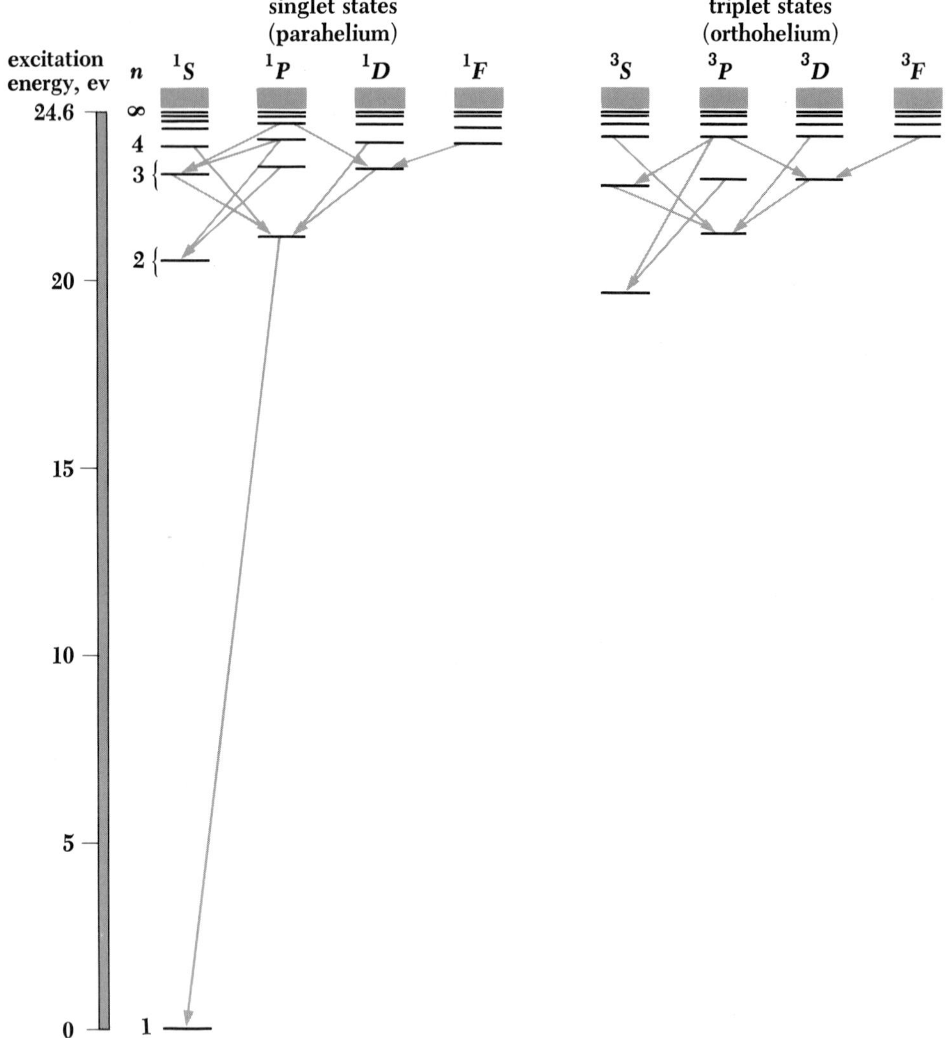

FIGURE 8-11 Energy-level diagram for helium showing the division into singlet (parahelium) and triplet (orthohelium) states. There is no 1^3S_1 state.

can occur between singlet states and triplet states, and the helium spectrum arises from transitions in one set or the other. Helium atoms in singlet states (antiparallel spins) constitute *parahelium* and those in triplet states (parallel spins) constitute *orthohelium*. An orthohelium atom can lose excitation energy in a collision and become one of parahelium, while a parahelium atom can gain excitation energy in a collision and become one of orthohelium; ordinary liquid or gaseous helium is therefore a mixture of both. The lowest triplet states are called *metastable* because, in the absence of collisions, an atom in one of them can retain its excitation energy for a relatively long time (a second or more) before radiating.

The second obvious peculiarity in Fig. 8-11 is the absence of the 1^3S state. The lowest triplet state is 2^3S, although the lowest singlet state is 1^1S. The 1^3S state is missing as a consequence of the exclusion principle, since in this state the two electrons would have parallel spins and therefore identical sets of quantum numbers. Third, the energy difference between the ground state and the lowest excited state is relatively large, which reflects the tight binding of closed-shell electrons discussed earlier in this chapter. The ionization energy of helium—the work that must be done to remove an electron from a helium atom— is 24.6 ev, the highest of any element.

The last energy-level diagram we shall consider is that of mercury, which has two electrons outside an inner core of 78 electrons in closed shells or subshells (Table 8.2). We expect a division into singlet and triplet states as in helium, but because the atom is so heavy we might also expect signs of a breakdown in the *LS* coupling of angular momenta. As Fig. 8-12 reveals, both of these expectations are realized, and several prominent lines in the mercury spectrum arise from transitions that violate the $\Delta S = 0$ selection rule. The transition $^3P_1 \rightarrow {}^1S_0$ is an example, and is responsible for the strong 2,537-A line in the ultraviolet. To be sure, this does not mean that the transition probability is necessarily very high, since the three $^3P_1$ states are the lowest of the triplet set and therefore tend to be highly populated in excited mercury vapor. The $^3P_0 \rightarrow {}^1S_0$ and $^3P_2 \rightarrow {}^1S_0$ transitions, respectively, violate the rules that forbid transitions from $J = 0$ to $J = 0$ and that limit ΔJ to 0 or ± 1, as well as violating $\Delta S = 0$, and hence are considerably less likely to occur than the $^3P_1 \rightarrow {}^1S_0$ transition. The $^3P_0$ and $^3P_2$ states are therefore metastable and, in the absence of collisions, an atom can persist in either of them for a relatively long time. The strong spin-orbit interaction in mercury that leads to the partial failure of *LS* coupling is also responsible for the wide spacing of the elements of the ^3P triplets.

8.7 X-ray Spectra

In Chap. 2 we learned that the X-ray spectra of targets bombarded by fast electrons exhibit narrow spikes at wavelengths characteristic of the target material in addition to a continuous distribution of wavelengths down to a minimum wavelength inversely proportional to the electron energy. The continuous X-ray spectrum is the result of the inverse photoelectric effect, with electron kinetic energy being transformed into photon energy $h\nu$. The discrete spectrum, on the other hand, has its origin in electronic transitions within atoms that have been disturbed by the incident electrons.

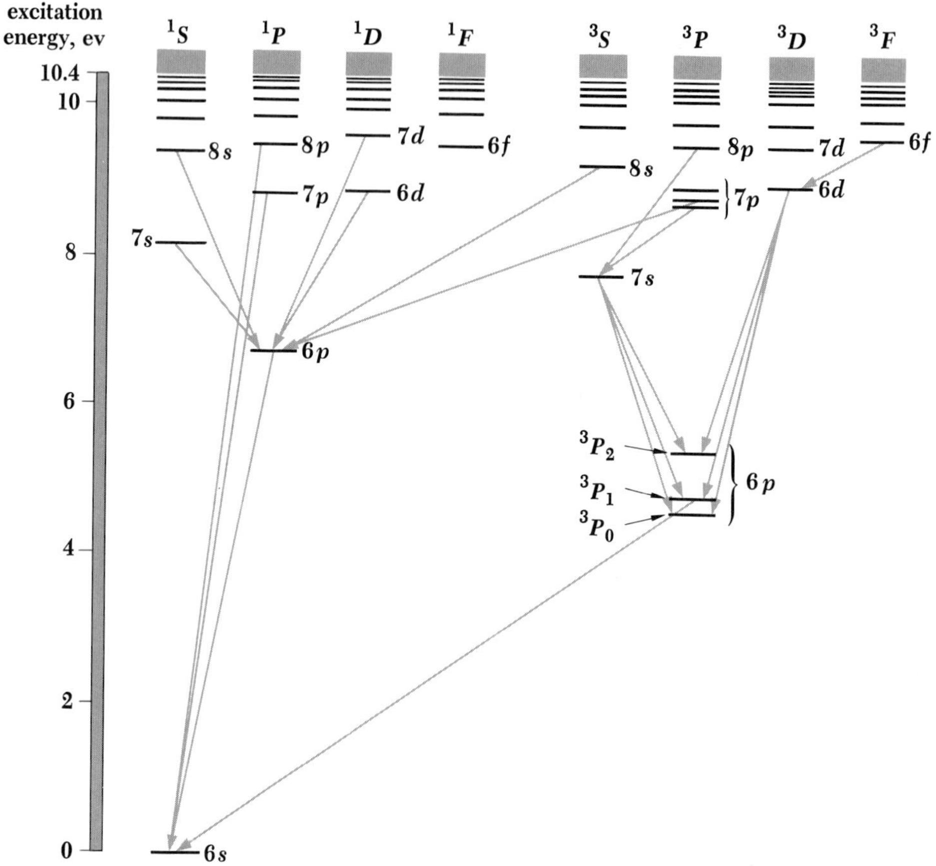

FIGURE 8-12 Energy-level diagram for mercury. In each excited level one outer electron is in the ground state, and the designation of the levels in the diagram corresponds to the state of the other electron.

Transitions involving the outer electrons of an atom usually involve only a few electron volts of energy, and even removing an outer electron requires at most 24.6 ev (for helium). These transitions accordingly are associated with photons whose wavelengths lie in or near the visible part of the electromagnetic spectrum, as is evident from the diagram in the back endpapers of this book. The inner electrons of heavier elements are a quite different matter, because these electrons experience all or much of the full nuclear charge without nearly complete shielding by intervening electron shells and in consequence are very tightly bound. In sodium, for example, only 5.13 ev is needed to remove the outermost 3*s* electron, while the corresponding figures for the inner ones are 31 ev for each 2*p* electron, 63 ev for each 2*s* electron, and 1,041 ev for each 1*s* electron. Transitions that involve the inner electrons in an atom are what give rise to discrete X-ray spectra because of the high photon energies involved.

Figure 8-13 shows the energy levels (not to scale) of a heavy atom classed by total quantum number n; energy differences between angular-momentum states within a shell are minor compared with the energy differences between shells. Let us consider what happens when an energetic electron strikes the atom and knocks out one of the K-shell electrons. (The K electron could also be elevated to one of the unfilled upper quantum states of the atom, but

FIGURE 8-13. The origin of X-ray spectra.

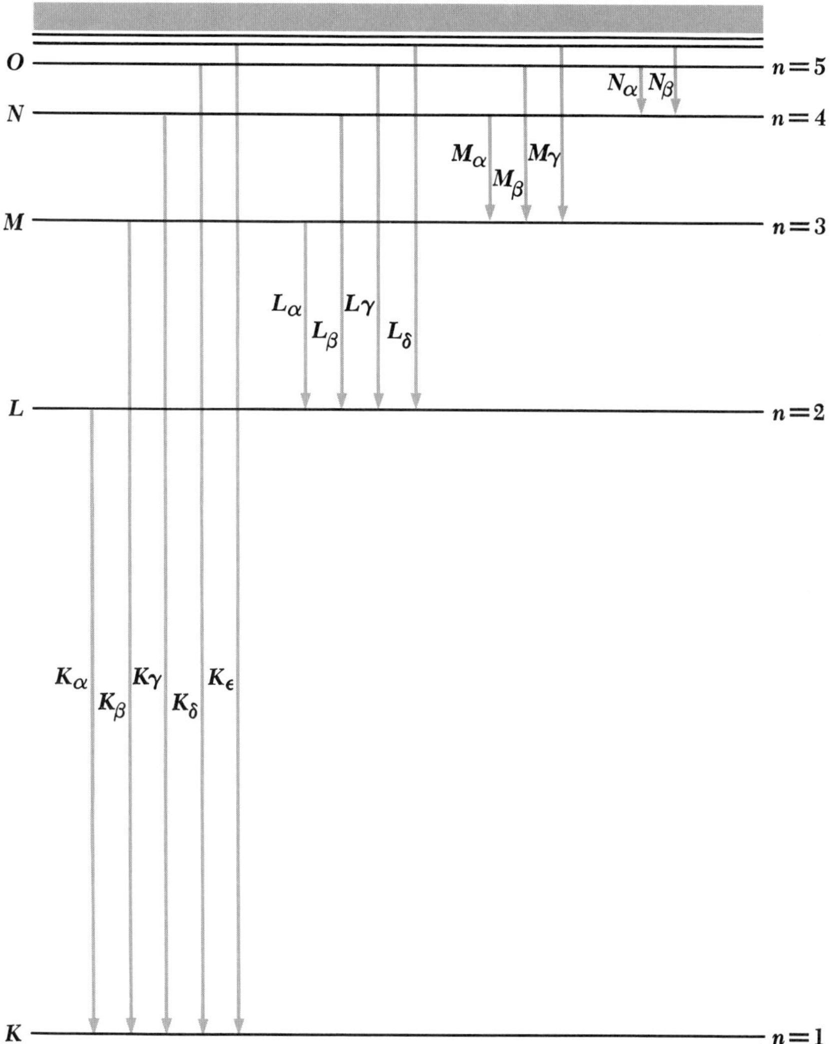

the difference between the energy needed to do this and that needed to remove the electron completely is insignificant, only 0.2 per cent in sodium and still less in heavier atoms.) An atom with a missing K electron gives up most of its considerable excitation energy in the form of an X-ray photon when an electron from an outer shell drops into the "hole" in the K shell. As indicated in Fig. 8-13, the K *series* of lines in the X-ray spectrum of an element consists of wavelengths arising in transitions from the $L, M, N, \ldots$ levels to the K level. Similarly the longer-wavelength L series originates when an L electron is knocked out of the atom, the M series when an M electron is knocked out, and so on. The two spikes in the X-ray spectrum of molybdenum in Fig. 2-5 are the K_α and K_β lines of its K series.

An atom with a missing inner electron can also lose excitation energy by the *Auger effect* without emitting an X-ray photon. In the Auger effect an outer-shell electron is ejected from the atom at the same time that another outer-shell electron drops to the incomplete inner shell; the ejected electron carries off the atom's excitation energy instead of a photon doing this. In a sense the Auger effect represents an internal photoelectric effect, although the photon never actually comes into being within the atom. The Auger process is competitive with X-ray emission in most atoms, but the resulting electrons are usually absorbed in the target material while the X rays readily emerge to be detected.

Problems

1. If atoms could contain electrons with principal quantum numbers up to and including $n = 6$, how many elements would there be?

2. The ionization energies of the elements of atomic numbers 20 through 29 are very nearly equal. Why should this be so when considerable variations exist in the ionization energies of other consecutive sequences of elements?

3. Prove that the maximum number of electrons in an atom that can have the same principal quantum number n is $2n^2$.

4. Many years ago it was pointed out that the atomic numbers of the rare gases are given by the following scheme:

$$
\begin{aligned}
Z(\mathrm{He}) &= 2(1^2) = 2 \\
Z(\mathrm{Ne}) &= 2(1^2 + 2^2) = 10 \\
Z(\mathrm{Ar}) &= 2(1^2 + 2^2 + 2^2) = 18 \\
Z(\mathrm{Kr}) &= 2(1^2 + 2^2 + 2^2 + 3^2) = 36
\end{aligned}
$$

$$Z(\mathrm{Xe}) = 2(1^2 + 2^2 + 2^2 + 3^2 + 3^2) = 54$$
$$Z(\mathrm{Rn}) = 2(1^2 + 2^2 + 2^2 + 3^2 + 3^2 + 4^2) = 86$$

Explain the origin of this scheme in terms of atomic theory.

5. A beam of electrons enters a uniform magnetic field of flux density 1.2 webers/m^2. Find the energy difference between electrons whose spins are parallel and antiparallel to the field.

6. How does the agreement between observations of the normal Zeeman effect and the theory of this effect tend to confirm the existence of electrons as independent entities within atoms?

7. A sample of a certain element is placed in a magnetic field of flux density 0.3 weber/m^2. How far apart are the Zeeman components of a spectral line of wavelength 4,500 A?

8. Why does the normal Zeeman effect occur only in atoms with an even number of electrons?

9. The observed frequencies of the K_α lines in the X-ray spectra of the heavy elements are observed to be proportional to the squares of the respective atomic numbers. Calculate the constant of proportionality.

10. Show that, if the angle between the directions of **L** and **S** in Fig. 8-5 is θ,

$$\cos\theta = \frac{j(j+1) - l(l+1) - s(s+1)}{2\sqrt{l(l+1)s(s+1)}}$$

11. The spin-orbit effect splits the $3P \rightarrow 3S$ transition in sodium (which gives rise to the yellow light of sodium-vapor highway lamps) into two lines, 5,890 A corresponding to $3P_{3/2} \rightarrow 3S_{1/2}$ and 5,896 A corresponding to $3P_{1/2} \rightarrow 3S_{1/2}$. Use these wavelengths to calculate the effective magnetic induction experienced by the outer electron in the sodium atom as a result of its orbital motion.

12. The magnetic moment $\boldsymbol{\mu}_J$ of an atom in which LS coupling holds has the magnitude

$$\mu_J = \sqrt{J(J+1)}\,g_J\mu_B$$

where $\mu_B = e\hbar/2m$ is the Bohr magneton and

$$g_J = 1 + \frac{J(J+1) - L(L+1) + S(S+1)}{2J(J+1)}$$

is the Landé g factor. (a) Derive this result with the help of the law of cosines starting from the fact that, averaged over time, only the components of $\boldsymbol{\mu}_L$ and $\boldsymbol{\mu}_S$ parallel to **J** contribute to $\boldsymbol{\mu}_J$. (*b*) Consider an atom that obeys *LS* coupling that is in a weak magnetic field **B** in which the coupling is preserved. How many substates are there for a given value of J? What is the energy difference between different substates?

13. The ground state of chlorine is $^2P_{3/2}$. Find its magnetic moment (see previous problem). Into how many substates will the ground state split in a weak magnetic field?

PROPERTIES OF MATTER

THE PHYSICS OF MOLECULES 9

What is the nature of the forces that bind atoms together to form molecules? This question, of fundamental importance to the chemist, is hardly less important to the physicist, whose theory of the atom cannot be correct unless it provides a satisfactory answer. The ability of the quantum theory of the atom not only to explain chemical binding but to do so partly in terms of a force that has no classical analog is further testimony to the power of this approach.

9.1 The Heteropolar Bond

Two kinds of interaction are responsible for chemical bonds, *heteropolar* (or ionic) and *homopolar* (or covalent). In most molecules neither kind acts exclusively, an intermediate situation generally being the case. We shall first look into the origin and properties of the heteropolar bond.

As we learned in the previous chapter, the most stable electron configuration of an atom consists of closed subshells. Atoms with incomplete outer subshells tend to gain or lose electrons in order to attain stable configurations, becoming negative or positive ions in the process. The *ionization energy* of an element is the energy needed to remove an electron from one of its atoms; it is therefore a measure of how tightly bound its outermost electron is. Table 9.1 shows the ionization energies of the elements in group I and period 2 of the periodic table. An atom of any of the alkali metals of group I has a single s electron outside a closed subshell. The electrons in the inner shells partially shield the outer electron from the nuclear charge $+Ze$, so that the effective charge holding the outer electron to the atom is just $+e$ rather than $+Ze$. Relatively little work must be done to detach an electron from such an atom, and the alkali metals form positive ions readily. The larger the atom, the farther the outer electron is from the nucleus and the weaker is the electrostatic force on it; this is why the ionization energy decreases as we go down group I. The increase in ionization energy from left to right across

Table 9.1

SOME IONIZATION ENERGIES (in ev)

	I	II	III	IV	V	VI	VII	VIII
1								
2	Li 5.4	Be 9.3	B 8.3	C 11.3	N 14.5	O 13.6	F 17.4	Ne 21.6
3	Na 5.1							
4	K 4.3							
5	Rb 4.2							
6	Cs 3.9							

any period is accounted for by the increase in nuclear charge while the number of inner shielding electrons stays constant. There are two electrons in the common inner shell of period 2 elements, and the effective nuclear charge acting on the outer electrons of these atoms is therefore $+(z-2)e$. The outer electron in a lithium atom is held to the atom by an effective charge of $+e$, while each outer electron in beryllium, boron, carbon, etc., atoms is held to its parent atom by effective charges of $+2e$, $+3e$, $+4e$, etc. The anomalously high ionization energies of beryllium and nitrogen are explained by the additional stability exhibited by filled subshells (the $2s$ subshell of beryllium is complete) and half-filled subshells (the $2p$ subshell of nitrogen is half filled).

At the other extreme from alkali metal atoms, which tend to lose their outermost electrons, are halogen atoms, which tend to complete their outer p subshells by picking up an additional electron each. The *electron affinity* of an element is defined as the energy released when an electron is added to an atom of each element. The greater the electron affinity, the more tightly bound is the added electron. Table 9.2 shows the electron affinities of the halogens. In general, electron affinities decrease going down any group of the periodic table and increase going from left to right across any period. The experimental determination of electron affinities is quite difficult, and those for only a few elements are accurately known.

Table 9.2

ELECTRON AFFINITIES OF THE HALOGENS (in ev)

Fluorine	3.6
Chlorine	3.8
Bromine	3.5
Iodine	3.2

A heteropolar bond between two atoms can occur when one of them has a low ionization energy, and hence a tendency to become a positive ion, while the other has a high electron affinity, and hence a tendency to become a negative ion. Sodium, with an ionization energy of 5.1 ev, is an example of the former and chlorine, with an electron affinity of 3.8 ev, is an example of the latter. When a Na^+ ion and a Cl^- ion are in the same vicinity and are free to move, the attractive electrostatic force between them brings them together. The condition that a stable molecule of NaCl result is simply that the total energy of the system of the two ions be less than the total energy of a system of two atoms of the same elements; otherwise the surplus electron on the Cl^- ion would transfer to the Na^+ ion, and the neutral Na and Cl atoms would no longer be bound together. Let us see how this criterion is met by NaCl.

Let us consider a Na atom and a Cl atom infinitely far apart. In order to remove the outer electron from the Na atom, leaving it a Na^+ ion, 5.1 ev of work must be done. That is,

9.1 $$\text{Na} + 5.1 \text{ ev} \rightarrow \text{Na}^+ + e^-$$

When this electron is brought to the Cl atom, the latter absorbs it to complete its outer electron subshell and thereby becomes a Cl^- ion. Since the electron affinity of chlorine is 3.8 ev, signifying that this amount of work must be done to liberate the odd electron from a Cl^- ion, the formation of Cl^- *evolves* 3.8 ev of energy. Hence

9.2 $$\text{Cl} + e^- \rightarrow \text{Cl}^- + 3.8 \text{ ev}$$

The net result of these two events is the sum of Eqs. 9.1 and 9.2:

9.3 $$\text{Na} + \text{Cl} + 1.3 \text{ ev} \rightarrow \text{Na}^+ + \text{Cl}^-$$

Our net expenditure of energy to form Na^+ and Cl^- ions from Na and Cl atoms has been only 1.3 ev.

What happens when the electrostatic attraction between the Na^+ and Cl^- ions brings them together, let us say to 4 A (4×10^{-10} m) of each other? The energy given off when this occurs is equal to the potential energy of the system of a $+e$ charge a distance of 4 A from a $-e$ charge. This energy is

$$\begin{aligned} V &= -\frac{e^2}{4\pi\varepsilon_0 r} \\ &= -\frac{(1.6 \times 10^{-19} \text{ coulomb})^2}{4\pi \times 8.85 \times 10^{-12} \text{ f/m} \times 4 \times 10^{-10} \text{ m}} \\ &= -5.8 \times 10^{-19} \text{ joule} \\ &= -3.6 \text{ ev} \end{aligned}$$

Schematically,

9.4 $$\underset{\longleftarrow\infty\longrightarrow}{Na^+ \quad + \quad Cl^-} \rightarrow \underset{\leftarrow 4\,A \rightarrow}{Na^+ + Cl^-} + 3.6 \text{ ev}$$

If we shift an electron from a Na atom to an infinitely distant Cl atom and allow the resulting ions to come together, then, the entire process evolves an energy of 3.6 ev − 1.3 ev = 2.3 ev:

9.5 $$\underset{\longleftarrow\infty\longrightarrow}{Na \quad + \quad Cl} \rightarrow \underset{\leftarrow 4\,A \rightarrow}{Na^+ + Cl^-} + 2.3 \text{ ev}$$

Evidently the NaCl molecule formed by the electrostatic attraction of Na^+ and Cl^- ions is stable, since work must be done on such a molecule in order to separate it into Na and Cl atoms.

We may well ask what stops the two ions from continuing to approach each other until their electron structures mesh together. It is easy to see that, if such a meshing were to take place, the positively charged nuclei of the ions would no longer be completely shielded by their surrounding electrons and would simply repel each other electrostatically. There is another phenomenon that is even more effective in keeping the ions apart. According to the Pauli exclusion principle, no two electrons in the same atomic system can exist in the same quantum state. If the electron structures of Na^+ and Cl^- overlap, they constitute a single atomic system rather than separate, independent systems. If such a system is to obey the exclusion principle, some electrons will have to go to higher quantum states than they would otherwise occupy. These states have more energy than those corresponding to the normal electron configuration of the ions, and so the total energy of the NaCl molecule increases when the Na^+ and Cl^- ions approach each other too closely.

An increasing potential energy when two bodies are brought together means that the force between them is repulsive, just as a decreasing potential energy under the same circumstances means an attractive force. Figure 9-1 contains a curve showing how the potential energy V of the system of a Na^+ ion and a Cl^- ion varies with separation distance r. At $r = \infty$, $V = +1.3$ ev. When the ions are closer together, the attractive electrostatic force between them takes effect, and the potential energy drops. Finally, when they are about 4 A apart, their electron structures begin to interact, a repulsive force comes into play, and the potential energy soon starts to rise. The minimum in the potential-energy curve occurs at $r = 2.4$ A, where $V = -4.2$ ev. At this separation distance the mutually attractive and repulsive forces on the ions exactly balance, and the system is in equilibrium. To *dissociate* a NaCl molecule into Na and Cl atoms requires an energy of 4.2 ev.

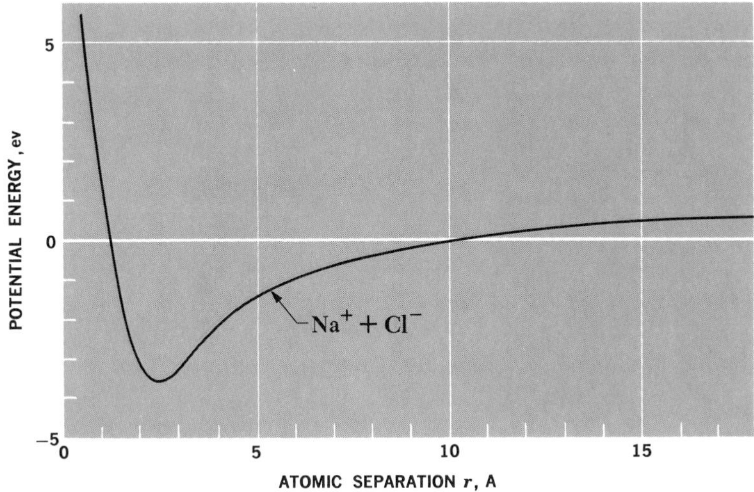

FIGURE 9-1 The variation of the potential energy of the system Na^+ and Cl^- with their distance apart.

9.2 The Covalent Bond: the Hydrogen Molecule

The theory of the ionic bond between atoms of opposite valence is unable to cope with such molecules as H_2 or Cl_2, whose constituent atoms are identical. Pure ionic binding involves the transfer of one or more electrons from those atoms in a molecule with positive valences to those with negative valences, and the resulting electrostatic forces are responsible for the stability of the molecule. In *covalent binding*, on the other hand, electrons may be thought of as being *shared* by the atoms composing a molecule. In the course of circulating among these atoms, the electrons spend more of their time between atoms than on the outside of the molecule; this leads to a net attractive electrostatic force that holds the molecule together. We shall consider the covalent bond between the hydrogen atoms in a hydrogen molecule in some detail.

A convenient way to investigate the force between two atoms is to determine their total energy E, including the energy of their interaction, as a function of the distance between them. If E increases when the atoms are brought together, the force between them is repulsive, while if E decreases, the force is attractive. We must solve Schrödinger's equation for the pair of hydrogen atoms shown in Fig. 9-2 in order to determine E as a function of R. We shall consider an electron in the atom containing nucleus a to be in quantum state a and an electron in the atom containing nucleus b to be in quantum state b; the electrons themselves we shall call 1 and 2. In terms of

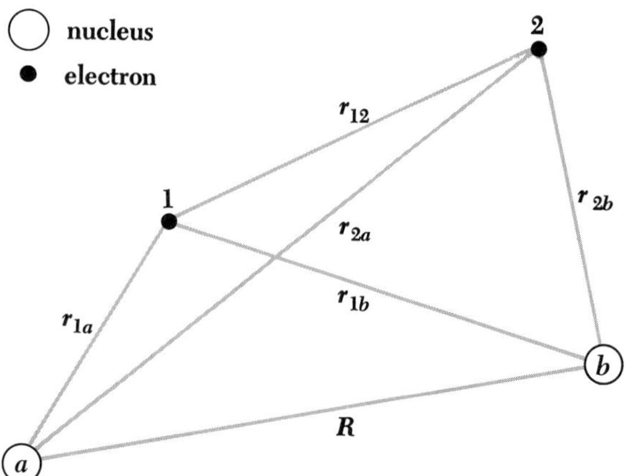

FIGURE 9-2 A system of two hydrogen atoms.

the distances between the various nuclei and electrons defined in Fig. 9-2, the total potential energy V of the system of two atoms is

9.6 $$V = \frac{e^2}{4\pi\varepsilon_0}\left(\frac{1}{R} + \frac{1}{r_{12}} - \frac{1}{r_{1a}} - \frac{1}{r_{1b}} - \frac{1}{r_{2a}} - \frac{1}{r_{2b}}\right)$$

The total potential energy of these atoms when infinitely separated is just

9.7 $$V_\infty = \frac{e^2}{4\pi\varepsilon_0}\left(-\frac{1}{r_{1a}} - \frac{1}{r_{2b}}\right)$$

and so the interaction adds four terms to the equation for V.

If electron 1 is in quantum state a and electron 2 in quantum state b, the wave function ψ (1, 2) of the system is

$$\psi_a(1)\psi_b(2)$$

while if electron 2 is in state a and electron 1 in state b, the wave function of the system is

$$\psi_a(2)\psi_b(1)$$

Since we cannot distinguish one electron from the other, the wave function $\psi(1, 2)$ that we must use should reflect the equal likelihood for either configuration to exist at any instant. As we know, the two possibilities for $\psi(1, 2)$ are the symmetric combination

9.8 $$\psi_S = \psi_a(1)\psi_b(2) + \psi_a(2)\psi_b(1)$$

and the antisymmetric combination

9.9 $$\psi_A = \psi_a(1)\psi_b(2) - \psi_a(2)\psi_b(1)$$

In order to determine when ψ_S must be used and when ψ_A, we make use of two fundamental principles: first, the *complete* wave function of a system is the product of a wave function $u(1, 2)$ describing the coordinates of the two electrons and a spin function $s(1, 2)$ describing the orientations of their spins; second, the complete wave function must be antisymmetric to an exchange of both coordinates and spins. The latter is just the exclusion principle. An antisymmetric complete wave function

$$\psi(1, 2) = u(1, 2)s(1, 2)$$

can result from the combination of a symmetric coordinate wave function u_S and an antisymmetric spin function s_A or from the combination of an antisymmetric coordinate wave function u_A and a symmetric spin function s_S. That is, only

$$\psi = u_S s_A$$

and

$$\psi = u_A s_S$$

are acceptable. If the spins of the two electrons are parallel, their spin function is symmetric since it does not change sign when the electrons are exchanged. Hence the coordinate wave function u for two electrons whose spins are parallel must be antisymmetric; we may express this by writing

9.10 $$u_{\uparrow\uparrow} = u_A$$

On the other hand, if the spins of the two electrons are antiparallel, their spin function is antisymmetric since it reverses sign when the electrons are exchanged. Hence the coordinate wave function u for two electrons whose spins are antiparallel must be symmetric, and we may express this by writing

9.11 $$u_{\uparrow\downarrow} = u_S$$

Now we can proceed to solve Schrödinger's equation for two hydrogen atoms, using the potential energy of Eq. 9.6 and investigating both symmetric (electron spins antiparallel) and antisymmetric (electron spins parallel) forms for the wave function u of the system. When this is done, it is found that the two types of wave function result in two different formulas for the energy of the system, namely,

9.12 $$E_S = E_0 + E_{e1} - E_{ex}$$

9.13 $$E_A = E_0 + E_{e1} + E_{ex}$$

The energy term E_0 is that of two isolated hydrogen atoms, corresponding to the potential-energy function of Eq. 9.7, while the other two terms arise from the interaction between the atoms. The term E_{e1} is given by

9.14 $$E_{e1} = \frac{\alpha e^2}{4\pi\varepsilon_0} \int \left(\frac{1}{R} + \frac{1}{r_{12}} - \frac{1}{r_{1b}} - \frac{1}{r_{2a}} \right) u_a{}^2(1) u_b{}^2(2)\, dV$$

where α is a normalization factor. Since $u_a{}^2(1)$ and $u_b{}^2(2)$ are the probability densities of the electrons, this term evidently equals the contributions to the total energy of the electrostatic potential energy of the two protons, the two electrons, proton a and electron 2, and proton b and electron 1. The magnitude of the term E_{ex} is given by

9.15 $$E_{ex} = \frac{\beta e^2}{4\pi\varepsilon_0} \int \left(\frac{1}{R} + \frac{1}{r_{12}} - \frac{1}{r_{1b}} - \frac{1}{r_{2a}} \right) u_a(1) u_b(2) u_a(2) u_b(1)\, dV$$

where β is another normalization factor, and arises from the exchange of electrons between the atoms. Instead of probability densities, Eq. 9.15 contains mixed products of wave functions, and the exchange energy E_{ex} cannot be directly interpreted on the basis of classical notions. The exchange energy has a negative sign (Eq. 9.12) when the system wave function is symmetric but a positive sign (Eq. 9.13) when the wave function is antisymmetric. Hence the energy of a pair of hydrogen atoms is least when their spins are antiparallel, most when their spins are parallel.

FIGURE 9-3 The variation of the potential energy of the system H + H with their distance apart when the electron spins are parallel and antiparallel.

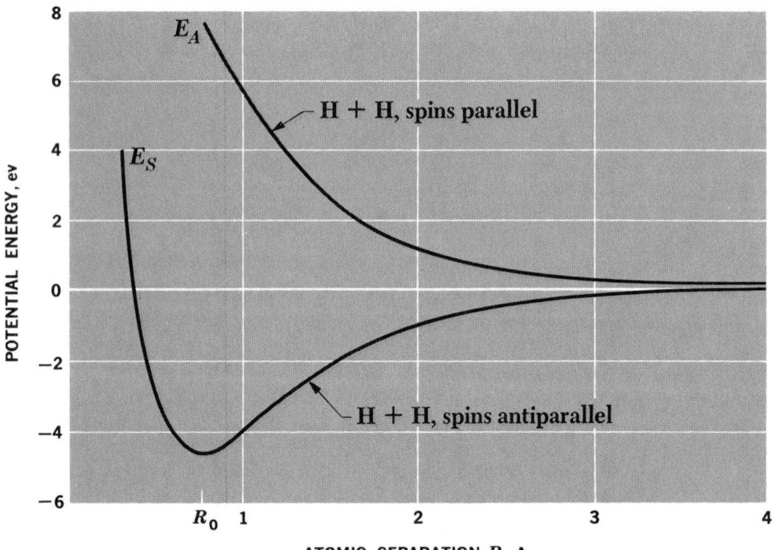

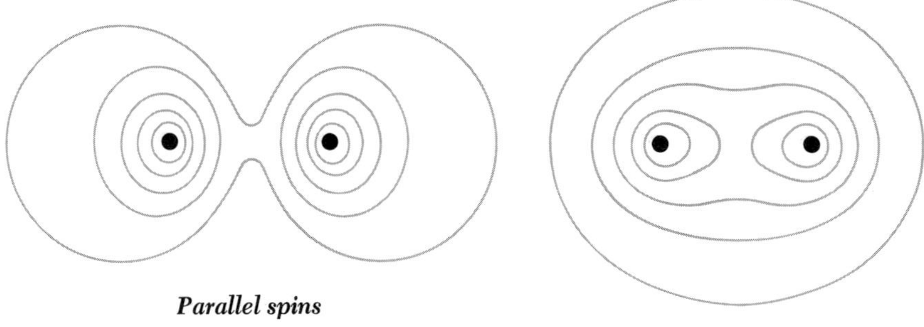

FIGURE 9-4 **Contours of electron probability for two adjacent hydrogen atoms whose electron spins are parallel and antiparallel.**

Figure 9-3 shows how E_S and E_A vary with separation distance R. When the electron spins in the two atoms are parallel, the total energy increases as R decreases, and the atoms repel each other for all separations R. On the other hand, when the electron spins in the two atoms are antiparallel, the total energy decreases with R down to $R_0 = 0.74$ A, and the atoms attract each other except when they are closer together than 0.74 A. In the former case a stable molecule cannot exist, while in the latter the atoms can form a stable molecule with a binding energy of 4.5 ev.

As we indicated at the start of this section, there is a simple intuitive interpretation of the exchange force between two adjacent hydrogen atoms. When their electrons have opposite spins, they are not affected by the exclusion principle, and we may imagine them as being shared by both atoms. In this sharing process the electrons spend on the average more time between the nuclei than on the outside (Fig. 9-4), and the resulting concentration of negative charge between the two positive nuclei keeps them together. When the atomic electrons have like spins, however, the exclusion principle tends to keep them away from the region between the nuclei where there is a common field of force and both cannot simultaneously be present in their lowest energy states. Hence these electrons spend on the average more time on the outside of the pair of nuclei, where they act to pull the nuclei apart in addition to permitting the nuclei themselves to repel each other. Binding is impossible in this situation.

This interpretation is supported by an inspection of the relevant wave functions. An electron is described by the wave function u_a when it is in state a (that is, belonging to atom a) and by the wave function u_b when in state b (belonging to atom b). These wave functions are sketched in Fig. 9-5 for various values of the interatomic separation R. A symmetric combination u_S of them leads to a high value of u^2 between the nuclei, indicating a high

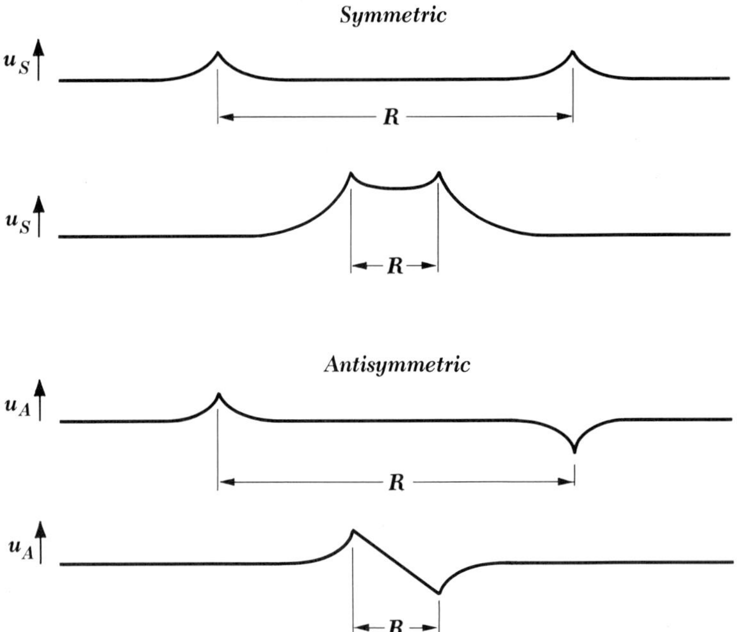

FIGURE 9-5 Symmetric and antisymmetric wave functions of the electrons in two adjacent hydrogen atoms.

probability of finding an electron there. An antisymmetric combination u_A of these functions leads to a low value of u^2 between the nuclei; in fact, ${u_A}^2 = 0$ in the exact middle.

9.3 Valence

We have seen that two H atoms can combine to form a H_2 molecule, and, indeed, hydrogen molecules in nature always consist of two H atoms. Let us now look into how the exclusion principle prevents molecules such as He_2 and H_3 from existing, while permitting such other molecules as H_2O to be stable.

Every He atom in its ground state has a 1*s* electron of each spin. If it is to join with another He atom by exchanging electrons, each atom will have two electrons with the same spin for part of the time. That is, one atom will have both electron spins up (↑↑) and the other will have both spins down (↓↓). The exclusion principle, of course, prohibits two 1*s* electrons in an atom from having the same spins, which is manifested in a repulsion between He atoms. Hence the He_2 molecule cannot exist. A similar argument holds in the case of H_3. An H_2 molecule contains two 1*s* electrons whose spins are antiparallel (↑↓). Should another H atom approach whose electron spin is, say, up, the resulting molecule would have two spins parallel (↑↑↓), and this is impossible if all three electrons are to be in 1*s* states. Hence the existing H_2 molecule repels the additional H atom. The exclusion-principle argument does not apply if one of the three electrons in H_3 is in an excited state. All such states are of higher energy than the 1*s* state, however, and the resulting configuration therefore has more energy than $H_2 + H$ and so will decay rapidly to $H_2 + H$.

The water molecule H_2O achieves stability because the O atom lacks two $2p$ electrons of having a completed outer electron shell. This lack is remedied when the O atom forms covalent bonds with two H atoms so that the latter's electrons are shared by the O atom without violating the exclusion principle. The H_2O structure has less energy than the separate atoms owing to the electron affinity of O, and so is favored.

Evidently the interatomic forces that lead to chemical bonds, covalent as well as ionic, exhibit *saturation:* an atom joins with only a limited number of other atoms. The chemist describes saturation in terms of *valence.* An atom that has only a few electrons in its outermost subshell tends to combine with one or more other atoms whose outermost subshells lack a total of the same number of electrons. An atom that donates n electrons to a bond is said to have a valence of $+n$, while one that receives m electrons from a bond is said to have a valence of $-m$. For stability the total numbers of positive and negative valences in a molecule must be equal. Thus H has a valence of $+1$ and O a valence of -2, and H_2O has a net valence of $1 + 1 - 2 = 0$. In the case of H_2, we may think of one H atom as having a valence of $+1$ and the other of -1. Atoms such as carbon, which form bonds by sharing four electrons, may be regarded as having valences of either $+4$ or -4, depending upon the atoms they combine with. For example, in CH_4 (methane) carbon has a valence of -4, while in CO_2 (carbon dioxide) its valence is $+4$. To be sure, in both cases the bonds are largely covalent and electrons resonate between the atoms of each molecule instead of being transferred as in purely ionic bonds, but the notion of valence is so convenient that it is retained whenever possible.

9.4 Directed Bonds

In Sec. 7.7 we mentioned the fact that the electron probability densities in the hydrogen atom for all but *s* states are not spherically symmetric, but exhibit definite maxima in certain directions. Similar lobed patterns are found in the electron probability densities of more complex atoms. Covalent binding occurs when the appropriate probability lobes of adjacent atoms overlap. When an atom combines with two or more other atoms, the latter

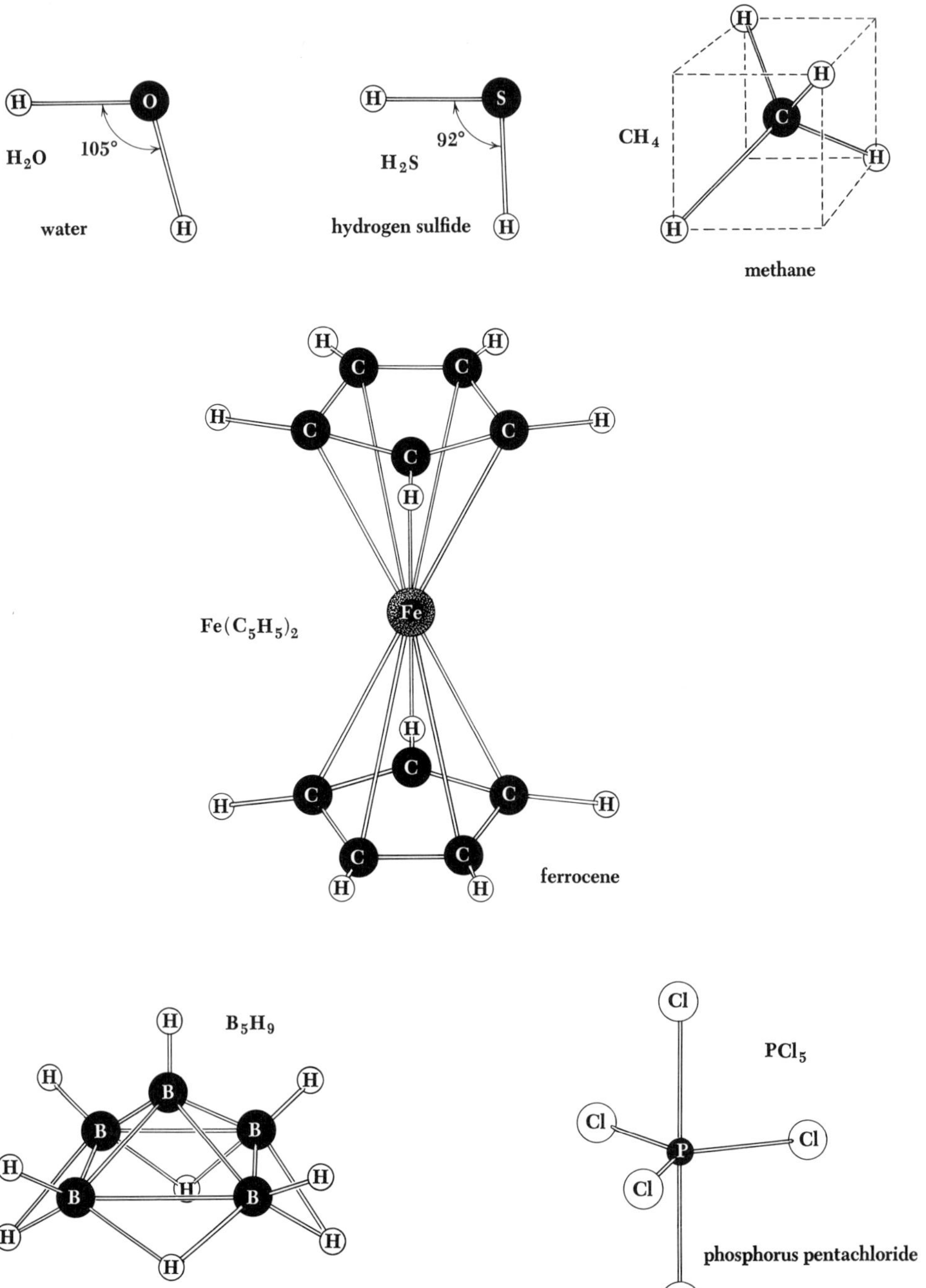

FIGURE 9-6 Directed bonds in complex molecules.

are generally "attached" at specific places on the atom corresponding to the lobes of greatest magnitude, instead of being located at random or in some universal configuration. Figure 9-6 shows some examples of directed bonds in complex molecules.

9.5 Molecular Spectra

Molecular energy states may arise from the rotation of a molecule as a whole and from the vibrations of its constituent atoms as well as from changes in its electronic configuration. The spectra arising from transitions among this multiplicity of levels are accordingly complex, but the difficulty of interpreting them is compensated for by the detailed picture of molecular structure that emerges.

The lowest energy levels of a diatomic molecule arise from rotation about its center of mass. We may picture such a molecule as consisting of atoms of masses m_1 and m_2 a distance r apart, as in Fig. 9-7. The moment of inertia of this molecule about an axis passing through its center of mass and perpendicular to a line joining the atoms is

9.16 $$I = m_1 r_1^2 + m_2 r_2^2$$

where r_1 and r_2 are the distances of atoms 1 and 2 respectively from the center of mass. Since

9.17 $$m_1 r_1 = m_2 r_2$$

by definition, the moment of inertia may be written

$$I = \left(\frac{m_1 m_2}{m_1 + m_2}\right)(r_1 + r_2)^2$$

9.18 $$= m' r^2$$

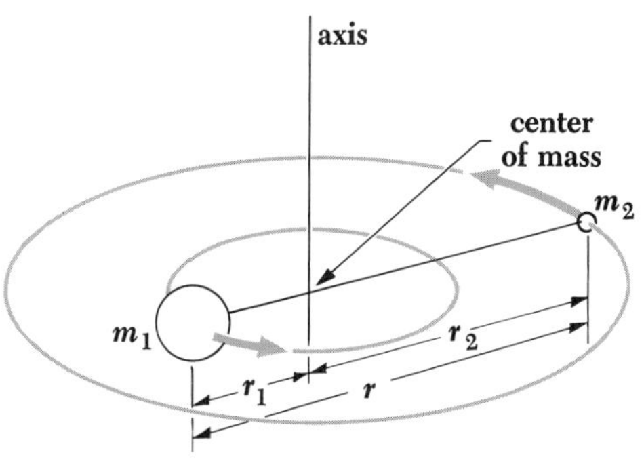

FIGURE 9-7 A diatomic molecule can rotate about its center of mass.

where

9.19 $$m' = \frac{m_1 m_2}{m_1 + m_2}$$

is the reduced mass of the molecule as discussed in Sec. 5.7. Equation 9.18 states that the rotation of a diatomic molecule is equivalent to the rotation of a single particle of mass m' about an axis located a distance of r away.

The angular momentum L_r of the molecule is

9.20 $$L_r = I\omega$$

where ω is its angular frequency. In Chap. 7 we learned that the orbital angular momentum of the electron in a hydrogen atom can have only the values

9.21 $$L = \sqrt{l(l+1)}\,\hbar$$

and the angular momentum of a rotating molecule turns out to be similarly quantized. That is,

9.22 $$L_r = \sqrt{K(K+1)}\,\hbar$$

where K, the *rotational quantum number*, is restricted to

$$K = 0, 1, 2, 3, \ldots$$

We are now in a position to relate the frequencies of the spectral lines that result from transitions between rotational energy levels with the molecular moment of inertia. The energy of a rotating molecule is $\frac{1}{2}I\omega^2$, and so its energy levels are specified by

$$E_r = \tfrac{1}{2}I\omega^2$$

$$= \frac{L_r^2}{2I}$$

9.23 $$= \frac{K(K+1)\hbar^2}{2I}$$ **Rotational states**

Only transitions involving a change in quantum number K of ± 1 can occur, with an increase in K corresponding to the absorption of energy and a decrease in K to the emission of energy. Only molecules that have electric dipole moments, that is, polar molecules such as HCl, can interact with electromagnetic photons, and rotational spectra accordingly exist only for them and not for such nonpolar molecules as H_2.

A transition between a rotational state of quantum number $K + 1$ and a state of quantum number K means an energy change of

$$\Delta E = E_{K+1} - E_K$$

9.24 $$= \frac{(K+1)\hbar^2}{I}$$

Since

$$\Delta E = h\nu$$

the photon involved in the transition has the frequency

9.25 $$\nu = \frac{\hbar}{2\pi I}(K+1)$$ **Rotational spectra**

Rotational spectra evidently consist of equally spaced lines, as in Fig. 9-8. These lines lie in the extreme infrared and microwave regions, between about 10^{-4} and 10^{-1} m in wavelength. From the rotational spectrum of a molecule its moment of inertia can be obtained, and, since the masses of its constituent atoms are known, the interatomic separation r can be calculated. Typical separations are 1.1 A in CO, 1.3 A in HCl, and 2.2 A in HgCl.

When sufficiently excited, a molecule can vibrate as well as rotate. Since

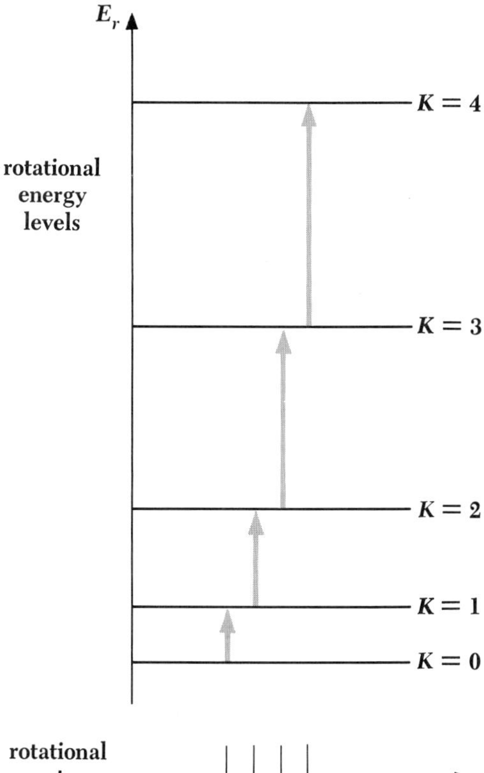

FIGURE 9-8 Energy levels and spectrum of molecular rotation.

the excitation energy involved in molecular vibration is greater than that involved in rotation, rotation always accompanies vibration. Vibration-rotation spectra are characterized by the presence of shorter wavelengths than pure rotational spectra and are generally in the near infrared region of the electromagnetic spectrum. Vibrational transitions occur most readily in polar molecules, as in the case of rotational transitions.

The potential energy of a diatomic molecule is plotted versus interatomic distance in Fig. 9-9. In the neighborhood of the minimum of this curve, which corresponds to the normal configuration of the molecule, the shape of the curve is very nearly a parabola. In this region, then,

$$V = V_0 + \tfrac{1}{2}k(r - r_e)^2 \qquad (9.26)$$

where r_e is the equilibrium separation of the atoms. The interatomic force that gives rise to this potential energy may be found by differentiating Eq. 9.26:

FIGURE 9-9 The potential energy of a diatomic molecule as a function of interatomic distance, showing vibrational and rotational energy levels.

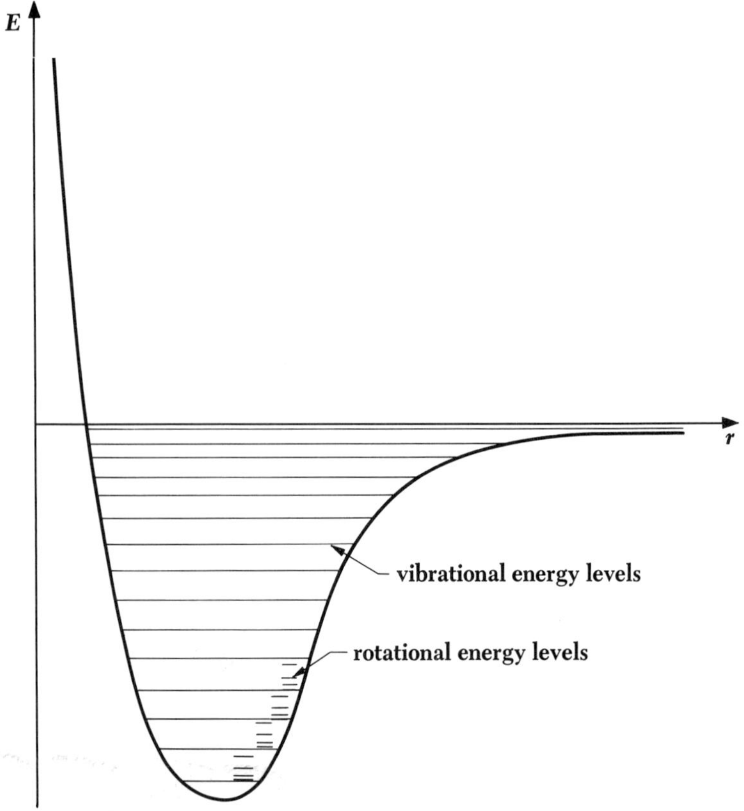

$$F = -\frac{dV}{dr}$$

9.27 $$= -k(r - r_e)$$

The force is just the restoring force that a stretched or compressed spring exerts, and, as with a spring, a molecule suitably excited can undergo simple harmonic oscillations. Classically, the frequency of a vibrating body of mass m connected to a spring of force constant k is

9.28 $$\nu' = \frac{1}{2\pi}\sqrt{\frac{k}{m}}$$

When the harmonic-oscillator problem is solved quantum-mechanically, as was done in Chap. 6, the allowed frequencies are instead specified by

$$\nu = (\upsilon + \tfrac{1}{2})\frac{1}{2\pi}\sqrt{\frac{k}{m}}$$ Vibrational spectra

9.29 $$= (\upsilon + \tfrac{1}{2})\,\nu'$$

where υ, the *vibrational quantum number,* may have the values

$$\upsilon = 0, 1, 2, 3, \ldots$$

The energies corresponding to these frequencies are

$$E_\upsilon = h\nu$$ Vibrational states

9.30 $$= (\upsilon + \tfrac{1}{2})h\nu'$$

Evidently the lowest vibrational state ($\upsilon = 0$) has the finite energy $\tfrac{1}{2}h\nu'$, and not the classical value of zero. In this connection we recall from Sec. 6.5 that a particle confined to a box, where the potential well has vertical rather than curved walls, also cannot have zero energy.

The higher vibrational states of a molecule do not obey Eq. 9.30 because the parabolic approximation to its potential-energy curve becomes less and less valid with increasing energy. The spacing between adjacent energy levels of high υ is less than the spacing between adjacent levels of low υ because of this effect, which is shown in Fig. 9-9. This diagram also shows the fine structure in the vibrational levels caused by the simultaneous excitation of rotational levels.

The energies of rotation and vibration in a molecule are due to the motion of its atomic nuclei, since the nuclei contain essentially all of the molecule's mass. The molecular electrons also can be excited to higher energy levels, though the spacing of these levels is much greater than that due to rotation or vibration. For the latter reason electronic transitions involve radiation in the visible or ultraviolet parts of the electromagnetic spectrum, with each

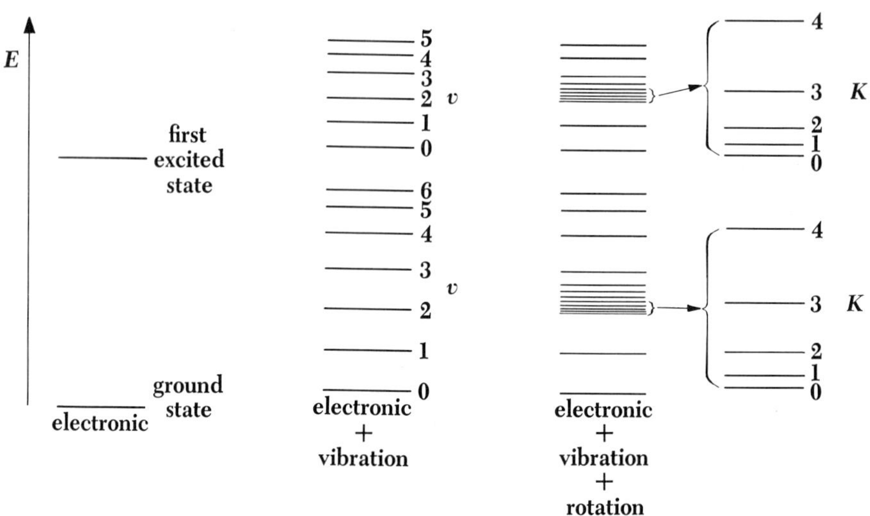

FIGURE 9-10 Bands in molecular spectra actually consist of closely spaced lines due to rotational and vibrational transitions.

purely electronic transition appearing as a series of closely spaced lines, called *bands,* because of superimposed rotational and vibrational effects (Fig. 9-10).

Problems

1. Sodium atoms exhibit more pronounced chemical activity than sodium ions. Why?

2. Chlorine atoms exhibit more pronounced chemical activity than chlorine ions. Why?

3. Why are lithium and sodium so similar chemically?

4. Why are fluorine and chlorine so similar chemically?

5. (*a*) How much energy is required to form a K^+ and Cl^- ion pair from a pair of these atoms? (*b*) What must the separation be between a K^+ and a Cl^- ion if their total energy is to be zero?

6. (*a*) How much energy is required to form a K^+ and I^- ion pair from a pair of these atoms? (*b*) What must the separation be between a K^+ and an I^- ion if their total energy is to be zero?

7. (*a*) How much energy is required to form a Li^+ and Br^- ion pair from a pair of these atoms? (*b*) What must the separation be between a Li^+ and a Br^- ion if their total energy is to be zero?

8. Calculate the energies of the four lowest rotational energy states of the H_2 and D_2 molecules, where D represents the deuterium atom ${}_1H^2$.

9. The rotational spectrum of HCl contains the following wavelengths:

12.03×10^{-5} m
9.60×10^{-5} m
8.04×10^{-5} m
6.89×10^{-5} m
6.04×10^{-5} m

If the isotopes involved are ${}_1H^1$ and ${}_{17}Cl^{35}$, find the distance between the hydrogen and chlorine nuclei in an HCl molecule. (The mass of Cl^{35} is 5.81×10^{-26} kg.)

10. Calculate the classical angular velocity of a rigid body whose energy is given by Eq. 9.23 for states of $K = K$ and $K = K + 1$, and show that the frequency of the spectral line associated with a transition between these states is intermediate between the angular velocities of the states.

11. A $Hg^{200}Cl^{35}$ molecule emits a 4.4-cm photon when it undergoes a rotational transition from $K = 1$ to $K = 0$. Find the interatomic distance in this molecule. (The masses of Hg^{200} and Cl^{35} are, respectively, 3.32×10^{-25} kg and 5.81×10^{-26} kg.)

12. Assume that the H_2 molecule behaves exactly like a harmonic oscillator with a force constant of 573 n/m and find the vibrational quantum number corresponding to its 4.5-ev dissociation energy.

13. The bond between the hydrogen and chlorine atoms in a HCl molecule has a force constant of 470 n/m. Is it likely that a HCl molecule will be vibrating in its first excited vibrational state at room temperature?

14. The observed molar specific heat of hydrogen gas at constant volume is plotted in Fig. 9-11 versus absolute temperature. (The temperature scale is logarithmic.) Since each degree of freedom (that is, each mode of energy possession) in a gas molecule contributes ~1 kcal/kmole °K to the specific heat of the gas, this curve is interpreted as indicating that only translational motion, with three degrees of freedom, is possible for hydrogen molecules at very low temperatures. At higher temperatures the specific heat rises to ~5 kcal/kmole °K, indicating that two more degrees of freedom are available, and at still higher temperatures the specific heat is ~7 kcal/kmole °K, indicating two further degrees of freedom. The additional pairs of degrees

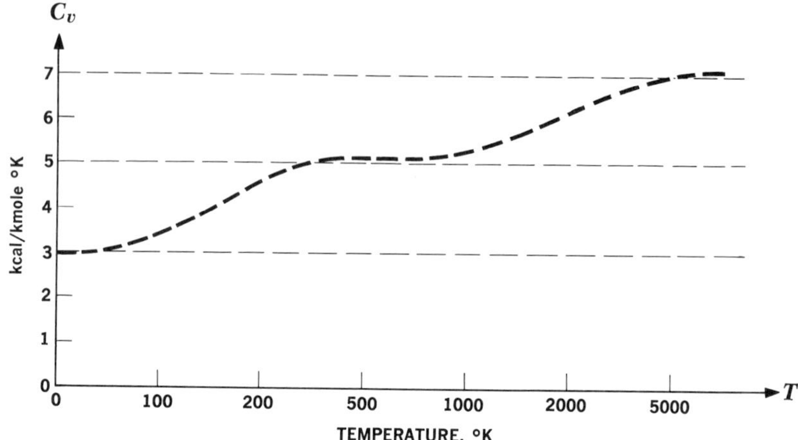

FIGURE 9-11 Molar specific heat of hydrogen at constant volume.

of freedom represent, respectively, rotation, which can take place about two independent axes perpendicular to the axis of symmetry of the H_2 molecule, and vibration, in which the two degrees of freedom correspond to the kinetic and potential modes of energy possession by the molecule. (*a*) Verify this interpretation of Fig. 9-11 by calculating the temperature at which kT is equal to the minimum rotational energy and to the minimum vibrational energy a H_2 molecule can have. Assume that the force constant of the bond in H_2 is 573 n/m and that the H atoms are 7.42×10^{-11} m apart. (At these temperatures, approximately half the molecules are rotating or vibrating, respectively, though in each case some are in higher states than $K = 1$ or $v = 1$.) (*b*) To justify considering only two degrees of rotational freedom in the H_2 molecule, calculate the temperature at which kT is equal to the minimum rotational energy a H_2 molecule can have for rotation about its axis of symmetry. (*c*) How many rotations does a H_2 molecule with $K = 1$ and $v = 1$ make per vibration?

STATISTICAL MECHANICS 10

The branch of physics known as *statistical mechanics* attempts to relate the macroscopic properties of an assembly of particles to the microscopic properties of the particles themselves. Statistical mechanics, as its name implies, is not concerned with the actual motions or interactions of individual particles, but investigates instead their *most probable* behavior. While statistical mechanics cannot help us determine the life history of a particular particle, it *is* able to inform us of the likelihood that a particle (exactly which one we cannot know in advance) has a certain position and momentum at a certain instant. Because so many phenomena in the physical world involve assemblies of particles, the value of a statistical rather than deterministic approach is clear. Owing to the generality of its arguments, statistical mechanics can be applied with equal facility to classical problems (such as that of molecules in a gas) and quantum-mechanical problems (such as those of free electrons in a metal or photons in a box), and it is one of the most powerful tools of the theoretical physicist.

10.1 Phase Space

The state of a system of particles is completely specified classically at a particular instant if the position and momentum of each of its constituent particles are known. Since position and momentum are vectors with three components apiece, we must know six quantities,

$$x, y, z, p_x, p_y, p_z$$

for each particle.

The position of a particle is a point having the coordinates x, y, z in ordinary three-dimensional space. It is convenient to generalize this conception by imagining a six-dimensional space in which a point has the six coordinates x, y, z, p_x, p_y, p_z. This combined position and momentum space is called *phase space.* The notion of phase space is introduced to enable us to develop statistical mechanics in a geometrical framework, thereby permitting a simpler and more straightforward method of analysis than an equivalent one wholly abstract in character. A point in phase space corresponds to a particular position *and* momentum, while a point in ordinary space corresponds to a particular position only. Thus every particle is completely specified by a point in phase space, and the state of a system of particles corresponds to a certain distribution of points in phase space.

The uncertainty principle compels us to elaborate what we mean by a "point" in phase space. Let us divide phase space into tiny six-dimensional cells whose sides are dx, dy, dz, dp_x, dp_y, dp_z. As we reduce the size of the cells, we approach more and more closely to the limit of a point in phase space. However, the volume of each of these cells is

$$\tau = dx\,dy\,dz\,dp_x\,dp_y\,dp_z$$

and, according to the uncertainty principle,

$$dx\,dp_x \geqslant h$$
$$dy\,dp_y \geqslant h$$
$$dz\,dp_z \geqslant h$$

Hence we see that

$$\tau \geqslant h^3$$

A "point" in phase space is actually a cell whose minimum volume is of the order of h^3. We must think of a particle in phase space as being located somewhere in such a cell centered at some location x, y, z, p_x, p_y, p_z instead of being precisely at the point itself.

While the notion of a point of infinitesimal size in phase space can have no physical significance, since it violates the uncertainty principle, the notion of a point of infinitesimal size in either position space or momentum space alone is perfectly acceptable: we *can* in principle determine the position of a particle with as much precision as we like merely by accepting an unlimited uncertainty in our knowledge of its momentum, and vice versa.

It is the task of statistical mechanics to determine the state of a system by investigating how the particles constituting the system distribute themselves in phase space. If we can find the probabilities of occurrence w of all possible distributions that are permitted by the nature of the system, we can immediately select the most probable one and assert that the system tends to behave according to this distribution of particle positions and momenta. That is, we assert that the state of a system when it is in thermal equilibrium corresponds to the most probable distribution of particles in phase space.

10.2 The Probability of a Distribution

We shall now consider an elementary example in order to introduce the mathematical methods that will be necessary. Let us suppose that we have a large box divided into k cells whose areas are $a_1, a_2, a_3, \ldots, a_k$, as in Fig. 10-1. We proceed to throw N balls into the box in a completely random manner, so that no part of the box is favored. We note how many balls fall in each cell, and then repeat the experiment. After a great many determinations of this kind we will find that a certain distribution of balls among the various cells occurs more often than any other; we call this the most probable distribution. While the most probable distribution may or may not actually be found after any particular throw of the balls, it is the one *most likely* to be found. Clearly the most probable distribution is that in which the number of balls in each cell is proportional to the size of the cell: a large cell is more apt to be hit than a small cell. Let us obtain this result by actually calculating the probabilities of the various possible distributions. Our procedure will resemble the use of a steamroller to crack a peanut, but for the moment our interest is in the operation of the steamroller.

The probability w that the balls be distributed in a certain way among the cells depends upon two factors: (1) the *a priori probability* of the distribution, which is based upon the properties of each cell, and (2) the *thermodynamic probability* of the distribution, which is the number of different ways the balls may be redistributed among the cells without changing the number in each cell. We thus explicitly assume that the balls are identical but distinguishable.

Here the a priori probability g_i that a ball fall into the ith cell is the ratio

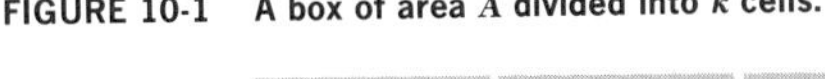

FIGURE 10-1 **A box of area A divided into k cells.**

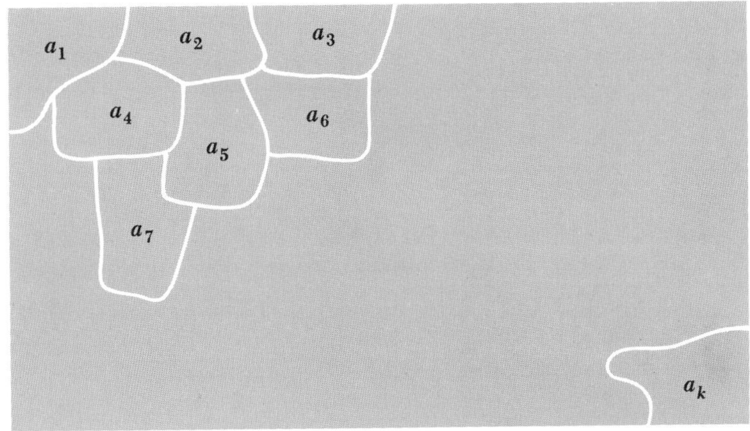

between the area a_i of the cell and the total area A of the entire box. That is,

10.1 $$g_i = \frac{a_i}{A}$$

where

10.2 $$A = a_1 + a_2 + \ldots + a_k$$

The sum of the a priori probabilities for all of the cells must be 1, since the ball, by hypothesis, falls into the box somewhere:

10.3 $$\Sigma g_i = g_1 + g_2 + \ldots + g_k = 1$$

The probability that *two* balls fall in the *i*th cell is g_i^2, the product of the probabilities g_i and g_i that each one separately fall in. (If there is a 10 per cent chance that one ball will fall in a given cell, there is only a 1 per cent chance that two will fall in.) The a priori probability for n_i balls to fall in the *i*th cell is $(g_i)^{n_i}$. The a priori probability of any particular *distribution* of the N balls among the k cells is therefore the product of k probabilities of the form $(g_i)^{n_i}$, namely,

10.4 $$(g_1)^{n_1}(g_2)^{n_2} \ldots (g_k)^{n_k}$$

subject to the condition that the total number of balls equal N:

10.5 $$\Sigma n_i = n_1 + n_2 + \ldots + n_k = N$$

The total number of permutations possible for N balls is $N!$; in other words, N balls can be arranged in $N!$ different sequences. As an example, we might have four balls, *a, b, c,* and *d*. The value of 4! is

$$4! = 4 \times 3 \times 2 \times 1 = 24$$

and there are indeed 24 ways of arranging them:

abcd	*bacd*	*cabd*	*dabc*
abdc	*badc*	*cadb*	*dacb*
acbd	*bcad*	*cbad*	*dbac*
acdb	*bcda*	*cbda*	*dbca*
adbc	*bdac*	*cdab*	*dcab*
adcb	*bdca*	*cdba*	*dcba*

When more than one ball is in a cell, however, permuting them among themselves has no significance in this situation. Thus n_i balls in the *i*th cell contribute $n_i!$ irrelevant permutations. If there are n_1 balls in cell 1, n_2 balls in cell 2, and so on through n_k balls in cell k, there are $n_1!n_2!n_3! \ldots n_k!$ irrelevant permutations. Therefore the thermodynamic probability of the distribution is the total number of possible permutations $N!$ divided by the total number of irrelevant permutations, or

$$\frac{N!}{n_1!n_2! \ldots n_k!} \tag{10.6}$$

The total probability W of the distribution is the product of the a priori probability (Eq. 10.4) and the thermodynamic probability (Eq. 10.6):

$$W = \frac{N!}{n_1!n_2! \ldots n_k!}(g_1)^{n_1}(g_2)^{n_2} \ldots (g_k)^{n_k} \tag{10.7}$$

To verify that Eq. 10.7 is correct, we add together the probabilities for all possible distributions and obtain

$$\begin{aligned}\Sigma W &= \Sigma \frac{N!}{n_1!n_2! \ldots n_k!}(g_1)^{n_1}(g_2)^{n_2} \ldots (g_k)^{n_k} \\ &= (g_1 + g_2 + \ldots + g_k)^N\end{aligned}$$

with the help of the multinomial theorem of algebra. (This theorem is a generalization of the binomial theorem.) Since, from Eq. 10.3, the sum of the a priori probabilities g_i is 1 and $1^N = 1$, the sum of the probabilities for all possible distributions is 1. The meaning of

$$\Sigma W = 1$$

is that all N balls are certain to fall in the box somewhere, which agrees with our initial hypothesis and confirms the correctness of Eq. 10.7.

What we now must do is determine just which distribution of the balls is most probable, that is, which distribution yields the largest value of W. Our first step is to obtain a suitable analytic approximation for the factorial of a large number. We note that, since

$$n! = n(n-1)(n-2) \ldots (4)(3)(2)$$

the natural logarithm of $n!$ is

$$\ln n! = \ln 2 + \ln 3 + \ln 4 + \ldots + \ln (n-1) + \ln n$$

Figure 10-2 is a plot of $\ln n$ versus n. The area under the stepped curve is $\ln n!$ When n is very large, the stepped curve and the smooth curve of $\ln n$ become indistinguishable, and we can find $\ln n!$ by merely integrating $\ln n$ from $n = 1$ to $n = n$:

$$\begin{aligned}\ln n! &= \int_1^n \ln n \, dn \\ &= n \ln n - n + 1\end{aligned}$$

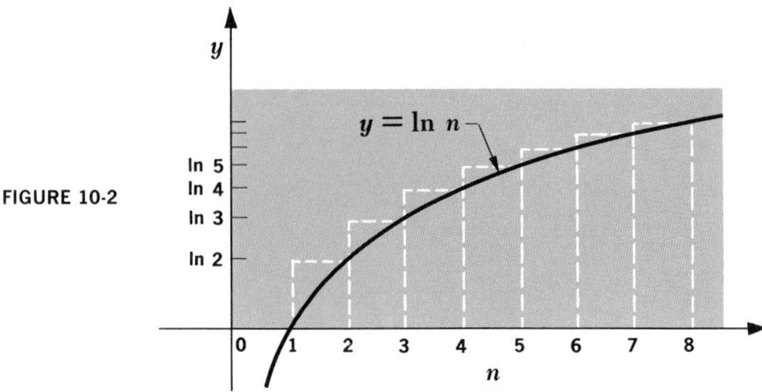

FIGURE 10-2

Because we are assuming that $n \gg 1$, we may neglect the 1 in the above result, and so we obtain

10.8 $$\ln n! = n \ln n - n \qquad n \gg 1$$

Equation 10.8 is known as *Stirling's formula.*

The natural logarithm of Eq. 10.7 is

$$\ln W = \ln N! - \Sigma \ln n_i! + \Sigma\, n_i \ln g_i$$

Stirling's formula enables us to write this expression as

$$\ln W = N \ln N - N - \Sigma\, n_i \ln n_i + \Sigma\, n_i + \Sigma\, n_i \ln g_i$$

Since $\Sigma\, n_i = N$,

10.9 $$\ln W = N \ln N - \Sigma\, n_i \ln n_i + \Sigma\, n_i \ln g_i$$

While we have an equation for ln W rather than for W itself, this is no handicap since

$$(\ln W)_{\max} = \ln W_{\max}$$

The condition for a distribution to be the most probable one is that small changes δn_i in any of the n_i's not affect the value of W. (If the n_i's were continuous variables instead of being restricted to integral values, we could express this condition in the usual way as $\partial W/\partial n_i = 0$.) If the change in ln W corresponding to a change in n_i of δn_i is $\delta \ln W$, from Eq. 10.9 we see that

10.10 $$\delta \ln W_{\max} = -\Sigma\, n_i \delta \ln n_i - \Sigma \ln n_i \delta n_i + \Sigma \ln g_i \delta n_i = 0$$

since $N \ln N$ is constant. Now

$$\delta \ln n_i = \frac{1}{n_i} \delta n_i$$

and so

$$\Sigma\, n_i \delta \ln n_i = \Sigma\, \delta n_i$$

Because the total number of balls is constant, the sum $\Sigma\, \delta n_i$ of all the changes in the number of balls in each cell must be 0, which means that

$$\Sigma\, n_i \delta \ln n_i = 0$$

Hence Eq. 10.10 becomes

10.11 $$-\Sigma \ln n_i \delta n_i + \Sigma \ln g_i \delta n_i = 0$$

While Eq. 10.11 must be fulfilled by the most probable distribution of the balls, it does not by itself completely specify this distribution. We must take account of the fact that the variations $\delta n_1, \delta n_2, \ldots$ in the number of balls in each of the k cells are not independent, since the total number of balls is fixed, but obey the relationship

10.12 $$\Sigma\, \delta n_i = \delta n_1 + \delta n_2 + \ldots + \delta n_k = 0$$

To do so, we make use of Lagrange's method of undetermined multipliers. We let α be a quantity that does not depend upon any of the n_i's and multiply Eq. 10.12 by α to give

10.13 $$\Sigma\, \alpha \delta n_i = 0$$

Now we add this equation to Eq. 10.11 and obtain

10.14 $$-\Sigma \ln n_i \delta n_i + \Sigma \ln g_i \delta n_i + \Sigma\, \alpha \delta n_i = \Sigma\, (-\ln n_i + \ln g_i + \alpha) \delta n_i = 0$$

In each of the k equations added together to give Eq. 10.14, the variation δn_i is effectively an independent variable. In order for Eq. 10.14 to be true, then, the quantity in parentheses must always be 0 regardless of the values given to the δn_i's. Thus we have

10.15 $$-\ln n_i + \ln g_i + \alpha = 0$$

$$n_i = g_i e^{\alpha}$$

Adding together the n_i's and noting that α does not depend upon i,

$$\Sigma\, n_i = e^{\alpha}\, \Sigma\, g_i$$

We recall that g_i is the a priori probability that a ball fall into the ith cell and, from Eq. 10.3,

$$\Sigma\, g_i = 1$$

Hence

$$\Sigma\, n_i = e^{\alpha}$$

and, since the total number of balls $\Sigma\, n_i$ is N, we see that e^α is just

$$e^\alpha = N$$

Equation 10.15 therefore becomes

$$n_i = Ng_i$$

The most probable number of balls in any cell is proportional both to the total number of balls N and to the a priori probability g_i, which is equal to the relative size of the cell. More precisely, we note from Eq. 10.1 that

$$g_i = \frac{a_i}{A}$$

and so

$$n_i = \frac{N}{A}\, a_i$$

The most probable number of balls in any cell is equal to the average density of balls N/A multiplied by the area of the cell. This is the result we set out to derive.

10.3 Maxwell-Boltzmann Statistics

We shall now apply the methods of statistical mechanics to determine how a fixed total amount of energy is distributed among the various members of an assembly of identical particles. That is, our problem is to find out how many particles (on the average) have the energy u_1, how many have the energy u_2, and so on. We shall consider assemblies of three kinds of particles:

A. Identical particles of any spin that are sufficiently widely separated to be distinguished. The molecules of a gas are particles of this kind.

B. Identical particles of 0 or integral spin that cannot be distinguished. These are Bose particles and do not obey the exclusion principle.

C. Identical particles of spin ½ that cannot be distinguished. These are Fermi particles and obey the exclusion principle.

In this section our efforts will be directed to obtaining the distribution law for particles of type A, which we shall refer to in general as *molecules* because gas molecules constitute the most important class of such particles.

Let us consider an assembly of N molecules whose energies are limited to the k values $u_1, u_2, \ldots u_k$, arranged in order of increasing energy. These energies may represent either discrete energy states or average energies within a sequence of energy intervals. If there are n_i molecules of energy u_i

and the total energy of the assembly is U, the most probable distribution of molecules among these k energies is subject to two conditions, namely, conservation of molecules,

$$\Sigma\, n_i = n_1 + n_2 + \ldots + n_k = N \tag{10.16}$$

and conservation of energy,

$$\Sigma\, n_i u_i = n_1 u_1 + n_2 u_2 + \ldots + n_k u_k = U \tag{10.17}$$

If the a priori probability for a molecule to have the energy u_i is g_i, the probability W for any distribution is given by Eq. 10.7. As before, the distribution of maximum probability must obey Eq. 10.11, but now there are *two* conditions the δn_i must fulfill. These conditions follow from Eq. 10.16 and 10.17; they are

$$\Sigma\, \delta n_i = \delta n_1 + \delta n_2 + \ldots + \delta n_k = 0 \tag{10.18}$$

$$\Sigma\, u_i \delta n_i = u_i \delta n_i + u_2 \delta n_2 + \ldots + u_k \delta n_k = 0 \tag{10.19}$$

To incorporate these conditions into Eq. 10.11, we multiply Eq. 10.18 by $-\alpha$ and Eq. 10.19 by $-\beta$, where α and β are quantities independent of the n_i's, and add these expressions to Eq. 10.11. We obtain

$$\Sigma\,(-\ln n_i + \ln g_i - \alpha - \beta u_i)\delta n_i = 0$$

Again, the δn_i's are now effectively independent, and the quantity in parentheses must be 0 for each value of i. Hence

$$-\ln n_i + \ln g_i - \alpha - \beta u_i = 0$$

$$n_i = g_i e^{-\alpha} e^{-\beta u_i} \tag{10.20}$$

Maxwell-Boltzmann distribution law

This result is known as the *Maxwell-Boltzmann distribution law.*

We shall now evaluate $e^{-\alpha}$ and β. To do so, it is convenient to consider a continuous distribution of molecular energies, rather than the discrete set $u_1, u_2, \ldots, u_k$, so that Eq. 10.20 becomes

$$n(u)\, du = g e^{-\alpha} e^{-\beta u}\, du \tag{10.21}$$

(This is a perfectly valid approximation for the molecules in a gas, where energy quantization is inconspicuous and the total number of molecules may be very large.) In Eq. 10.21 $n(u)\,du$ is interpreted as the number of molecules whose energies lie between u and $u + du$. In terms of molecular momentum, since

$$u = \frac{p^2}{2m}$$

we have

$$n(p)\,dp = ge^{-\alpha}e^{-\beta p^2/2m}\,dp$$

The a priori probability g that a molecule have a momentum between p and $p + dp$ is equal to the number of cells in phase space within which such a molecule may exist. If each cell has the infinitesimal volume h^3,

$$g\,dp = \frac{\iiiint\!\!\int dx\,dy\,dz\,dp_x\,dp_y\,dp_z}{h^3}$$

where the numerator is the phase-space volume occupied by particles with the specified momenta. Here

$$\iiint dx\,dy\,dz = V$$

where V is the volume occupied by the gas in ordinary position space, and

$$\iint dp_x\,dp_y\,dp_z = 4\pi p^2\,dp$$

where $4\pi p^2\,dp$ is the volume of a spherical shell of radius p and thickness dp in momentum space. Hence

10.22 $$g(p)\,dp = \frac{4\pi V p^2\,dp}{h^3}$$

and

10.23 $$n(p)\,dp = \frac{4\pi V p^2 e^{-\alpha}e^{-\beta p^2/2m}}{h^3}\,dp$$

We are now able to find $e^{-\alpha}$. Since

$$\int_0^\infty n(p)\,dp = N$$

we find by integrating Eq. 10.23 that

$$\begin{aligned} N &= \frac{4\pi e^{-\alpha}V}{h^3}\int_0^\infty p^2 e^{-\beta p^2/2m}\,dp \\ &= \frac{e^{-\alpha}V}{h^3}\left(\frac{2\pi m}{\beta}\right)^{3/2} \end{aligned}$$

where we have made use of the definite integral

$$\int_0^\infty x^2 e^{-ax^2}\,dx = \frac{1}{4}\sqrt{\frac{\pi}{a^3}}$$

Hence

$$e^{-\alpha} = \frac{Nh^3}{V}\left(\frac{\beta}{2\pi m}\right)^{3/2}$$

and

10.24 $$n(p)\,dp = 4\pi N\left(\frac{\beta}{2\pi m}\right)^{3/2} p^2 e^{-\beta p^2/2m}\,dp$$

To find β, we compute the total energy U of the assembly of molecules. Since

$$p^2 = 2mu \qquad \text{and} \qquad dp = \frac{m\,du}{\sqrt{2mu}}$$

we can rewrite Eq. 10.24 in the form

10.25 $$n(u)\,du = \frac{2N\beta^{3/2}}{\sqrt{\pi}}\sqrt{u}\,e^{-\beta u}\,du$$

The total energy is

$$U = \int_0^\infty u n(u)\,du$$

$$= \frac{2N\beta^{3/2}}{\sqrt{\pi}}\int_0^\infty u^{3/2}e^{-\beta u}\,du$$

10.26 $$= \frac{3}{2}\frac{N}{\beta}$$

where we have made use of the definite integral

$$\int_0^\infty x^{3/2}e^{-ax}\,dx = \frac{3}{4a^2}\sqrt{\frac{\pi}{a}}$$

According to the kinetic theory of gases, the total energy U of N molecules of an ideal gas (which is what we have been considering) at the absolute temperature T is

10.27 $$U = \frac{3}{2}NkT$$

where k is Boltzmann's constant

$$k = 1.380 \times 10^{-23}\ \text{joule/molecule-degree}$$

Equations 10.26 and 10.27 agree if

10.28 $$\beta = \frac{1}{kT}$$

Now that the parameters α and β have been evaluated, we can write the Boltzmann distribution law in its various final forms. Thus

10.29 $$n(u)\,du = \frac{2\pi N}{(\pi kT)^{3/2}}\sqrt{u}\,e^{-u/kT}\,du$$ **Boltzmann distribution of energies**

is the number of molecules having energies between u and $u + du$;

10.30 $$n(p)\,dp = \frac{\sqrt{2}\pi N}{(\pi mkT)^{3/2}}\,p^2 e^{-p^2/2mkT}\,dp$$ **Boltzmann distribution of momenta**

is the number of molecules having momenta between p and $p + dp$; and

10.31 $$n(v)\,dv = \frac{\sqrt{2}\pi N m^{3/2}}{(\pi kT)^{3/2}}\,v^2 e^{-mv^2/2kT}\,dv$$ **Boltzmann distribution of velocities**

is the number of molecules having velocities between v and $v + dv$. The last formula, which was first obtained by Maxwell in 1859, is plotted in Fig. 10-3. The corresponding energy distribution is plotted in Fig. 10-4.

According to Eq. 10.26, the total energy U of an assembly of N molecules is

$$U = \frac{3}{2}\frac{N}{\beta}$$

FIGURE 10-3 Maxwell-Boltzmann velocity distribution.

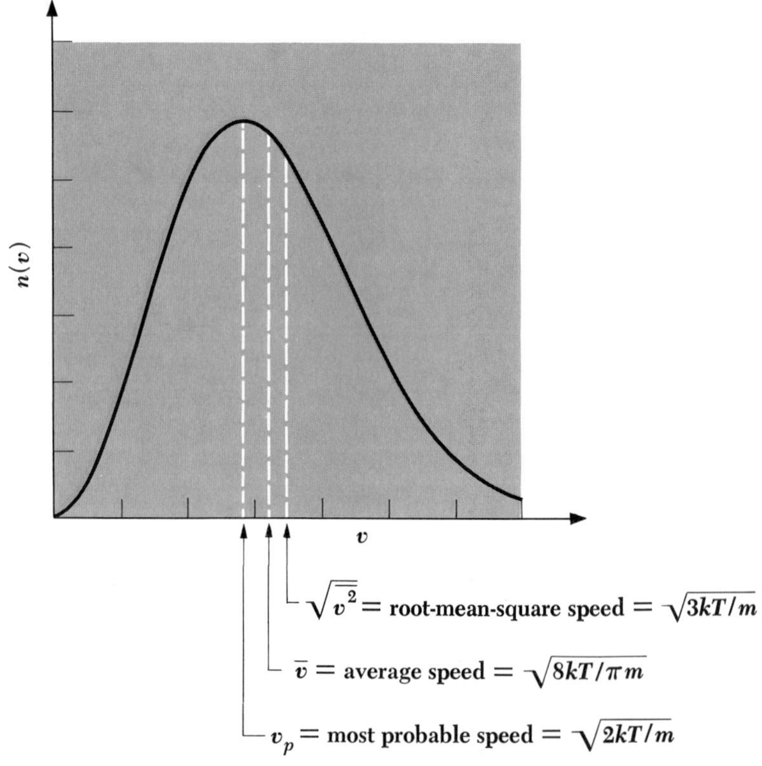

The Zartman-Ko Experiment

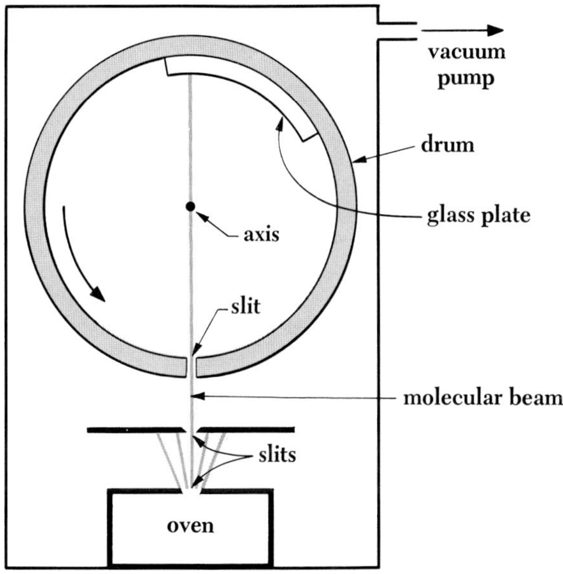

A beam of molecules that emerged from a slit in the wall of an oven provided the basis of a direct test of the Maxwell-Boltzmann velocity distribution. This experiment, performed in the early 1930s by I. F. Zartman and C. C. Ko, made use of the apparatus shown above. The oven contains bismuth vapor at about 800°C, some of which escapes through a slit and is collimated by another slit a short distance away. Above the second slit is a drum that rotates about a horizontal axis at 6,000 rpm. At those instants when the slit in the drum faces the bismuth beam, a burst of molecules enters the drum. These molecules reach the opposite face of the drum, where a glass plate is attached, at various times, depending upon their speeds. Because the drum is turning, the faster and slower molecules strike different parts of the plate. From the resulting distribution of deposited bismuth on the plate it is possible to infer the distribution of speeds in the beam, and this distribution agrees with the prediction of Maxwell-Boltzmann statistics.

The *average energy* $\overline{u}$ per molecule is U/N, so that

$$\overline{u} = 3/2\beta$$

10.32
$$= \frac{3}{2}kT$$

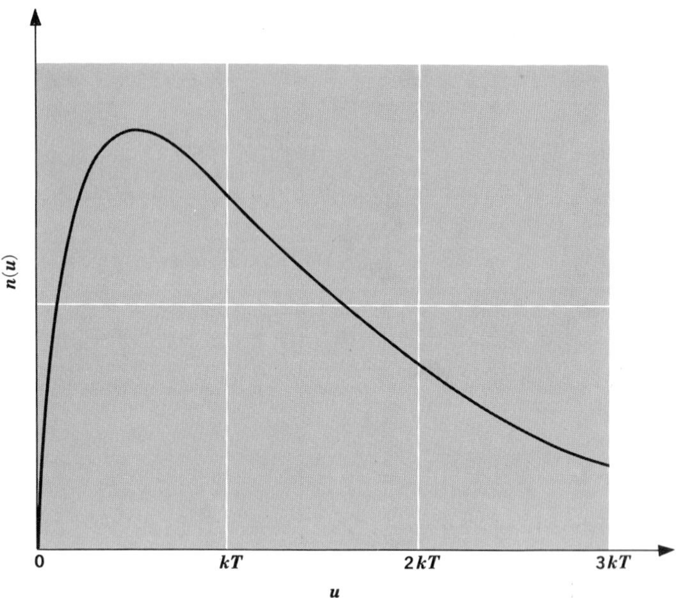

FIGURE 10-4 Maxwell-Boltzmann energy distribution.

At 300°K, which is approximately room temperature,

$$\begin{aligned}\bar{u} &= 6.21 \times 10^{-21} \text{ joule/molecule} \\ &\approx 1/25 \text{ ev/molecule}\end{aligned}$$

This average energy is the same for all molecules at 300°K, regardless of their mass.

10.4 Bose-Einstein Statistics

The basic difference between Maxwell-Boltzmann statistics and Bose-Einstein statistics is that the former applies to identical particles which can be distinguished from one another in some way, while the latter applies to identical particles which cannot be distinguished. In Bose-Einstein statistics we assume that all quantum states have equal a priori probabilities, so that g_i represents the number of states having the same energy u_i. Each quantum state corresponds to a cell in phase space, so our first step is to determine the number of ways in which n_i indistinguishable particles can be distributed in g_i cells. To do this, we consider a series of $n_i + g_i - 1$ objects placed in a line (Fig. 10-5). We note that $g_i - 1$ of the objects can be regarded as partitions separating a total of g_i intervals, with the entire series therefore representing n_i particles arranged in g_i cells. In the picture $g_i = 12$ and $n_i = 20$; 11 partitions separate the 20 particles into 12 cells. The first cell contains two particles, the second none, the third one particle, the fourth three particles, and so on. There are $(n_i + g_i - 1)!$ possible permutations among $n_i + g_i - 1$ objects, but of these the $n_i!$ permutations of the n_i particles among themselves and the $(g_i - 1)!$ permutations of the $g_i - 1$ partitions among themselves do not affect the distribution and are meaningless. Hence there are

$$\frac{(n_i + g_i - 1)!}{n_i!(g_i - 1)!}$$

possible distinguishably different arrangements of the n_i indistinguishable particles among the g_i cells.

The probability W of the entire distribution of N particles is the product

10.33 $$W = \prod \frac{(n_i + g_i - 1)!}{n_i!(g_i - 1)!}$$

of the numbers of distinct arrangements of particles among the states having each energy. We now assume that

$$(n_i + g_i) \gg 1$$

so that $(n_i + g_i - 1)$ can be replaced by $(n_i + g_i)$, and take the natural logarithm of both sides of Eq. 10.33 to give

$$\ln W = \Sigma\,[\ln\,(n_i + g_i)! - \ln n_i! - \ln\,(g_i - 1)!]$$

Stirling's formula

$$\ln n! = n \ln n - n$$

permits us to rewrite ln W as

10.34 $$\ln W = \Sigma\,[(n_i + g_i)\ln\,(n_i + g_i) - n_i \ln n_i - \ln\,(g_i - 1)! - g_i]$$

FIGURE 10-5 A series of n_i indistinguishable particles separated by g_i -1 partitions into g_i cells.

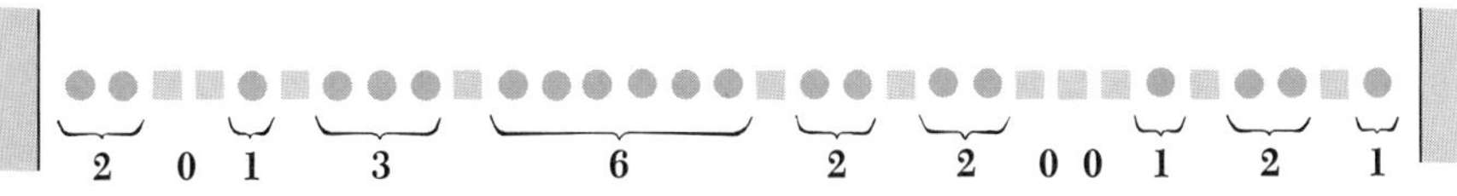

number of indistinguishable particles $= n_i = 20$
number of partitions $= g_i - 1 = 11$
number of cells $= g_i = 12$

As before, the condition that this distribution be the most probable one is that small changes δn_i in any of the individual n_i's not affect the value of W. If a change in ln W of δ ln W occurs when n_i changes by δn_i, the above condition may be written

$$\delta \ln W_{\text{max}} = 0$$

Hence, if the W of Eq. 10.34 represents a maximum,

10.35 $$\delta \ln W_{\text{max}} = \Sigma\,[\ln(n_i + g_i) - \ln n_i]\,\delta n_i = 0$$

where we have made use of the fact that

$$\delta \ln x = \frac{1}{x}\,\delta x$$

Again we incorporate the conservation of particles,

(10.18) $$\Sigma\,\delta n_i = 0$$

and the conservation of energy,

(10.19) $$\Sigma\,u_i\,\delta n_i = 0$$

by multiplying the former equation by $-\alpha$ and the latter by $-\beta$ and adding to Eq. 10.35. The result is

10.36 $$\Sigma\,[\ln(n_i + g_i) - \ln n_i - \alpha - \beta u_i]\,\delta n_i = 0$$

Since the δn_i's are independent, the quantity in brackets must vanish for each value of i. Hence

$$\ln \frac{n_i + g_i}{n_i} - \alpha - \beta u_i = 0$$

$$1 + \frac{g_i}{n_i} = e^{\alpha} e^{\beta u_i}$$

and

10.37 $$n_i = \frac{g_i}{e^{\alpha} e^{\beta u_i} - 1}$$

Substituting

(10.28) $$\beta = \frac{1}{kT}$$

we arrive at the *Bose-Einstein distribution law:*

10.38 $$n_i = \frac{g_i}{e^{\alpha} e^{u_i/kT} - 1}$$ **Bose-Einstein distribution law**

10.5 The Planck Radiation Formula

Every substance emits electromagnetic radiation, the character of which depends upon the nature and temperature of the substance. We have already discussed the discrete spectra of excited gases which arise from electronic transitions within isolated atoms. At the other extreme, dense bodies such as solids radiate continuous spectra in which all frequencies are present; the atoms in a solid are so close together that their mutual interactions result in a multitude of adjacent quantum states indistinguishable from a continuous band of permitted energies. The ability of a body to radiate is closely related to its ability to absorb radiation. This is to be expected, since a body at a constant temperature is in thermal equilibrium with its surroundings and must absorb energy from them at the same rate as it emits energy. It is convenient to consider as an ideal body one that absorbs *all* radiation incident upon it, regardless of frequency. Such a body is called a *black body.*

It is easy to show experimentally that a black body is a better emitter of radiation than anything else. The experiment, illustrated in Fig. 10-6, involves two identical pairs of dissimilar surfaces. No temperature difference is observed between surfaces I′ and II′. At a given temperature the surfaces I and I′ radiate at the rate of e_1 watts/m^2, while II and II′ radiate at the different rate e_2. The surfaces I and I′ absorb some fraction a_1 of the radiation falling on them, while II and II′ absorb some other fraction a_2. Hence I′ absorbs energy from II at a rate proportional to a_1e_2, and II′ absorbs energy from I at a rate proportional to a_2e_1. Because I′ and II′ remain at the same temperature, it must be true that

$$a_1e_2 = a_2e_1$$

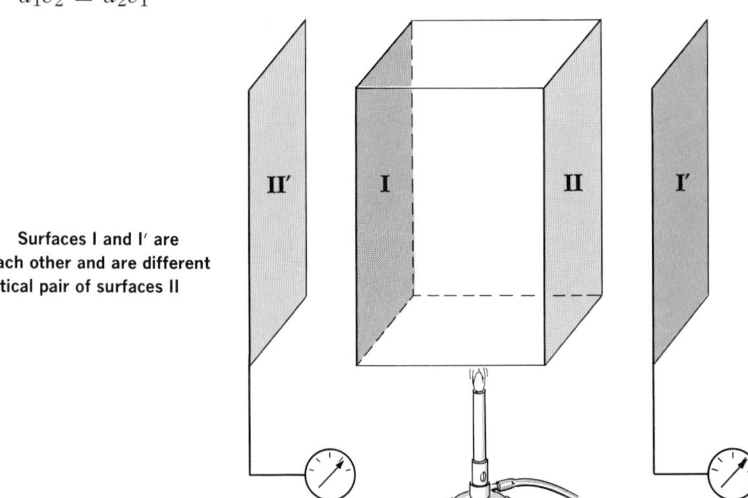

FIGURE 10-6 Surfaces I and I′ are identical to each other and are different from the identical pair of surfaces II and II′.

FIGURE 10-7 A hole in the wall of a hollow object is an excellent approximation of a black body.

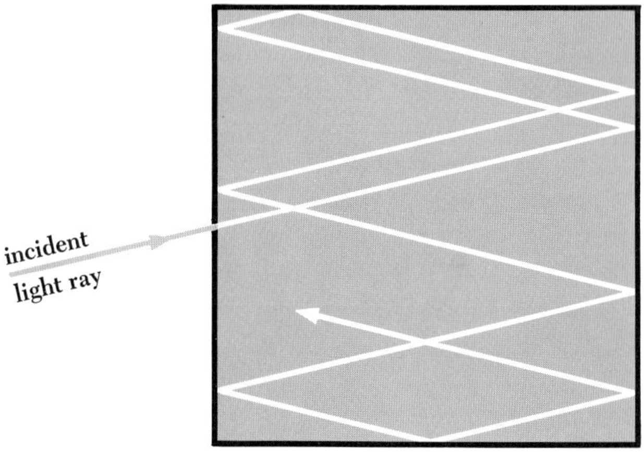

incident light ray

and

$$\frac{e_1}{a_1} = \frac{e_2}{a_2} \tag{10.39}$$

The ability of a body to emit radiation is proportional to its ability to absorb radiation. Let us suppose that I and I′ are black bodies, so that $a_1 = 1$. Now

$$e_1 = \frac{e_2}{a_2}$$

and, since $a_2 < 1$, $e_1 > e_2$. A black body at a given temperature radiates energy at a faster rate than any other body.

The point of introducing the idealized black body in a discussion of thermal radiation is that we can now disregard the precise nature of whatever is radiating, since all black bodies behave identically. In the laboratory a black body can be approximated by a hollow object with a very small hole leading to its interior (Fig. 10-7). Any radiation striking the hole enters the cavity, where it is trapped by reflection back and forth until it is absorbed. The cavity walls are constantly emitting and absorbing radiation, and it is in the properties of this radiation (*black-body radiation*) that we are interested. Experimentally we can sample black-body radiation simply by inspecting what emerges from the hole. The results agree with our everyday experience; a black body radiates more when it is hot than when it is cold, and the spectrum of a hot black body has its peak at a higher frequency than the peak in the spectrum of a cooler one. These facts correspond to the familiar behavior of an iron bar as it is heated to progressively higher temperatures: at first it glows dull red, then bright orange-red, and eventually becomes "white hot." The spectrum of black-body radiation is shown in Fig. 10-8 for two temperatures.

The principles of classical physics are unable to account for the observed black-body spectrum. In fact, it was this particular failure of classical physics that led Max Planck in 1900 to suggest that light is a quantum phenomenon. We shall use quantum-statistical mechanics to arrive at the Planck radiation formula, which predicts the same spectrum as that found by experiment.

Our theoretical model of a black body will be the same as the laboratory version, namely, a cavity in some opaque material. This cavity has some volume V, and it contains a large number of indistinguishable photons of various frequencies. Photons do not obey the exclusion principle, and so they are Bose particles that follow the Bose-Einstein distribution law. The a priori probability $g(p)$ that a photon have a momentum between p and $p + dp$ is equal to twice the number of cells in phase space within which such a photon may exist. The reason for the possible double occupancy of each cell is that photons of the same frequency can have two different directions of polarization (circularly clockwise and circularly counter-clockwise). Hence, using the argument that led to Eq. 10.22,

$$g(p)\,dp = \frac{8\pi V p^2\,dp}{h^3}$$

Since the momentum of a photon is $p = h\nu/c$,

$$p^2\,dp = \frac{h^3\nu^2\,d\nu}{c^3}$$

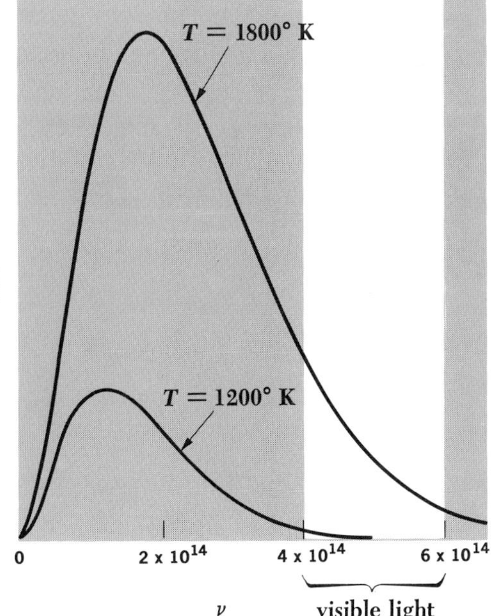

FIGURE 10-8 Black-body radiation.

and

10.40 $$g(\nu)\,d\nu = \frac{8\pi V}{c^3}\nu^2\,d\nu$$

We must now evaluate the Lagrangian multiplier α in Eq. 10.38. To do this, we note that the number of photons in the cavity need *not* be conserved. Unlike gas molecules or electrons, photons may be created and destroyed, and so, while the total radiant energy within the cavity must remain constant, the number of photons that incorporate this energy can change. For instance, two photons of energy $h\nu$ can be emitted simultaneously with the absorption of a single photon of energy $2h\nu$. Hence

$$\Sigma\,\delta n_i \neq 0$$

which we can express by setting α equal to zero since it multiplies Eq. 10.18.

Substituting Eq. 10.40 for g_i and $h\nu$ for u_i, and letting $\alpha = 0$ in the Bose-Einstein distribution law (Eq. 10.38), we find that the number of photons with frequencies between ν and $\nu + d\nu$ in the radiation within a cavity of volume V whose walls are at the absolute temperature T is

10.41 $$n(\nu)\,d\nu = \frac{8\pi V}{c^3}\frac{\nu^2\,d\nu}{e^{h\nu/kT} - 1}$$

The corresponding spectral energy density $u(\nu)\,d\nu$, which is the energy per unit volume in radiation between ν and $\nu + d\nu$ in frequency, is given by

$$u(\nu)\,d\nu = \frac{h\nu n(\nu)\,d\nu}{V}$$

10.42 $$= \frac{8\pi h}{c^3}\frac{\nu^3\,d\nu}{e^{h\nu/kT} - 1}$$ **Planck radiation formula**

Equation 10.42 is the *Planck radiation formula,* which agrees with experiment.

Two interesting results can be obtained from the Planck radiation formula. To find the wavelength whose energy density is greatest, we set

$$\frac{du(\lambda)}{d\lambda} = 0$$

and solve for $\lambda = \lambda_{max}$. We obtain

$$\frac{hc}{kT\lambda_{max}} = 4.965$$

which is more conveniently expressed as

$$\lambda_{max}T = \frac{hc}{4.965k}$$

10.43 $$= 2.898 \times 10^{-3}\ \text{m}^\circ\text{K}$$

Equation 10.43 is known as *Wien's displacement law.* It quantitatively expresses the empirical fact that the peak in the black-body spectrum shifts to progressively shorter wavelengths (higher frequencies) as the temperature is increased.

Another result we can obtain from Eq. 10.42 is the total energy u within the cavity. This is the integral of the energy density over all frequencies,

$$\begin{aligned} u &= \int_0^\infty u(\nu)\, d\nu \\ &= \frac{8\pi^5 k^4}{15 c^3 h^3} T^4 \\ &= aT^4 \end{aligned}$$

where a is a universal constant. The total energy is proportional to the fourth power of the absolute temperature of the cavity walls. We therefore expect that the energy e radiated by a black body per second per unit area is also proportional to T^4, a conclusion embodied in the *Stefan-Boltzmann law:*

10.44 $$e = \sigma T^4$$

The value of Stefan's constant σ is

$$\sigma = 5.67 \times 10^{-8} \text{ watt/m}^2\ {}^\circ\text{K}^4$$

Both Wien's displacement law and the Stefan-Boltzmann law are evident in qualitative fashion in Fig. 10-8; the maxima in the various curves shift to higher frequencies and the total areas underneath them increase rapidly with rising temperature.

10.6 Fermi-Dirac Statistics

Fermi-Dirac statistics apply to indistinguishable particles which are governed by the exclusion principle. Our derivation of the Fermi-Dirac distribution law will therefore parallel that of the Bose-Einstein distribution law except that now each cell (that is, quantum state) can be occupied by at most one particle. If there are g_i cells having the same energy u_i and n_i particles, n_i cells are filled and $(g_i - n_i)$ are vacant. The g_i cells can be rearranged in $g_i!$ different ways, but the $n_i!$ permutations of the filled cells among themselves are irrelevant since the particles are indistinguishable and the $(g_i - n_i)!$ permutations of the vacant cells among themselves are irrelevant since the cells are not occupied. The number of distinguishable arrangements of the particles among the cells is therefore

$$\frac{g_i!}{n_i!(g_i - n_i)!}$$

The probability W of the entire distribution of particles is the product

10.45 $$W = \prod \frac{g_i!}{n_i!(g_i - n_i)!}$$

Taking the natural logarithm of both sides,

$$\ln W = \Sigma\, [\ln g_i! - \ln n_i! - \ln (g_i - n_i)!]$$

which Stirling's formula

$$\ln n! = n \ln n - n$$

permits us to rewrite as

10.46 $$\ln W = \Sigma\, [g_i \ln g_i - n_i \ln n_i - (g_i - n_i) \ln (g_i - n_i)]$$

For this distribution to represent maximum probability, small changes δn_i in any of the individual n_i's must not alter W. Hence

10.47 $$\delta \ln W_{\max} = \Sigma\, [-\ln n_i + \ln (g_i - n_i)]\, \delta n_i = 0$$

As before, we take into account the conservation of particles and of energy by adding

$$-\alpha \Sigma\, \delta n_i = 0$$

and

$$-\beta \Sigma\, u_i \delta n_i = 0$$

to Eq. 10.47, with the result that

10.48 $$\Sigma\, [-\ln n_i + \ln (g_i - n_i) - \alpha - \beta u_i]\, \delta n_i = 0$$

Since the δn_i's are independent, the quantity in brackets must vanish for each value of i, and so

$$\ln \frac{g_i - n_i}{n_i} - \alpha - \beta u_i = 0$$

$$\frac{g_i}{n_i} - 1 = e^{\alpha} e^{\beta u_i}$$

10.49 $$n_i = \frac{g_i}{e^{\alpha} e^{\beta u_i} + 1}$$

Substituting

$$\beta = \frac{1}{kT}$$

yields the *Fermi-Dirac distribution law,*

10.50 $$n_i = \frac{g_i}{e^{\alpha}e^{u_i/kT} + 1}$$ **Fermi-Dirac distribution law**

The most important application of the Fermi-Dirac distribution law is in the free-electron theory of metals, which we shall examine in the next chapter.

10.7 Comparison

The three statistical distribution laws are as follows:

10.51 $$n_i = \frac{g_i}{e^{\alpha}e^{u_i/kT}}$$ **Maxwell-Boltzmann**

10.52 $$n_i = \frac{g_i}{e^{\alpha}e^{u_i/kT} - 1}$$ **Bose-Einstein**

10.53 $$n_i = \frac{g_i}{e^{\alpha}e^{u_i/kT} + 1}$$ **Fermi-Dirac**

In these formulas n_i is the number of particles whose energy is u_i and g_i is the number of states that have the same energy u_i. The quantity

10.54 $$f(u_i) = \frac{n_i}{g_i}$$ **Occupation index**

called the *occupation index* of a state of energy u_i, is therefore the average number of particles in each of the states of that energy. The occupation index does not depend upon how the energy levels of a system of particles are distributed, and for this reason it provides a convenient way of comparing the essential natures of the three distribution laws.

The Maxwell-Boltzmann occupation index is plotted in Fig. 10-9 for three different values of T and α. This index is a pure exponential, dropping by the factor $1/e$ for each increase in u_i of kT. While $f(u_i)$ depends upon the parameter α, the *ratio* between the occupation indices $f(u_i)$ and $f(u_j)$ of the two energy levels u_i and u_j does not;

10.55 $$\frac{f(u_i)}{f(u_j)} = e^{(u_j - u_i)/kT}$$ **Boltzmann factor**

This formula is useful because under certain circumstances the Bose-Einstein and Fermi-Dirac distributions resemble the Maxwell-Boltzmann distribution, and it then permits us to determine the relative degrees of occupancy of two energy states. For example, we can use Eq. 10.55 to compute the relative

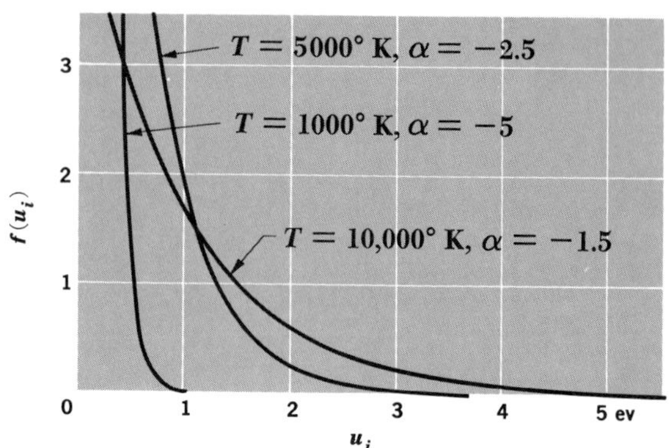

FIGURE 10-9 The occupation indexes for three Maxwell-Boltzmann distributions.

numbers of atoms in different quantum states in a gas at a particular temperature, from which we can deduce the relative brightness of the spectral lines arising from transitions involving these states.

The Bose-Einstein occupation index is plotted in Fig. 10-10 for temperatures of 1000°K, 5000°K, and 10,000°K, in each case for $\alpha = 0$ (corresponding to a "gas" of photons). When $u_i \gg kT$, the Bose-Einstein distribution approaches the Maxwell-Boltzmann distribution, while when $u_i \ll kT$, the -1 term in the denominator of Eq. 10.52 causes the occupation index of the former distribution to be much greater.

The Fermi-Dirac occupation index is plotted in Fig. 10-11 for four values of T and α. The occupation index never goes above 1, signifying one particle per state, which is a consequence of the obedience of Fermi particles to the exclusion principle. At low temperatures virtually all of the lower energy states are filled, with the occupation index dropping rapidly near a certain critical energy known as the *Fermi energy*. At high temperatures the

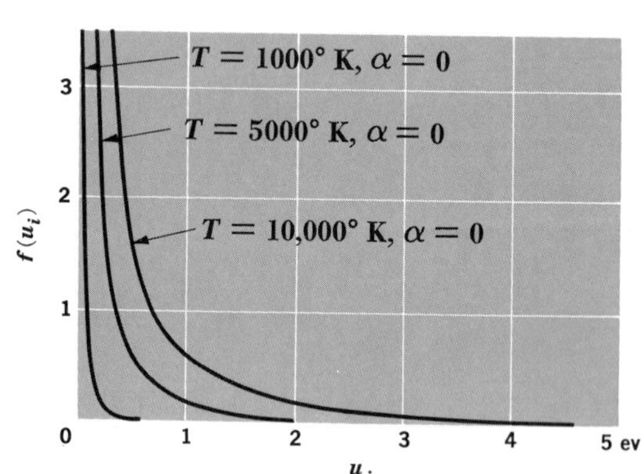

FIGURE 10-10 The occupation indexes for three Bose-Einstein distributions.

FIGURE 10-11 The occupation indexes for three Fermi-Dirac distributions.

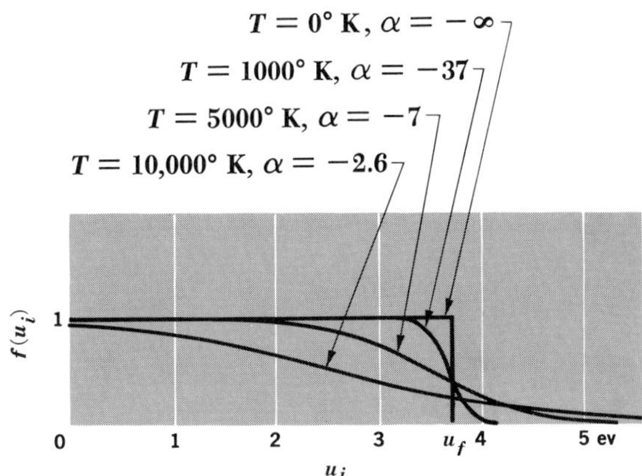

Fermi-Dirac statistics

occupation index is sufficiently small at all energies for the effects of the exclusion principle to be unimportant, and the Fermi-Dirac distribution becomes similar to the Maxwell-Boltzmann one.

10.8 Transitions between States

In this chapter we have been discussing particles of various kinds that can exist in different energy states and have considered the relative populations of these states in an assembly of particles of each kind. We shall now draw upon some of these ideas to look into the transitions between different energy states in the special case of atoms.

Let us consider two energy levels in a particular atom, a lower one i and an upper one j (Fig. 10-12). If the atom is initially in state i, it can be raised to state j by absorbing a photon of light whose frequency is

10.56 $$\nu = \frac{E_j - E_i}{h}$$

FIGURE 10-12 Transitions between two energy levels in an atom can occur by induced absorption, spontaneous emission, and induced emission.

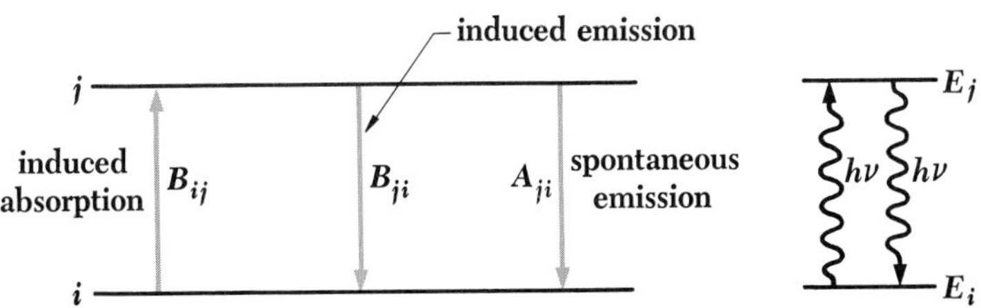

(We shall neglect recoil effects in this analysis.) The likelihood that the atom will actually undergo the transition is proportional to the rate at which photons of frequency ν fall on it and therefore to the spectral energy density $u(\nu)$. Of course, the transition probability also depends upon the properties of states i and j, as we saw in Secs. 7.8 and 7.9, but we can include this dependence for any specific pair of states in some constant B_{ij}. Hence if we shine light of frequency ν and energy density $u(\nu)$ on the atom when it is in the lower state i, the probability for it to go to the higher state j is

10.57 $$P_{i\to j} = B_{ij}u(\nu)$$

If the atom is initially in the upper state j, it has a certain probability A_{ji} to spontaneously drop to state i by emitting a photon of frequency ν. Let us also suppose that shining light of frequency ν on the atom when it is in the upper state somehow helps induce its transition to the lower state. A spectral energy density of $u(\nu)$ therefore means a probability for *induced emission* of $B_{ji}u(\nu)$, where B_{ji}, like B_{ij} and A_{ji}, depends upon the detailed properties of the states i and j. The total probability for an atom in state j to fall to the lower state i is therefore

10.58 $$P_{j\to i} = A_{ji} + B_{ji}u(\nu)$$

(We might remark that induced emission involves no novel concepts. Let us consider a harmonic oscillator, for instance a pendulum, which has a sinusoidal force applied to it whose period is the same as its natural period of vibration. If the applied force is exactly in phase with the pendulum swings, the amplitude of the latter increases; this corresponds to induced absorption of energy. However, if the applied force is 180° out of phase with the pendulum swings, the amplitude of the latter *decreases;* this corresponds to induced emission of energy. If the applied force is random in phase relative to the pendulum, the probabilities of induced absorption and induced emission are equal—that is, if a number of trials are made, on the average half the trials will result in absorption and half in emission of energy. This corresponds to the fact that $B_{ij} = B_{ji}$, which is demonstrated more formally in what follows. In any case, we have not assumed that induced emission *does* occur, but only that it *may* occur. If we have been mistaken, we should ultimately find merely that $B_{ji} = 0$.)

Next we consider an assembly of N_i atoms in state i and N_j atoms in state j, all in thermal equilibrium at the temperature T with light of frequency ν and energy density $u(\nu)$. The number of atoms in state i that absorb a photon and go to state j per second is

$$N_iP_{i\to j} = N_iB_{ij}u(\nu)$$

while the corresponding number in state j that drop to state i either by spontaneously emitting a photon or by being induced to do so is

$$N_j P_{j\to i} = N_j [A_{ji} + B_{ji} u(\nu)]$$

The opposite processes that participate in an equilibrium must occur equally often if that equilibrium is to continue, and so

$$N_i P_{i\to j} = N_j P_{j\to i}$$
$$N_i B_{ij} u(\nu) = N_j [A_{ji} + B_{ji} u(\nu)]$$

Dividing both sides of the latter equation by $N_j B_{ji}$ and solving it for $u(\nu)$, we obtain

$$\left(\frac{N_i}{N_j}\right)\left(\frac{B_{ij}}{B_{ji}}\right) u(\nu) = \frac{A_{ji}}{B_{ji}} + u(\nu)$$

10.59 $$u(\nu) = \frac{A_{ji}/B_{ji}}{\left(\frac{N_i}{N_j}\right)\left(\frac{B_{ij}}{B_{ji}}\right) - 1}$$

Finally we draw upon Eq. 10.55 for the ratio between the populations of states i and j to obtain

$$\frac{N_i}{N_j} = e^{(E_j - E_i)/kT} = e^{h\nu/kT}$$

with the result that

10.60 $$u(\nu) = \frac{A_{ji}/B_{ji}}{\left(\frac{B_{ij}}{B_{ji}}\right) e^{h\nu/kT} - 1}$$

Equation 10.60 is a formula for the energy density of photons of frequency ν in equilibrium at the temperature T with atoms whose possible energies are E_i and E_j. We can see at once that, if this formula is to be consistent with the Planck radiation formula of Eq. 10.42, it must be true that

10.61 $$B_{ij} = B_{ji}$$

and

10.62 $$\frac{A_{ji}}{B_{ji}} = \frac{8\pi h\nu^3}{c^3}$$

The preceding analysis, which was first carried out by Einstein in 1917, thus not only confirms that induced emission can occur, but also shows that its coefficient for a transition between two states is the same as the coefficient for induced absorption. In addition, there is a definite ratio between the spontaneous emission and induced emission coefficients that varies with ν^3, so that the relative likelihood of spontaneous emission increases rapidly with the energy difference between the two states. All we need know is one of the coefficients A_{ji}, B_{ij}, or B_{ji} to find the others.

From the discussion of Sec. 10.7 we note that, since $\alpha = 0$ for a photon gas, the average number N_ν of photons of frequency ν at the temperature T is

$$N_\nu = f(u_\nu) = \frac{1}{e^{h\nu/kT} - 1}$$

Average number of photons

Thus Planck's formula can be expressed in terms of N_ν as

$$u(\nu) = \frac{8\pi h\nu^3}{c^3} N_\nu$$

It is instructive to rewrite Eqs. 10.57 and 10.58 for the transition probabilities $P_{i\to j}$ and $P_{j\to i}$ in terms of N_ν and one of the transition coefficients, say B_{ij}. The result is

$$P_{i\to j} = \left(\frac{8\pi h\nu^3}{c^3} B_{ij}\right) N_\nu$$

$$P_{j\to i} = \left(\frac{8\pi h\nu^3}{c^3} B_{ij}\right) + \left(\frac{8\pi h\nu^3}{c^3} B_{ij}\right) N_\nu$$

$$= \left(\frac{8\pi h\nu^3}{c^3} B_{ij}\right)(N_\nu + 1)$$

If there are N_ν photons present, the probability that an atom in the lower state i will absorb a photon is proportional to N_ν, and the probability that an atom in the upper state j will emit a photon is proportional to $(N_\nu + 1)$. Evidently the process of spontaneous emission, which contributes the additive factor 1 in the latter case, is intimately related to the processes of induced emission and induced absorption. Induced emission and absorption can be understood in a straightforward way by considering the interaction between an atom and an electromagnetic wave of frequency ν, but spontaneous emission occurs in the absence of any such wave, apparently by a comparable interaction to judge by the above formulas. This paradox is removed by a quantum-theoretical treatment of the electromagnetic field, which shows that actual fields constantly fluctuate about what would be expected on classical grounds. The fluctuations occur even when electromagnetic waves are absent and when, classically, $\mathbf{E} = \mathbf{B} = 0$, and it is these fluctuations (often called "vacuum fluctuations" and analogous in a sense to the zero-point vibrations of a harmonic oscillator) that induce the "spontaneous" emission of photons by atoms in excited states.

Masers and Lasers

Since $h\nu$ is normally much greater than kT for atomic and molecular radiations, at thermal equilibrium the population of upper energy states in an atomic system is considerably smaller than that of the lowest state. Suppose we shine light of frequency ν upon a system in which the energy difference between the ground state E_0 and an excited state E_1 is $E_1 - E_0 = h\nu$. With the upper state largely unoccupied, there will be little stimulated emission, and the chief events that occur will be absorption of incident photons by atoms in the ground state and the subsequent spontaneous reradiation of photons of the same frequency. (Depending upon the system, a certain proportion of excited atoms will give up their energies in collisions.)

Certain atomic systems can sustain inverted energy populations, with an upper state occupied to a greater extent than the ground state. (This situation corresponds to a negative absolute temperature T in Eq. 10.55, which is an interesting notion.) The figure shows a three-level system in which the intermediate state 1 is metastable (Sec. 8-6), which means that the transition from it to the ground state is forbidden by selection rules. The system can be "pumped" to the upper state 2 by radiation of frequency $\nu' = (E_2 - E_0)/h$. Atoms in state 2 have lifetimes of about 10^{-8} sec against spontaneous emission via an allowed transition, so they fall to the metastable state 1 (or to the ground state) almost at once. Metastable states may have lifetimes of well over 1 sec against spontaneous emission, and it is therefore possible to continue pumping until there is a higher population in state 1 than there is in state 0. If now we direct radiation of frequency $\nu = (E_1 - E_0)/h$ on the system, the induced emission of photons of this frequency will exceed their absorption since more atoms are in the higher state, and the net result will be an output of radiation of frequency ν that exceeds the input. This is the principle of the *maser* (*m*icrowave *a*mplification by *s*timulated *e*mission of *r*adiation) and the laser (*l*ight *a*mplification by *s*timulated *e*mission of *r*adiation).

The radiated waves from spontaneous emission are, as might be expected, incoherent, with random phase relationships in space and time since there is no coordination among the atoms involved. The radiated waves from induced emission, however, are in phase with the inducing waves, which makes it possible for a maser or laser to produce a completely coherent beam. A typical laser is a gas-filled tube or a transparent solid that has mirrors at both ends, one of them partially transmitting to allow some of the light produced to emerge. The pumping light of frequency ν' is directed at the active medium from the sides of the tube, while the back-and-forth traversals of the trapped light stimulate emissions of frequency ν that maintain the emerging beam collimated. A wide variety of masers and lasers have been devised; in many of them the required inverted energy distribution is obtained less directly than by the straightforward mechanism described above.

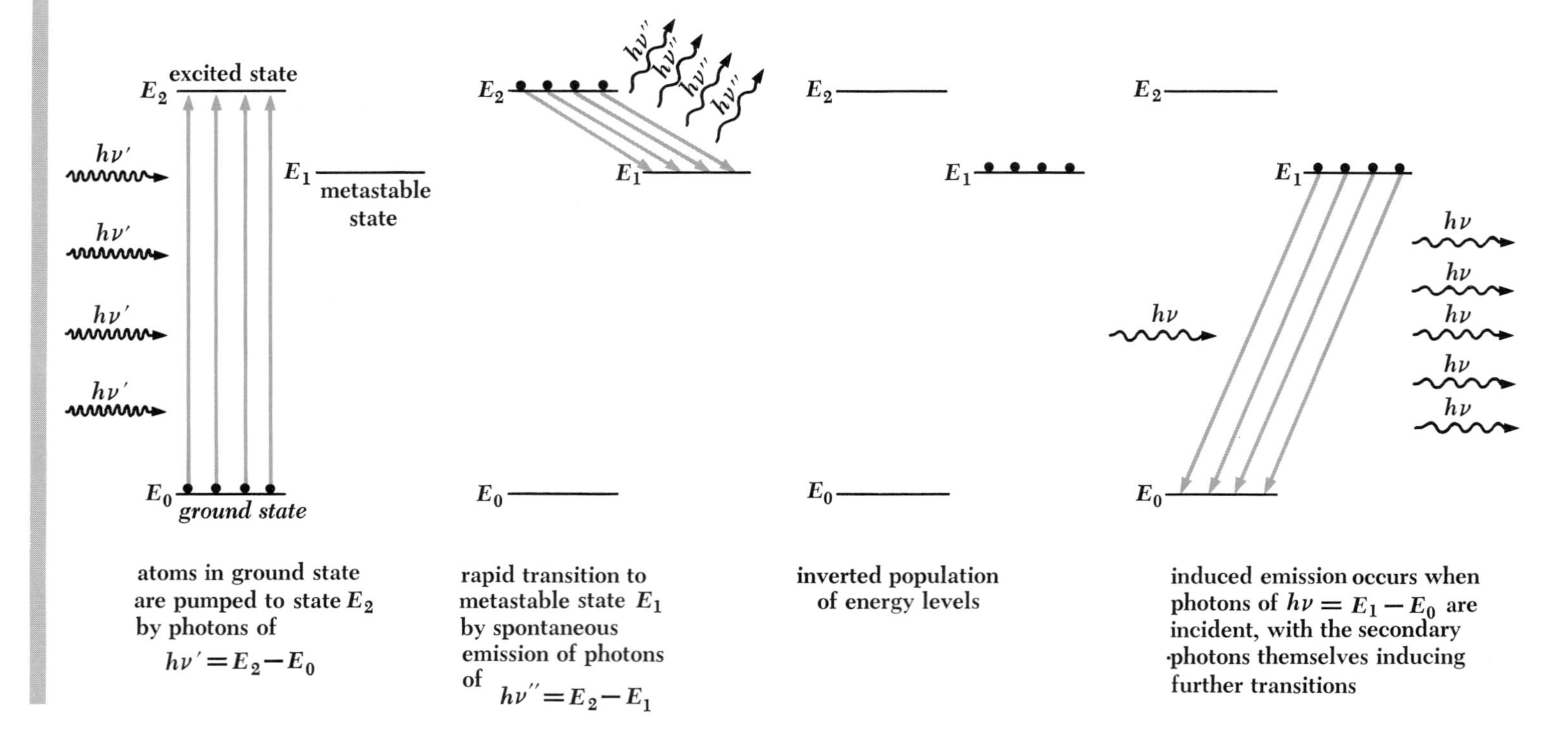

atoms in ground state are pumped to state E_2 by photons of $h\nu' = E_2 - E_0$

rapid transition to metastable state E_1 by spontaneous emission of photons of $h\nu'' = E_2 - E_1$

inverted population of energy levels

induced emission occurs when photons of $h\nu = E_1 - E_0$ are incident, with the secondary photons themselves inducing further transitions

Problems

1. Find the thermodynamic probability of the most probable distribution of 10^6 identical particles among 5×10^5 identical cells.

2. Find the thermodynamic probability of the least probable distribution of 10^6 identical particles among 5×10^5 identical cells.

3. Show that the most probable speed of a molecule of an ideal gas is equal to $\sqrt{2kT/m}$.

4. Show that the average speed of a molecule of an ideal gas is equal to $\sqrt{8kT/\pi m}$.

5. Show that the root-mean-square speed of a molecule of an ideal gas is equal to $\sqrt{3kT/m}$.

6. Is the most probable molecular energy in an ideal gas equal to $\frac{1}{2}mv_m^2$, where v_m is the most probable molecular speed?

7. Find the average value of $1/v$ in a gas obeying Maxwell-Boltzmann statistics.

8. What proportion of the molecules of an ideal gas have components of velocity in any particular direction greater than twice the most probable speed?

9. A flux of 10^{12} neutrons/m^2 emerges each second from a port in a nuclear reactor. If these neutrons have a Maxwell-Boltzmann energy distribution corresponding to $T = 300°\text{K}$, calculate the density of neutrons in the beam.

10. Express the Planck radiation law in terms of wavelength rather than frequency and show that, in the limit of $\lambda \to \infty$, it reduces to the Rayleigh-Jeans formula, $u(\lambda)\, d\lambda = 8\pi kT\, d\lambda/\lambda^4$.

11. Compare the relative occupancy of the rotational states $K = 1$ and $K = 0$ in HCl vapor at 288 and 2.88°K.

11 THE SOLID STATE

A solid consists of atoms packed closely together, and their proximity is responsible for the characteristic properties of this state of matter. Because most solids are crystalline in nature or nearly so, we shall consider them exclusively. The ionic and covalent bonds that are involved in the formation of molecules have important counterparts in the solid state. In addition, there are the *van der Waals* and *metallic* bonds that provide the cohesive forces in, respectively, molecular crystals and metals. All of these bonds are electrostatic in origin, so that the chief distinctions among them lie in the distribution of electrons around the atoms and molecules whose regular arrangement constitutes a crystal lattice. After a discussion of the various kinds of crystal structures found in solids, we shall look into the behavior of the electrons in them; such a study will permit us to understand how so remarkable a phenomenon as electrical conduction in metals arises.

11.1 Ionic and Covalent Crystals

Ionic bonds in crystals are very similar to heteropolar bonds in molecules. Such bonds come into being when atoms which have low ionization energies, and hence lose electrons readily, interact with other atoms which have high electron affinities. The former atoms give up electrons to the latter, and they thereupon become positive and negative ions respectively. In an ionic crystal these ions come together in an equilibrium configuration in which the attractive forces between positive and negative ions predominate over the repulsive forces between similar ions. As in the case of molecules, crystals of all types are prevented from collapsing under the influence of the cohesive forces present by the action of the exclusion principle, which requires the occupancy of higher energy states when electron shells of different atoms overlap and mesh together.

Figure 11-1 shows the arrangement of Na^+ and Cl^- ions in a sodium chloride crystal. The ions of either kind may be thought of as being located at the corners and at the centers of the faces of an assembly of cubes, with

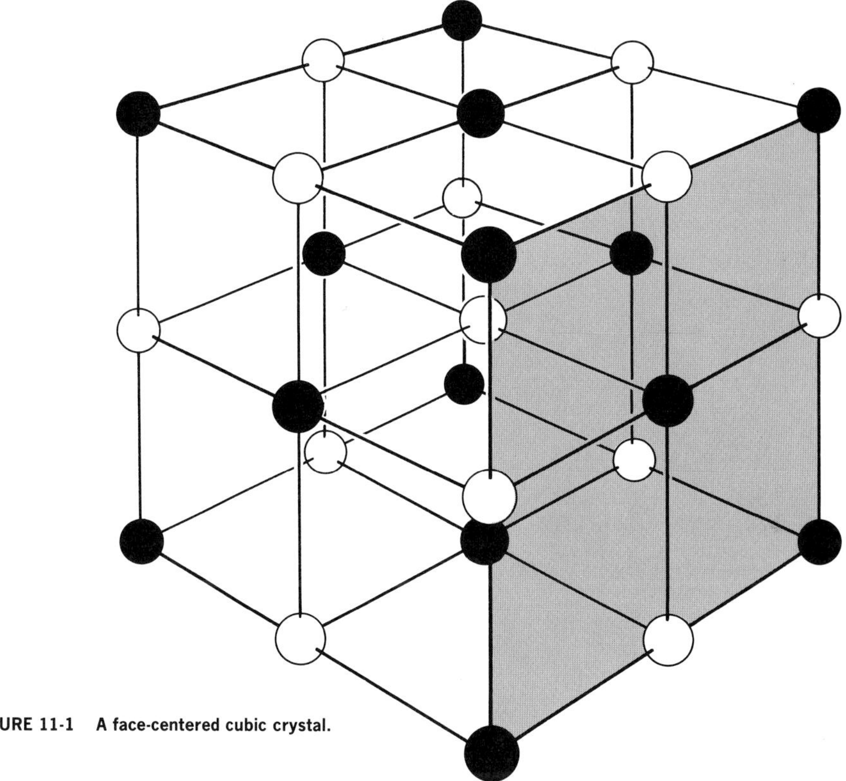

FIGURE 11-1 A face-centered cubic crystal.

the Na^+ and Cl^- assemblies interleaved. Each ion thus has six nearest neighbors of the other kind. Such a crystal structure is called *face-centered cubic.* In NaCl crystals the distance between like ions is 5.63 A.

A different structure is found in cesium chloride crystals, where each ion is located at the center of a cube at whose corners are ions of the other kind (Fig. 11-2). This structure is called *body-centered cubic,* and each ion has eight nearest neighbors of the other kind. In CsCl crystals the distance between like ions is 4.11 A.

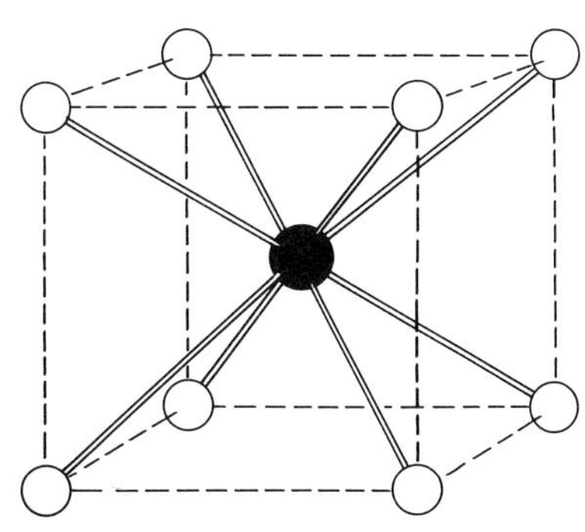

FIGURE 11-2 A body-centered cubic crystal.

The cohesive forces in covalent crystals arise from the presence of electrons between adjacent atoms. Each atom participating in a covalent bond contributes an electron to the bond, and these electrons are shared by both atoms rather than being the virtually exclusive property of one of them as in an ionic bond. Diamond is an example of a crystal whose atoms are linked by covalent bonds. Figure 11-3 shows the structure of a diamond crystal. Each carbon atom has four nearest neighbors and shares an electron pair with each of them. The length of each bond is 3.08 A. Silicon, germanium, and silicon carbide are among the substances with crystal structures the same as that of diamond.

As in the case of molecules, it is not always possible to classify a given crystal as being wholly ionic or covalent. Silicon dioxide (quartz) and tungsten carbide, for instance, contain bonds of mixed character. Both ionic and covalent crystals are held together with relatively strong binding forces: about 8 ev is required to remove each atom in NaCl and diamond crystals, compared with about 4 ev per atom in the metal iron and 0.1 ev per atom in solid methane (CH_4), which is held together by van der Waals bonds.

11.2 Van der Waals Forces

There are many substances whose molecules are so stable that, when brought together, they have no tendency to lose their individuality by joining in a collective lattice with multiple linkages like those found in ionic and covalent crystals. Most organic compounds are examples of such noninteracting sub-

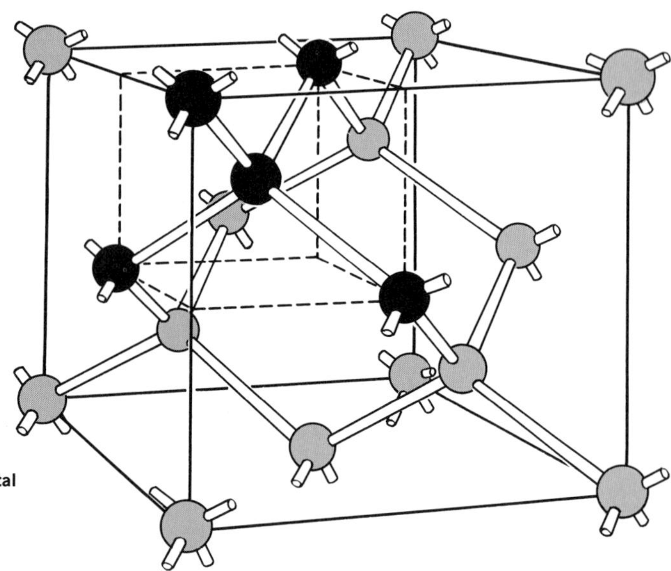

FIGURE 11-3 The crystal lattice of diamond.

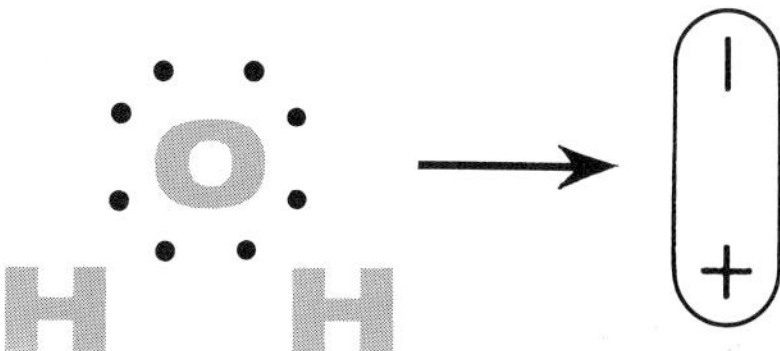

FIGURE 11-4 Water consists of polar molecules.

stances. Even they can exist as liquids and solids, however, through the action of the attractive *van der Waals* intermolecular forces. These forces are also responsible for many other aspects of the behavior of matter in bulk. Surface tension, friction, change of phase, adhesion, cohesion, viscosity, and so on all arise from van der Waals forces.

We begin by noting that many molecules (called *polar molecules*) possess permanent electric dipole moments. An example is the H_2O molecule, in which the concentration of electrons around the oxygen atom makes that end of the molecule more negative than the end where the hydrogen atoms are (Fig. 11-4). Such molecules tend to align themselves so that ends of opposite sign are adjacent, as in Fig. 11-5, and in this orientation the molecules strongly attract each other.

A polar molecule is also able to attract molecules which do not normally have a permanent dipole moment. The process is illustrated in Fig. 11-6: the electric field of the polar molecule causes a separation of charge in the other molecule, with the induced moment the same in direction as that of the polar molecule. The result is an attractive force. (The effect is the same as that involved in the attraction of an unmagnetized piece of iron by a magnet.) It is not difficult to determine the characteristics of this attractive force. The electric field **E** a distance r from a dipole of moment **p** is given by

11.1 $$\mathbf{E} = \frac{1}{4\pi\varepsilon_0}\left[\frac{\mathbf{p}}{r^3} - \frac{3(\mathbf{p}\cdot\mathbf{r})}{r^5}\mathbf{r}\right]$$

($\mathbf{p}\cdot\mathbf{r} = pr\cos\theta$, where θ is the angle between **p** and **r**.) The field **E** induces

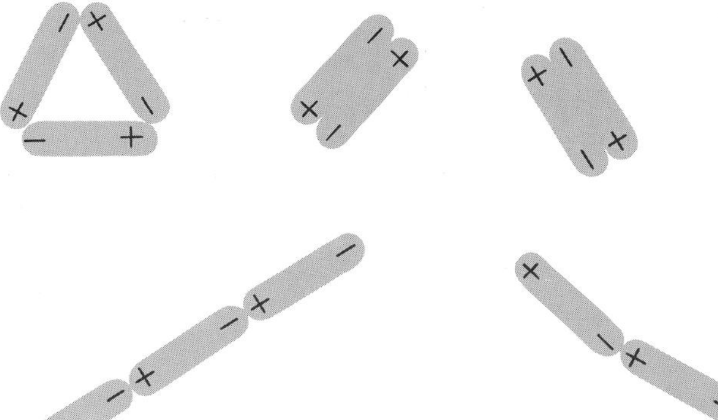

FIGURE 11-5 Polar molecules attract each other.

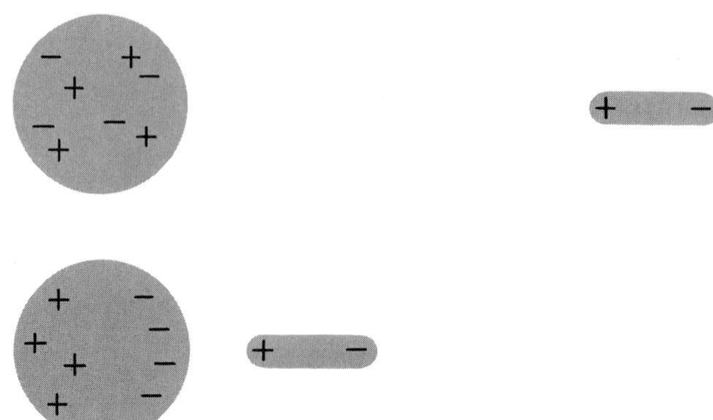

FIGURE 11-6 **Polar molecules attract polarizable molecules.**

in the other, normally nonpolar molecule an electric dipole moment $\mathbf{p}'$ proportional to $\mathbf{E}$ in magnitude and ideally in the same direction. Hence

11.2 $$\mathbf{p}' = \alpha \mathbf{E}$$

where α is a constant called the *polarizability* of the molecule. The energy of the induced dipole in the electric field $\mathbf{E}$ is

$$V = -\mathbf{p}' \cdot \mathbf{E}$$

$$= -\frac{\alpha}{(4\pi\varepsilon_0)^2}\left[\frac{p^2}{r^6} - \frac{3p^2}{r^6}\cos^2\theta - \frac{3p^2}{r^6}\cos^2\theta + \frac{9p^2}{r^6}\cos^2\theta\right]$$

11.3 $$= -\frac{\alpha}{(4\pi\varepsilon_0)^2}(1 + 3\cos^2\theta)\frac{p^2}{r^6}$$

The mutual energy of the two molecules that arises from their interaction is thus negative, signifying that the force between them is attractive, and is inversely proportional to r^6. The force itself is equal to $-dV/dr$ and so is proportional to r^{-7}, which means that it drops rapidly with increasing separation.

Only a limited number of molecules are polar, while van der Waals forces act between *all* molecules. There is no contradiction here, however, because the interaction energy V is proportional to p^2, the square of the dipole moment. Even though the electron distribution in a nonpolar molecule is symmetrical *on the average,* at any instant it is asymmetrical and so possesses an electric dipole moment. This instantaneous moment fluctuates in magnitude and direction, but while the average moment $\overline{p}$ is zero, the average of the square of the moment $\overline{p^2}$ is *not* zero but has some finite value. The forces arising from such instantaneous molecular moments according to the mechanism of the previous paragraph are present not only between all molecules but also between all atoms, including those of the rare gases which do not otherwise interact. The values of $\overline{p^2}$ and α are comparable for most molecules; this is part of the reason why the densities and heats of vaporization of liquids, properties that depend upon the strength of intermolecular forces, have a rather narrow range.

Van der Waals forces are much weaker than those found in ionic and covalent bonds, and as a result molecular crystals generally have low melting and boiling points and little mechanical strength.

11.3 The Metallic Bond

The basic concept that underlies the modern theory of metals is that the valence electrons of the atoms comprising a metal may be common to the entire atomic aggregate, so that a kind of "gas" of free electrons pervades it. The interaction between this gas and the positive metal ions constitutes a strong cohesive force. Further, the presence of such free electrons accounts very nicely for the high electrical conductivities and other unique properties of metals. To be sure, no electrons in any solid, even a metal, are able to move about its interior with complete freedom. All of them are influenced to some extent by the other particles present, and when we refine the theory of metals to include these perturbations, there emerges a comprehensive picture that is in excellent agreement with experiment.

A convenient way to regard the metallic bond is to view it as an unsaturated covalent bond. Let us compare the binding processes in hydrogen and in lithium, both members of group I of the periodic table. A H_2 molecule contains two $1s$ electrons with opposite spins, the maximum number of K electrons that can be present. The H_2 molecule is therefore saturated, since the exclusion principle requires that any additional electrons be in states of higher energy and the stable attachment of further H atoms cannot occur unless their electrons are in $1s$ states. Superficially lithium might seem obliged to behave in a similar way, having the electron configuration $1s^22s$. There are, however, six *unfilled* $2p$ states in every Li atom whose energies are only very slightly greater than those of the $2s$ states. When a Li atom comes near a Li_2 molecule, it readily becomes attached with a covalent bond without violating the exclusion principle, and the resulting Li_3 molecule is stable since all its valence electrons remain in L shells. There is no limit to the number of Li atoms that can join together in this way, since lithium forms body-centered cubic crystals (Fig. 11-2) in which each atom has eight nearest neighbors. With only one electron per atom available to enter into bonds, each bond involves one-fourth of an electron on the average instead of two electrons as in ordinary covalent bonds. Hence the bonds are far from being saturated; this is true of the bonds in other metals as well.

One consequence of the unsaturated nature of the metallic bond is the weakness of metals as compared with ionic and covalent crystals having saturated bonds. Another is the ease with which metals can be deformed. With so many vacancies in their outermost electron shells, metal atoms do not exhibit directional preferences in the locations of their bonds, and the atoms accordingly can be rearranged in position without loss of strength by the metal as a whole. A third consequence is the most striking, however: there are so many unoccupied states in each atom of a metal that an electron can wander from atom to atom without being limited by the exclusion principle, instead of being compelled to remain attached to a particular atom, as in an ionic crystal, or to resonate between a particular pair of atoms, as in a covalent crystal. Thus the valence electrons in a metal behave in a manner similar to that of molecules in a gas. Before we go into the properties of this electron gas, let us see how the simple fact of its existence leads to a theory of electrical conduction that agrees with experience.

11.4 Ohm's Law

The current i that flows through a metallic conductor when the potential difference V is established across its ends is, within wide limits, directly proportional to V. This empirical relationship, called *Ohm's law*, is usually expressed as

11.4 $$i = \frac{V}{R}$$

where R, the *resistance* of the conductor, is a function of its dimensions, composition, and temperature but is independent of V. To derive Ohm's law, we begin by assuming that the free electrons in a metal, like the molecules in a gas, move in random directions and constantly collide with other electrons. Between collisions there is some average time interval τ which depends upon the detailed characteristics of the electron gas. When we apply a potential difference across the ends of a conductor, there will be an electric field E within it. The force this field exerts on an electron is eE and is opposite in direction to E, since electrons are negatively charged. The second law of motion yields

$$F = ma$$

$$eE = m\frac{\Delta v}{\tau}$$

where Δv is the change in speed the electron experiences in the time τ. Hence

$$\Delta v = \frac{eE\tau}{m}$$

Each time an electron undergoes a collision, it rebounds in a random direction and, on the average, no longer has any component of motion parallel to the field. Hence at the start of each time interval τ we may imagine the electron at rest in the field direction, and at the end of the interval to have the speed Δv. Hence the average drift speed v_d through the conductor is

$$v_d = \frac{\Delta v}{2} = \frac{eE\tau}{2m}$$

The name "drift speed" is apt since v_d is very small relative to the average speed $\bar{v}$ of the random electron motion. The effect of imposing the electric field E on the free-electron gas in a metal, then, is to superimpose a general drift on the more pronounced but random motions of the electrons (Fig. 11-7).

We can now compute the total current i flowing in a wire of length L and cross-sectional area A that contains η free electrons per unit volume (Fig. 11-8). The current is

Total current = free-electron density × cross-sectional area × charge per electron × electron drift speed

$$i = \eta A e v_d = \frac{\eta A e^2 E\tau}{2m}$$

Since the electric field intensity E in the wire is just

$$E = \frac{V}{L}$$

the current is

11.5 $$i = \frac{\eta e^2 \tau}{2m}\frac{A}{L}V$$

Equation 11.5 becomes Ohm's law if we set

11.6 $$R = \left(\frac{2m}{\eta e^2 \tau}\right)\frac{L}{A}$$

FIGURE 11-7 An electric field produces a general drift superimposed on the random motion of a free electron.

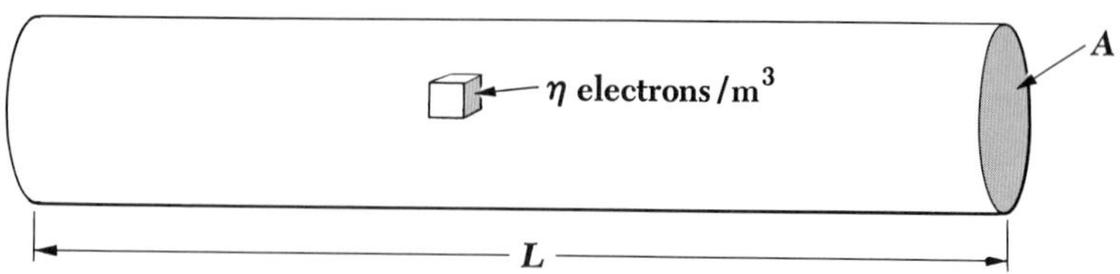

FIGURE 11-8 A model for deriving Ohm's law.

This formula for the resistance of a conductor indicates that R should be proportional to the length of the wire and inversely proportional to its cross-sectional area; this agrees with experience. At a given temperature the quantity in parentheses in Eq. 11.6, the *resistivity* of the metal, is a constant that depends solely upon the temperature and the properties of the metal. Our hypothesis that electrical conduction in metals occurs because of the presence of a gas of free electrons in them is evidently on the right track.

11.5 The Fermi Energy

Granting the initial assumption of an electron gas in metals, we next inquire into the nature of this gas. We might suspect from the discussions of the previous chapter that it behaves like a system of Fermi particles and hence obeys Fermi-Dirac statistics, and this is indeed the case. The Fermi-Dirac distribution law for the number of electrons n_i with the energy u_i is

11.7 $$n_i = \frac{g_i}{e^{\alpha}e^{u_i/kT} + 1}$$

It is more convenient to consider a continuous distribution of electron energies than the discrete distribution of Eq. 11.7, so that the distribution law becomes

11.8 $$n(u)\,du = \frac{g(u)\,du}{e^{\alpha}e^{u/kT} + 1}$$

To find $g(u)\,du$, the number of quantum states available to electrons with energies between u and $u + du$, we use the same reasoning as for the photon gas involved in black-body radiation. The correspondence is exact because there are two possible spin states, $s = +\frac{1}{2}$ and $s = -\frac{1}{2}$, for electrons, thus doubling the number of available phase-space cells just as the existence of two possible directions of polarization for otherwise identical photons doubles the number of cells for a photon gas. In terms of momentum we found that

$$g(p)\,dp = \frac{8\pi V p^2\,dp}{h^3}$$

For nonrelativistic electrons

$$p^2\,dp = (2m^3u)^{1/2}\,du$$

with the result that

11.9 $$g(u)\,du = \frac{8\sqrt{2}\pi V m^{3/2}}{h^3}\,u^{1/2}\,du$$

The next step is to evaluate the parameter α. In order to do this, we consider the condition of the electron gas at low temperatures. As observed in the previous chapter, the occupation index when T is small is 1 from $u = 0$ until near the Fermi energy u_F, where it drops rapidly to 0. This situation reflects the effect of the exclusion principle: no states can contain more than one electron, and so the minimum energy configuration of an electron gas is one in which the lowest states are filled and the remaining ones empty. If we set

11.10 $$\alpha = -\frac{u_F}{kT}$$

the occupation index becomes

11.11 $$f(u) = \frac{n(u)}{g(u)} = \frac{1}{e^{(u-u_F)/kT} + 1}$$

The formula is in accord with the exclusion principle. At $T = 0°\text{K}$,

$$\begin{aligned} f(u) &= 1 \text{ when } u < u_F \\ &= 0 \text{ when } u > u_F \end{aligned}$$

As the temperature increases, the occupation index changes from 1 to 0 more and more gradually, as in Fig. 11-9. At all temperatures

$$f(u) = \tfrac{1}{2} \text{ when } u = u_F$$

If a particular metal sample contains N free electrons, we can calculate its Fermi energy u_F by filling up its energy states with these electrons in order of

FIGURE 11-9 **The occupation index for a Fermi-Dirac distribution at absolute zero and at a higher temperature.**

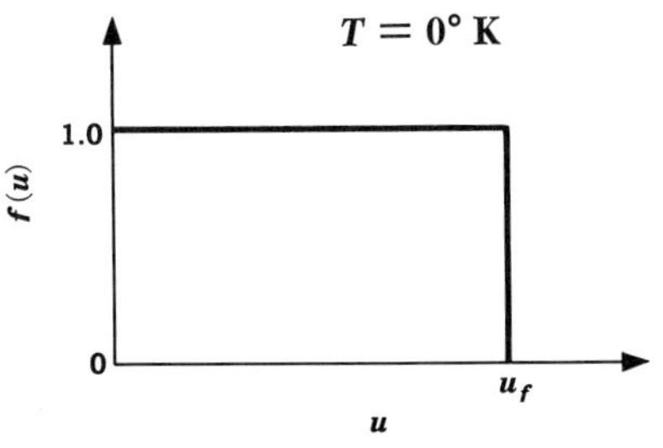

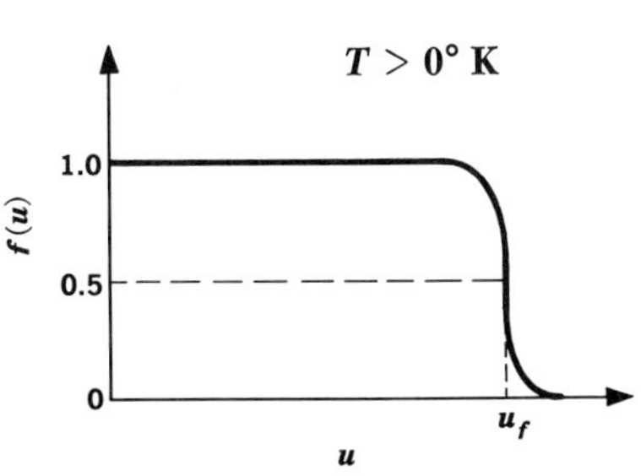

increasing energy starting from $u = 0$. The highest energy state to be filled will then have the energy $u = u_F$ by definition. The number of electrons that can have the same energy u is equal to the number of states $g(u)$ that have this energy, since each state is limited to one electron. Hence

$$\int_0^{u_F} g(u)\, du = N \tag{11.12}$$

Substituting Eq. 11.9 for $g(u)\, du$ yields

$$N = \frac{8\sqrt{2}\pi V m^{3/2}}{h^3} \int_0^{u_F} u^{1/2}/du$$

$$= \frac{16\sqrt{2}\pi V m^{3/2}}{3h^3} u_F{}^{3/2}$$

and

$$u_F = \frac{h^2}{2m}\left(\frac{3N}{8\pi V}\right)^{2/3} \tag{11.13}$$ Fermi energy

The quantity N/V is the density of free electrons; hence u_F is independent of the dimensions of whatever metal sample is being considered.

Let us use Eq. 11.13 to calculate the Fermi energy in copper. The electron configuration of the ground state of the copper atom is $1s^2 2s^2 2p^6 3s^2$-$3p^6 3d^{10} 4s$; that is, each atom has a single $4s$ electron outside closed inner shells. It is therefore reasonable to assume that each copper atom contributes one free electron to the electron gas. The electron density $\eta = N/V$ is accordingly equal to the number of copper atoms per unit volume, which is given by

$$\frac{\text{Atoms}}{\text{Volume}} = \frac{(\text{atoms/kmole}) \times (\text{mass/volume})}{\text{mass/kmole}} \tag{11.14}$$

$$= \frac{N_0 \rho}{w}$$

Here

$$N_0 = \text{Avogadro's number} = 6.02 \times 10^{26} \text{ atoms/kmole}$$
$$\rho = \text{density of copper} = 8.94 \times 10^3 \text{ kg/m}^3$$
$$w = \text{atomic weight of copper} = 63.5 \text{ kg/kmole}$$

so that

$$\eta = \frac{6.02 \times 10^{26} \text{ atoms/kmole} \times 8.94 \times 10^3 \text{ kg/m}^3}{63.5 \text{ kg/kmole}}$$

$$= 8.5 \times 10^{28} \text{ atoms/m}^3$$

$$= 8.5 \times 10^{28} \text{ electrons/m}^3$$

The corresponding Fermi energy is, from Eq. 11.13,

$$u_F = \frac{(6.63 \times 10^{-34}\text{ joule-sec})^2}{2 \times 9.11 \times 10^{-31}\text{ kg/electron}} \left(\frac{3 \times 8.5 \times 10^{28}\text{ electrons}/m^3}{8\pi}\right)^{2/3}$$
$$= 1.13 \times 10^{-18}\text{ joule}$$
$$= 7.04\text{ ev}$$

At absolute zero, $T = 0°$K, there would be electrons with energies of up to 7.04 ev in copper. By contrast, *all* of the molecules in an ideal gas at absolute zero would have zero energy. Because of its decidedly nonclassical behavior, the electron gas in a metal is said to be *degenerate*. Table 11.1 gives the Fermi energies of several common metals.

11.6 Electron-energy Distribution

We may now substitute for α and $g(u)\,du$ in Eq. 11.8 to obtain a formula for the number of electrons in an electron gas having energies between some value u and $u + du$. This formula is

11.15
$$n(u)\,du = \frac{(8\sqrt{2}\pi V m^{3/2}/h^3)u^{1/2}\,du}{e^{(u-u_F)/kT} + 1}$$

If we express the numerator of Eq. 11.15 in terms of the Fermi energy u_F, we obtain

11.16
$$n(u)\,du = \frac{(3N/2){u_F}^{-3/2}u^{1/2}\,du}{e^{(u-u_F)/kT} + 1}$$

Equation 11.16 is plotted in Fig. 11-10 for the temperatures $T = 0$, 300, and 1200°K.

It is interesting to determine the average electron energy at absolute zero. To do this, we first obtain the total energy U_0 at 0°K, which is

$$U_0 = \int_0^{u_F} u n(u)\,du$$

Table 11.1

SOME FERMI ENERGIES

Metal	Fermi energy, ev
Lithium (Li)	4.72
Sodium (Na)	3.12
Potassium (K)	2.14
Cesium (Cs)	1.53
Copper (Cu)	7.04
Silver (Ag)	5.51
Gold (Au)	5.54

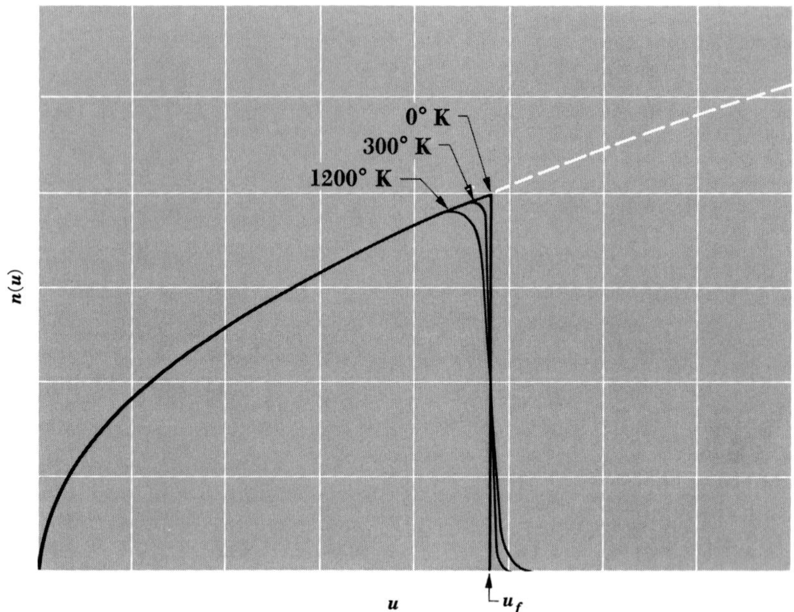

FIGURE 11-10 Distribution of electron energies in a metal at various temperatures.

Since at $T = 0°\text{K}$ all of the electrons have energies less than or equal to the Fermi energy u_F, we may let

$$e^{(u-u_F)/kT} = 0$$

and

11.17 $$U_0 = \frac{3N}{2} u_F^{-3/2} \int_0^{u_F} u^{3/2}\, du = \frac{3}{5} N u_F$$

The average electron energy $\bar{u}_0$ is this total energy divided by the number of electrons present N, which yields

11.18 $$\bar{u}_0 = \frac{3}{5} u_F$$

Since Fermi energies for metals are usually several electron volts, the average electron energy in them at 0°K will also be of this order of magnitude. The temperature of an ideal gas whose molecules have an average kinetic energy of 1 ev is 11,600°K; this means that, if free electrons behaved classically, a sample of copper would have to be at a temperature of about 50,000°K for its electrons to have the same average energy they actually have at 0°K.

The considerable amount of kinetic energy possessed by the valence electrons in the electron gas of a metal represents a repulsive influence. The act of assembling a group of metal atoms into a solid requires that additional energy be given to the valence electrons in order to elevate them to the higher energy states required by the exclusion principle. The atoms in a metallic solid, however, are closer together because of their bonds than they would be otherwise. As a result the valence electrons are, on the average, closer to an atomic nucleus in a metallic solid than they are in an isolated metal atom. These electrons accordingly have lower potential energies in the former case than in the latter, sufficiently lower to lead to a net cohesive force even when the added electron kinetic energy is taken into account. We shall go into this effect in more detail in the next section.

11.7 The Band Theory of Solids

The atoms in almost every crystalline solid, whether a metal or not, are so close together that their valence electrons constitute a single system of electrons common to the entire crystal. The exclusion principle is obeyed by such an electron system because the energy states of the outer electron shells of the atoms are all altered somewhat by their mutual interactions. In place of each precisely defined characteristic energy level of an individual atom, the entire crystal possesses an *energy band* composed of myriad separate levels very close together. Since there are as many of these separate levels as there are atoms in the crystal, the band cannot be distinguished from a continuous spread of permitted energies.

Figure 11-11 shows the energy bands in solid sodium plotted versus internuclear distance. The $3p$ band is the first to broaden, closely followed by the $3s$ band; the $2p$ band does not begin to spread until a quite small internuclear separation. This behavior reflects the order in which the electron subshells of sodium atoms interact as the atoms are brought together. The average energies in the $3p$ and $3s$ bands drop at first, implying attractive forces. The actual internuclear distance in solid sodium is indicated, and it corresponds, as it should, to a situation of minimum average energy.

The energy bands in a solid correspond to the energy levels in an atom, and an electron in a solid can possess only those energies that fall within these energy bands. The various energy bands in a solid may overlap, as in Fig. 11-12, in which case its electrons have a continuous distribution of permitted energies. In other solids the bands may *not* overlap (Fig. 11-13), and the intervals between them represent energies which their electrons cannot possess. Such intervals are called *forbidden bands.* The electrical behavior of a crystalline solid is determined both by its energy-band structure and by how these bands are normally filled by electrons.

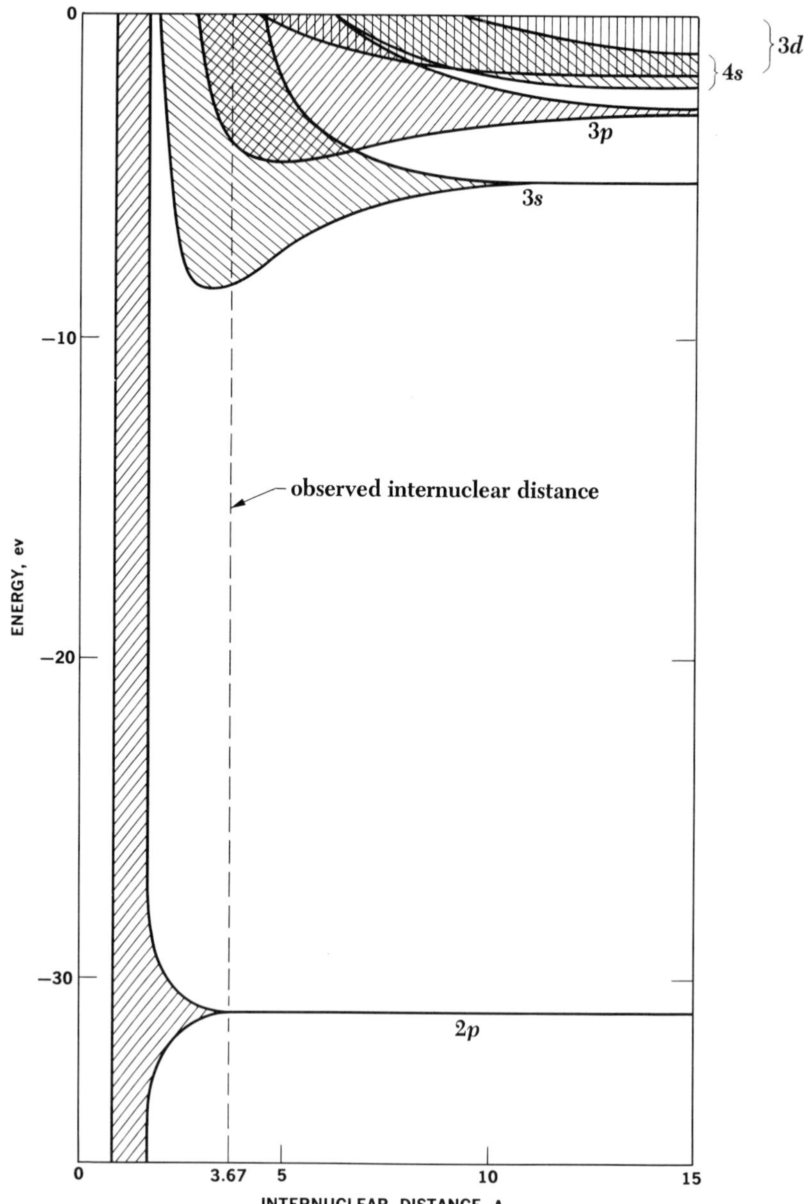

FIGURE 11-11 The energy levels of sodium atoms become bands as their internuclear distance decreases. The observed internuclear distance in solid sodium is 3.67 A.

Figure 11-14 is a simplified diagram of the energy levels of a sodium atom and the energy bands of solid sodium. A sodium atom has a single 3*s* electron in its outer shell. This means that the lowest energy band in a sodium crystal is only half occupied, since each level in the band, like each level in

FIGURE 11-12 The energy bands in a solid may overlap.

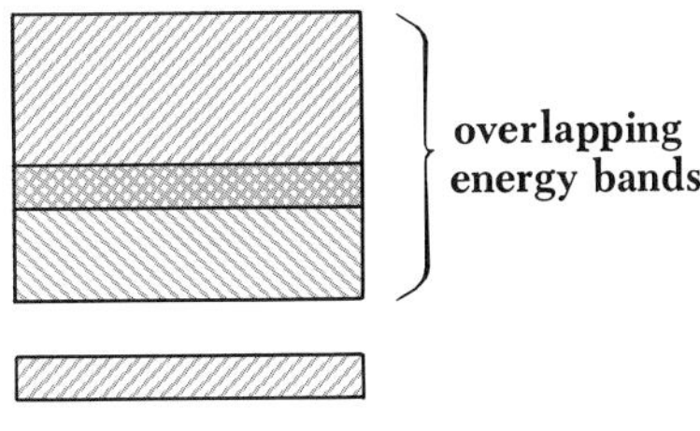

the atom, is able to contain *two* electrons. When an electric field is set up across a piece of solid sodium, electrons easily acquire additional energy while remaining in their original energy band. The additional energy is in the form of kinetic energy, and the moving electrons contribute to an electric current by the mechanism discussed earlier. Sodium is therefore a good conductor of electricity, as are other crystalline solids with energy bands that are only partially filled.

Figure 11-15 is a simplified diagram of the energy bands of diamond. The two lower energy bands are completely filled with electrons, and there is a gap of 6 ev between the top of the higher of these bands and the empty band above it. This means that at least 6 ev of additional energy must be provided an electron in a diamond crystal if it is to have any kinetic energy, since it cannot have an energy lying in the forbidden band. An energy increment of this magnitude cannot readily be given to an electron in a crystal by an electric field. An electron moving through a crystal collides with another electron after an average of perhaps 10^{-8} m, and it loses much of the energy it gained from any electric field in the collision. An electric field intensity of 6×10^8 volts/m is necessary if an electron is to gain 6 ev in a path length of 10^{-8} m, well over 10^{10} times greater than the electric intensity needed to cause a current to flow in sodium. Diamond is

FIGURE 11-13 A forbidden band separates nonoverlapping energy bands.

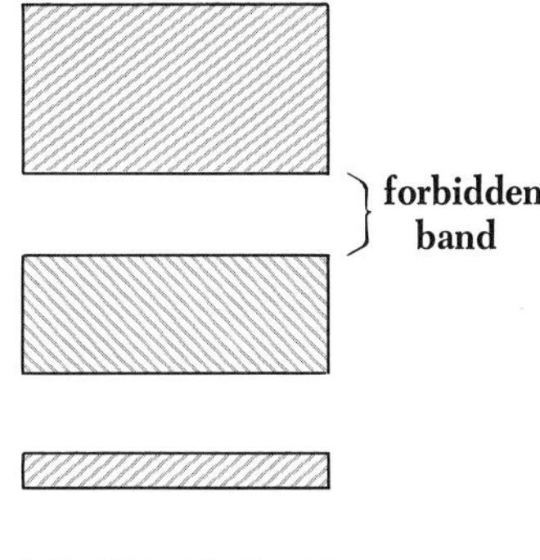

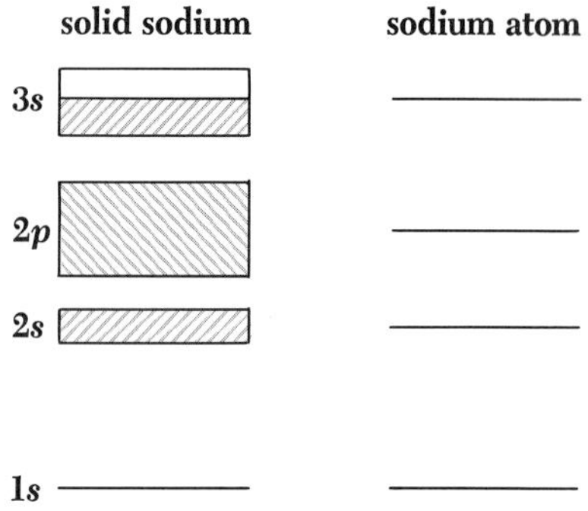

FIGURE 11-14 Energy levels in the sodium atom and the corresponding bands in solid sodium (not to scale).

therefore a very poor conductor of electricity and is accordingly classed as an insulator.

Silicon has a crystal structure resembling that of diamond, and, as in diamond, a gap separates the top of a filled energy band from a vacant higher band. The forbidden band in silicon, however, is only 1.1 ev wide. At low temperatures silicon is little better than diamond as a conductor, but at room temperature a small proportion of its electrons have sufficient kinetic energy of thermal origin to jump the forbidden band and enter the energy band above it. These electrons are sufficient to permit a limited amount of current to flow when an electric field is applied. Thus silicon has an electrical resistivity intermediate between those of conductors and those of insulators; it is termed a *semiconductor.*

11.8 Impurity Semiconductors

The resistivity of semiconductors can be significantly affected by small amounts of impurity. Let us incorporate a few arsenic atoms in a silicon crystal. Arsenic atoms have five electrons in their outermost shells, while silicon atoms have four. (These shells have the configurations $4s^24p^3$ and $3s^23p^2$ respectively.) When an arsenic atom replaces a silicon atom in a silicon crystal, four of its electrons are incorporated in covalent bonds with its

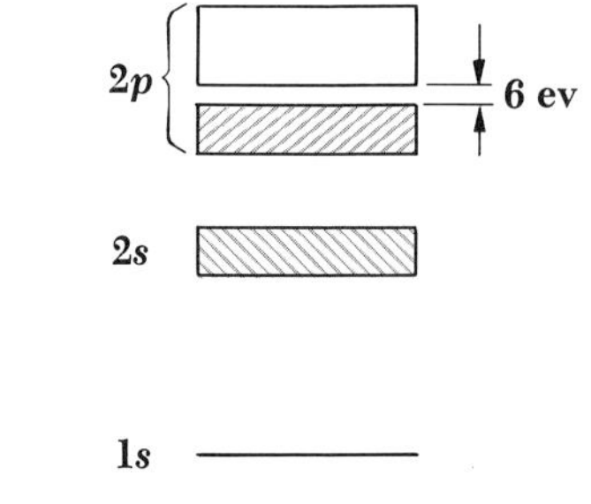

FIGURE 11-15 Energy bands in diamond (not to scale).

The Hall Effect

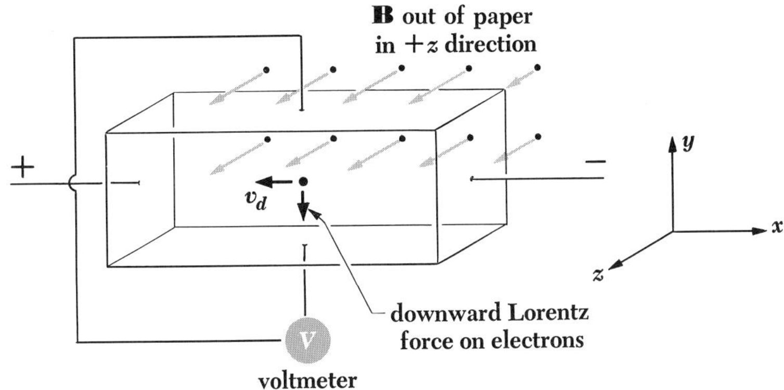

The Hall effect makes it possible to determine experimentally several of the microscopic properties of conductors and semiconductors. As in the figure, a block of a solid is placed in perpendicular electric and magnetic fields. If the electric field is in the $+x$ direction, the electrons in the solid experience the force $-eE_x$ and drift in the $-x$ direction with the speed v_d. The magnetic field exerts the Lorentz force $\mathbf{F} = -e\mathbf{v} \times \mathbf{B}$ on moving electrons; this force averages out to zero on the random motions of the electrons, but it has a resultant component in the $-y$ direction because of the drift v_d. This component has the magnitude $F_y = -ev_dB$, and it causes the electrons to collect at the lower face of the block until there is built up between the upper and lower faces an electric field E_y that opposes the further transverse migration of electrons. The condition for force balance on an electron is thus

$$\begin{aligned} F_y &= eE_y \\ -ev_dB &= eE_y \\ E_y &= -v_dB \end{aligned}$$

The potential difference between the upper and lower faces of the block can be measured; from it E_y can be determined and, since B is known, v_d can be found. By measuring the resistivity of the material at the same time, it is possible to find the number of electrons per unit volume participating in the conduction process. In the case of semiconductors, where conduction may take place by the motion of "holes" as well as by electron motion, the direction of E_y indicates whether the material is an n- or p-type semiconductor.

nearest neighbors. The fifth electron requires little energy to be detached and move about in the crystal. As shown in Fig. 11-16, the presence of arsenic as an impurity provides energy levels just below the band which electrons must occupy for conduction to take place. Such levels are termed

Semiconductor Rectifiers

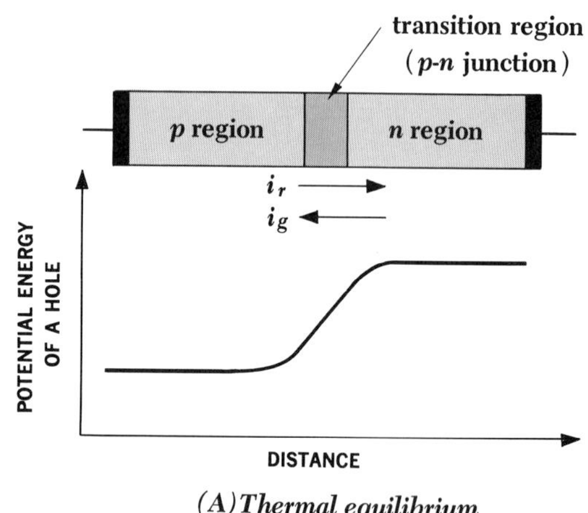

(A) Thermal equilibrium

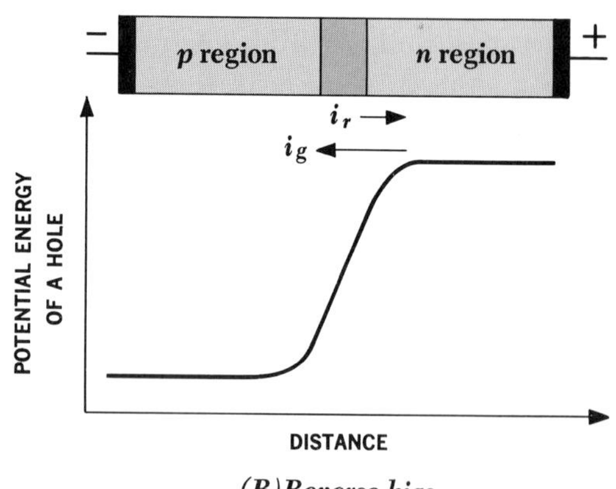

(B) Reverse bias

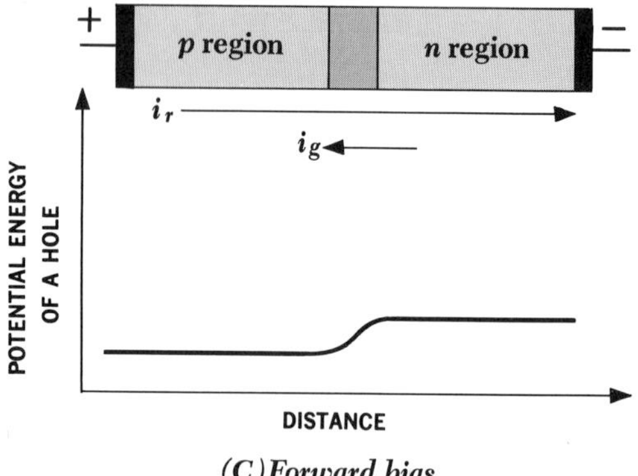

(C) Forward bias

The addition of impurities to germanium and silicon crystals can be so controlled that part of a crystal is n-type and the rest p-type, with only a thin intermediate region. The simplest of several methods of producing such crystals is to gradually pull out a crystal that is growing in, say, molten germanium that contains a donor impurity, and quickly add an acceptor impurity to the melt in the middle of the process. The first portion of the resulting crystal to solidify will be n-type and the remainder will be p-type if the proportion of acceptor impurity exceeds that of donor impurity in the last stage of the crystal growth.

An important property of a p-n junction in a semiconductor crystal is that it conducts electric current much more readily in one direction than in the other. In the absence of any applied voltage across the ends of a crystal containing a p-n junction, there normally are small flows of electrons and holes in both directions across the junction. Thus some holes have enough energy to diffuse from the p region into the n region, where they recombine with conduction electrons, while at the same time thermal fluctuations lead to the spontaneous generation of holes in the n region that diffuse across the junction into the p region. At equilibrium the two currents, i_r and i_g respectively, are equal, as is shown above. The curve shows how the potential energy of a hole varies through the crystal. At B a voltage is applied across the crystal with the p end negative and the n end positive. In this situation, called *reverse bias,* the potential-energy difference across the junction for a hole is increased over that in A, which impedes the recombination current i_r without affecting i_g appreciably. Hence the limit of the reverse hole current is i_g. At C a forward bias is applied, with the p end of the crystal positive and the n end negative. Now the diffusion of holes across the junction into the n region of the crystal is enhanced, and i_r increases exponentially with the applied voltage. The behavior of the electron current corresponds exactly to that of the hole current except for a change in sign, so that under reverse bias few electrons flow from the n to the p region, while under forward bias a high density of electrons can surmount the potential barrier to complement the hole current in the opposite direction, yielding a high resultant electric current. More complex semiconductor devices, notably transistors, are in wide use in modern electronic circuits.

donor levels, and the substance is called an *n-type* semiconductor because electric current in it is carried by negative charges.

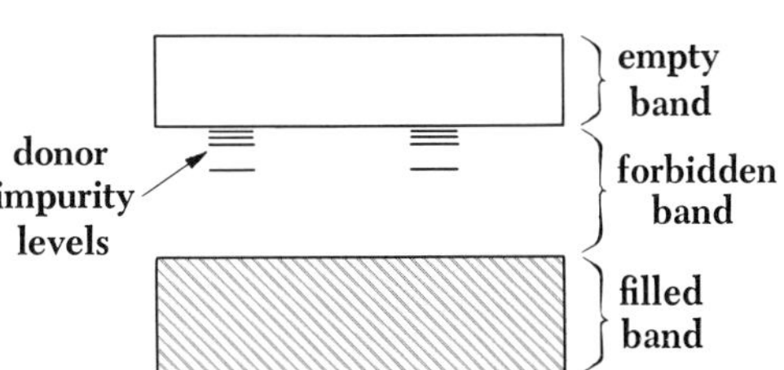

FIGURE 11-16 A trace of arsenic in a silicon crystal provides donor levels in the normally forbidden band, producing an n-type semiconductor.

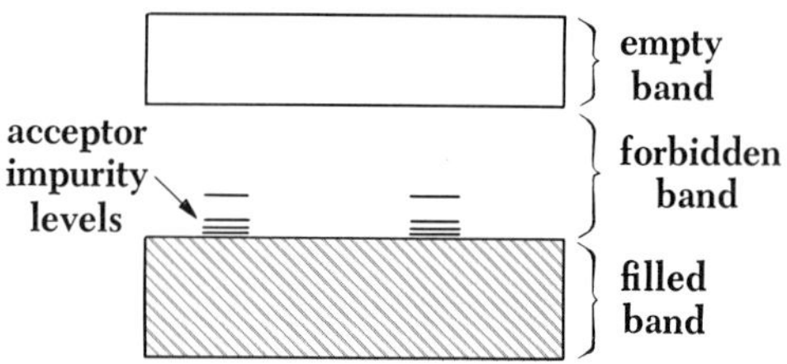

FIGURE 11-17 A trace of gallium in a silicon crystal provides acceptor levels in the normally forbidden band, producing a p-type semiconductor.

If we alternatively incorporate gallium atoms in a silicon crystal, a different effect occurs. Gallium atoms have only three electrons in their outer shells, whose configuration is $4s^24p$, and their presence leaves vacancies called *holes* in the electron structure of the crystal. An electron needs relatively little energy to enter a hole, but as it does so, it leaves a new hole in its former location. When an electric field is applied across a silicon crystal containing a trace of gallium, electrons move toward the anode by successively filling holes. The flow of current here is conveniently described with reference to the holes, whose behavior is like that of positive charges since they move toward the negative electrode. A substance of this kind is called a *p-type* semiconductor. In the energy-band diagram of Fig. 11-17 we see that the presence of gallium provides energy levels, termed *acceptor levels*, just above the highest filled band. Any electrons that occupy these levels leave behind them, in the formerly filled band, vacancies which permit electric current to flow.

Problems

1. Find the Fermi energy in calcium under the assumption that there are two free electrons per atom.

2. Find the Fermi energy in aluminum under the assumption that there are three free electrons per atom.

3. Compare the Fermi energies of Table 11.1 with kT at room temperature. How would the degree of occupancy of each quantum state change if the electrons in a metal were to have a Maxwell-Boltzmann energy distribution instead of a Fermi-Dirac distribution?

4. The binding energy of the hydrogen molecule is relatively high. What bearing do you think this has on the fact that hydrogen is not a metal?

5. Find the mean kinetic energy of the conduction electrons in a metal in terms of the Fermi energy.

6. Explain why a solid whose energy bands are filled cannot be a metal.

7. A small proportion of indium is incorporated in a germanium crystal. Is the crystal an n-type or a p-type semiconductor?

8. What is the connection between the fact that the free electrons in a metal obey Fermi statistics and the fact that the photoelectric effect is virtually temperature-independent?

9. Explain why the free electrons in a metal make only a minor contribution to its specific heat.

10. The potential energy $V(x)$ of a pair of atoms in a solid that are displaced by x from their equilibrium separation at $0°K$ may be written $V(x) = ax^2 - bx^3 - cx^4$, where the anharmonic terms $-bx^3$ and $-cx^4$ represent, respectively, the asymmetry introduced by the repulsive forces between the atoms and the leveling off of the attractive forces at large displacements (see Fig. 9-1). According to Eq. 10.55, at a temperature T the likelihood that a displacement x will occur relative to the likelihood of no displacement is $e^{-V/kT}$, so that the average displacement x at this temperature is

$$\overline{x} = \frac{\int_{-\infty}^{\infty} xe^{-V/kT}\, dx}{\int_{-\infty}^{\infty} e^{-V/kT}\, dx}$$

Show that, for small displacements, $\overline{x} \approx 3bkT/4a^2$. (This is the reason that the change in length of a solid when its temperature changes is proportional to ΔT.)

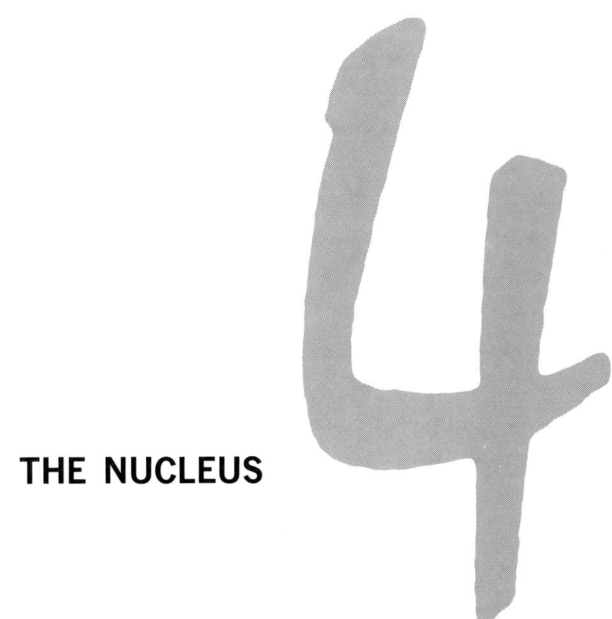

THE NUCLEUS

THE ATOMIC NUCLEUS 12

Until now we have been able to regard the atomic nucleus solely as a point mass possessing positive charge. Actually, of course, nuclei are quite complex entities, so complex that serious problems still remain in understanding their properties and behavior. In this chapter we shall consider some fundamental information about nuclei together with several of the more successful attempts at understanding their structure.

12.1 Atomic Masses

Table 12.1 is a list of the atomic masses of the various elements expressed in *atomic mass units* (amu), where, by definition,

$$1 \text{ amu} = 1.66 \times 10^{-27} \text{ kg}$$

We note at once that many of these masses are very close to being integral multiples of the mass of the hydrogen atom, though none is an exact multiple of it. Not long after the development of methods for determining atomic masses early in this century, it was discovered that not all of the atoms of a particular element have the same mass; the different varieties of the same element are called its *isotopes*. While the *average* atomic mass of an element, which is listed in Table 12.1, may not be nearly an integral multiple of the hydrogen mass, the atomic masses of its isotopes always are. Thus the average atomic mass of chlorine is 35.46, but closer study shows that chlorine is composed of two isotopes of masses 34.98 and 36.98 amu. The presence of these isotopes in about a 3:1 ratio respectively in natural chlorine accounts for the average figure of 35.46.

Even hydrogen is found to have isotopes, though the two heavier ones are not at all abundant. Their atomic masses are 1.0078, 2.0141, and 3.0160 amu. The nucleus of the lightest isotope is the *proton*, whose mass of

$$m_p = 1.0073 \text{ amu}$$

Table 12.1

ATOMIC MASSES OF THE ELEMENTS

Atomic number	Element	Symbol	Atomic mass*
1	Hydrogen	H	1.008
2	Helium	He	4.003
3	Lithium	Li	6.939
4	Beryllium	Be	9.012
5	Boron	B	10.81
6	Carbon	C	12.01
7	Nitrogen	N	14.01
8	Oxygen	O	16.00
9	Fluorine	F	19.00
10	Neon	Ne	20.18
11	Sodium	Na	22.99
12	Magnesium	Mg	24.31
13	Aluminum	Al	26.98
14	Silicon	Si	28.09
15	Phosphorus	P	30.97
16	Sulfur	S	32.06
17	Chlorine	Cl	35.45
18	Argon	Ar	39.95
19	Potassium	K	39.10
20	Calcium	Ca	40.08
21	Scandium	Sc	44.96
22	Titanium	Ti	47.90
23	Vanadium	V	50.94
24	Chromium	Cr	52.00
25	Manganese	Mn	54.94
26	Iron	Fe	55.85
27	Cobalt	Co	58.93
28	Nickel	Ni	58.71
29	Copper	Cu	63.54
30	Zinc	Zn	65.37
31	Gallium	Ga	69.72
32	Germanium	Ge	72.59
33	Arsenic	As	74.92
34	Selenium	Se	78.96
35	Bromine	Br	79.91
36	Krypton	Kr	83.80
37	Rubidium	Rb	85.47
38	Strontium	Sr	87.62
39	Yttrium	Y	88.91
40	Zirconium	Zr	91.22
41	Niobium	Nb	92.91
42	Molybdenum	Mo	95.94
43	Technetium	Tc	(99)
44	Ruthenium	Ru	101.1
45	Rhodium	Rh	102.9
46	Palladium	Pd	106.4
47	Silver	Ag	107.9
48	Cadmium	Cd	112.4
49	Indium	In	114.8
50	Tin	Sn	118.7
51	Antimony	Sb	121.8
52	Tellurium	Te	127.6

Table 12.1

(Continued)

Atomic number	Element	Symbol	Atomic mass*
53	Iodine	I	126.9
54	Xenon	Xe	131.3
55	Cesium	Cs	132.9
56	Barium	Ba	137.4
57	Lanthanum	La	138.9
58	Cerium	Ce	140.1
59	Praseodymium	Pr	140.9
60	Neodymium	Nd	144.2
61	Promethium	Pm	(147)
62	Samarium	Sm	150.4
63	Europium	Eu	152.0
64	Gadolinium	Gd	157.3
65	Terbium	Tb	158.9
66	Dysprosium	Dy	162.5
67	Holmium	Ho	164.9
68	Erbium	Er	167.3
69	Thulium	Tm	168.9
70	Ytterbium	Yb	173.0
71	Lutetium	Lu	175.0
72	Hafnium	Hf	178.5
73	Tantalum	Ta	181.0
74	Tungsten	W	183.9
75	Rhenium	Re	186.2
76	Osmium	Os	190.2
77	Iridium	Ir	192.2
78	Platinum	Pt	195.1
79	Gold	Au	197.0
80	Mercury	Hg	200.6
81	Thallium	Tl	204.4
82	Lead	Pb	207.2
83	Bismuth	Bi	209.0
84	Polonium	Po	(210)
85	Astatine	At	(210)
86	Radon	Rn	(222)
87	Francium	Fr	(223)
88	Radium	Ra	(226)
89	Actinium	Ac	(227)
90	Thorium	Th	232.0
91	Protactinium	Pa	(231)
92	Uranium	U	238.1
93	Neptunium	Np	(237)
94	Plutonium	Pu	(242)
95	Americium	Am	(243)
96	Curium	Cm	(247)
97	Berkelium	Bk	(249)
98	Californium	Cf	(251)
99	Einsteinium	Es	(254)
100	Fermium	Fm	(253)
101	Mendelevium	Md	(256)
102	Nobelium	No	(254)
103	Lawrencium	Lw	(257)

* Masses in parentheses are those of the most stable isotopes of the elements.

is the 1.0078-amu mass of the entire atom minus the 0.00055-amu mass of the electron it also contains. The proton, like the electron, is an *elementary particle* rather than a composite of other particles. (We shall consider the notion of elementary particle in some detail in Chap. 15.)

12.2 Nuclear Electrons

While the masses of nuclei heavier than hydrogen are very nearly integral multiples of the proton mass, they are always greater than (often more than double) the mass of a number of protons equal to their atomic numbers. An explanation that comes to mind immediately is that electrons may be present in nuclei which neutralize the positive charge of some of the protons. Thus the helium nucleus, whose mass is four times that of the proton though its charge is only $+2e$, would be regarded as being composed of four protons and two electrons. This explanation is buttressed by the fact that certain radioactive nuclei spontaneously emit electrons, a phenomenon called *beta decay,* whose occurrence is not easy to account for if electrons are absent from nuclei. Despite the superficial attractiveness of the hypothesis of nuclear electrons, however, there are a number of strong arguments against it:

1. Nuclear size. Nuclei are only 10^{-15} to 10^{-14} m across, and to confine an electron to so small a volume requires, by the uncertainty principle, an uncertainty in electron momentum equivalent to an energy uncertainty of the order of 10^2 Mev. Hence the energy of a nuclear electron must be of this magnitude. This figure may also be obtained by calculating the lowest energy level of an electron in a box of nuclear size (this calculation must be made relativistically). However, the electrons emitted during beta decay have energies of only 2 or 3 Mev—a major discrepancy.

2. Nuclear spin. Protons and electrons are Fermi particles with spins of $\frac{1}{2}$, that is, angular momenta of $\frac{1}{2}\hbar$. Thus nuclei with an even number of protons plus electrons should have integral spins, while those with an odd number of protons plus electrons should have half-integral spins. This prediction is not obeyed. The fact that the deuteron, which is the nucleus of an isotope of hydrogen, has an atomic number of 1 and a mass number of 2, would be interpreted as implying the presence of two protons and one electron. Depending upon the orientations of the particles, the nuclear spin of $_1H^2$ should therefore be $\frac{3}{2}$, $\frac{1}{2}$, $-\frac{1}{2}$, or $-\frac{3}{2}$. However, the observed spin of the deuteron is 1, something that cannot be reconciled with the hypothesis of nuclear electrons.

3. Magnetic moment. A proton with a given angular momentum has a much smaller ($\sim 10^{-3}$) magnetic moment than an electron with the same angular momentum because it is so much heavier and therefore spins more slowly. However, the magnetic moments of nuclei are of the order of magnitude of that of the proton, not of that of the electron; this cannot be understood if electrons are nuclear constituents.

4. Electron-nuclear interaction. It is observed that the forces that act between nuclear particles lead to binding energies of the order of 10 Mev per particle. It is therefore hard to see why, if electrons interact strongly with protons to form nuclei, the orbital electrons in an atom interact only electrostatically with its nucleus. Furthermore, when fast electrons are scattered by nuclei, they behave as though acted upon solely by electrostatic forces, while the nuclear scattering of fast protons reveals departures from electrostatic influences that can be ascribed only to a specifically nuclear force.

The difficulties of the nuclear electron hypothesis were known for some time before the correct explanation for nuclear masses came to light, but there seemed to be no serious alternative. The problem of the mysterious ingredient besides the proton in atomic nuclei was not solved until 1932.

12.3 The Neutron

The composition of atomic nuclei was finally understood in 1932. Two years earlier the German physicists W. Bothe and H. Becker had bombarded beryllium with alpha particles from a sample of polonium and found that radiation was emitted which was able to penetrate matter readily. Bothe and Becker ascertained that the radiation did not consist of charged particles and assumed, quite naturally, that it consisted of gamma rays. (Gamma rays are electromagnetic waves of extremely short wavelength.) The ability of the radiation to pass through as much as several centimeters of lead without being absorbed suggested gamma rays of unprecedentedly short wavelength. Other physicists became interested in this radiation, and a number of experiments were performed to determine its properties in detail. In one such experiment Irène Curie and F. Joliot observed that when the radiation struck a slab of paraffin, a hydrogen-rich substance, protons were knocked out. At first glance this is not very surprising: X rays can give energy to electrons in Compton collisions, and there is no reason why shorter-wavelength gamma rays cannot give energy to protons in similar processes.

Curie and Joliot found proton recoil energies of up to about 5.3 Mev. From Eq. 2.15 for the Compton effect we can calculate the minimum gamma-ray photon energy $E = h\nu$ needed to transfer the kinetic energy T to a proton. Equation 2.15 states that

$$2m_0c^2(h\nu - h\nu') = 2(h\nu)(h\nu')(1 - \cos\phi)$$

where ϕ is the angle through which the photon is scattered. Maximum energy transfer occurs when $\phi = 180°$, corresponding to a head-on collision, so that $(1 - \cos \phi) = 2$. Here m_0 is the proton rest mass and $T = h\nu - h\nu'$. Hence

$$m_0c^2T = 2E(E - T)$$

and

12.1 $$E = \tfrac{1}{2}[T + (T^2 + 2m_0c^2T)^{1/2}]$$

The proton rest energy m_0c^2 is 938 Mev and the observed maximum proton recoil energy T is 5.3 Mev. Substituting these values in Eq. 12.1 yields a minimum initial gamma-ray photon energy of 53 Mev!

This result seemed peculiar because no nuclear radiation known at the time had more than a small fraction of this considerable energy. The peculiarity became even more striking when it was calculated that the presumed reaction of an alpha particle and a beryllium nucleus to yield a $_6C^{13}$ nucleus would result in a mass decrease of 0.01144 amu, which is equivalent to only 10.7 Mev—one-fifth the energy needed by a gamma-ray photon if it is to knock 5.3 Mev protons out of paraffin.

In 1932 James Chadwick, an associate of Rutherford, proposed an alternative hypothesis for the now-mysterious radiation emitted by beryllium when bombarded by alpha particles. He assumed that the radiation consisted of neutral *particles* whose mass is approximately the same as that of the proton. The electrical neutrality of these particles, which were called *neutrons*, accounted for their ability to penetrate matter readily. Their mass accounted nicely for the observed proton recoil energies: a moving particle colliding head-on with one at rest whose mass is the same can transfer all of its kinetic energy to the latter. A maximum proton energy of 5.3 Mev thus implies a neutron energy of 5.3 Mev, not the 53 Mev required by a gamma ray to cause the same effect. Other experiments had shown that such light nuclei as those of helium, carbon, and nitrogen could also be knocked out of appropriate absorbers by the beryllium radiation, and the measurements made of the energies of these nuclei fit in well with the neutron hypothesis. In fact, Chadwick arrived at the neutron-mass figure of $m_n \approx m_p$ from an analysis of observed proton and nitrogen nuclei recoil energies; no other mass gave as good agreement with the experimental data.

Before we consider the role of the neutron in nuclear structure, we should note that it is not a stable particle outside nuclei. The free neutron decays radioactively into a proton, an electron, and an antineutrino after a mean lifetime of about 10^3 sec. We shall discuss the decay of the neutron in the next chapter.

Immediately after its discovery the neutron was recognized as the missing ingredient in atomic nuclei. Its mass of slightly more than that of the proton, its electrical neutrality, and its spin of ½ all fit in perfectly with the observed properties of nuclei when it is assumed that nuclei are composed solely of neutrons and protons.

The following terms and symbols are widely used to describe a nucleus:

Z = atomic number = number of protons
N = neutron number = number of neutrons
$A = Z + N$ = mass number = total number of neutrons and protons

The term *nucleon* refers to both protons and neutrons, so that the mass number A is the number of nucleons in a particular nucleus. Nuclear species (sometimes called *nuclides*) are identified according to the scheme

$$_ZX^A$$

where X is the chemical symbol of the species. Thus the arsenic isotope of mass number 75 is denoted by

$$_{33}\mathrm{As}^{75}$$

since the atomic number of arsenic is 33. Similarly a nucleus of ordinary hydrogen, which is a proton, is denoted by

$$_1\mathrm{H}^1$$

Here the atomic and mass numbers are the same because no neutrons are present.

The fact that nuclei are composed of neutrons as well as protons immediately explains the existence of isotopes: the isotopes of an element all contain the same numbers of protons but have different numbers of neutrons. Since its nuclear charge is what is ultimately responsible for the characteristic properties of an atom, the isotopes of an element all have identical chemical behavior and differ physically only in mass.

12.4 Nuclear Size

The Rutherford scattering experiment provided the first evidence that nuclei are of finite size. In that experiment, as we saw in Chap. 4, an incident alpha particle is deflected by a target nucleus in a manner consistent with Coulomb's law provided the distance between them exceeds about 10^{-14} m. For smaller separations the predictions of Coulomb's law are not obeyed because the nucleus no longer appears as a point charge to the alpha particle.

Since Rutherford's time a variety of experiments have been performed to

The Mass Spectrometer

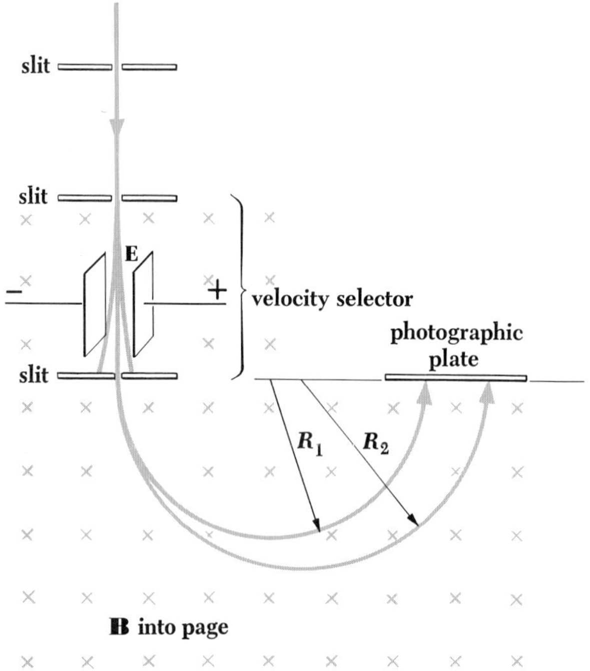

An instrument used to measure atomic masses is called a *mass spectrometer.* A variety of mass spectrometers have been devised, one of which is shown schematically above. The first step in its operation is the production of ions of the substance under study. If the substance is a gas, ions can readily be formed by electron bombardment, while if it is a solid, it is often convenient to incorporate it into an electrode used as one terminal of an arc discharge. The ions emerge from their source through a slit with the charge $+e$ and are then accelerated by an electric field. (The presence of ions with other charges is easily taken into account; we consider only those with single charges for simplicity.) When they enter the spectrometer itself the ions as a rule are traveling in slightly different directions with slightly different speeds. A pair of slits serves to collimate the ion beam, which then passes through a *velocity selector.* The velocity selector consists of uniform electric and magnetic fields that are perpendicular to each other and to the beam. The electric field exerts the force

$$F_e = eE$$

on the ions, while the magnetic field exerts the force

$$F_m = Bev$$

on them in the opposite direction. In order for an ion to reach the slit at the far end of the velocity selector, it must suffer no deflection within the selector, which means that only those ions escape for which

$$F_e = F_m$$

$$v = \frac{E}{B}$$

The ions in the beam are now all moving in the same direction with the same velocity. Once past the velocity selector, they enter a uniform magnetic field and follow circular paths whose radii R may be found by equating the magnetic force Bev on them with the centripetal force mv^2/R:

$$Bev = \frac{mv^2}{R}$$

$$R = \frac{mv}{eB}$$

Since v, e, and B are known, a measurement of R yields a value for the ion mass m. In some spectrometer designs the ions fall upon a photographic plate, permitting R to be determined by the position of their image, while in others B is varied to bring the ion beam to a fixed detector for which R is known.

---

determine nuclear dimensions, with particle scattering still a favored technique. In every case it is found that the volume of a nucleus is directly proportional to the number of nucleons it contains, which is its mass number A. If a nuclear radius is R, the corresponding volume is $\frac{4}{3}\pi R^3$, and so R^3 is proportional to A. This relationship is usually expressed in inverse form as

12.2 $$R = R_0 A^{1/3}$$

The value of R_0 is about

$$R_0 \approx 1.3 \times 10^{-15} \text{ m}$$

The uncertainty in R_0 is a consequence, not just of experimental error, but of the different character of the various experiments. For instance, an incident neutron interacts with a nucleus only through nuclear forces, while an incident electron interacts only through electrical forces. Interestingly enough the value of R_0 is smaller when it is deduced from electron scattering than from neutron scattering; this implies that nuclear matter and nuclear charge are not identically distributed throughout a nucleus.

Now that we know nuclear sizes as well as masses, we can compute the density of nuclear matter. This is very nearly 2×10^{17} kg/m^3 for all nuclei—a billion tons per cubic inch! Certain stars, called *white dwarfs,* are composed of atoms whose electron shells have "collapsed" owing to enormous pressure, and the densities of such stars approach that of pure nuclear matter.

12.5 Binding Energy

Stable nuclei invariably have smaller masses than the combined masses of their constituent particles. The nucleus $_3Li^6$, for example, has a mass of 6.0135 amu, while the three protons and three neutrons of which it is composed have a total mass (as free particles) of 6.0473 amu. The "missing" mass Δm is therefore 0.0338 amu, which is equivalent to 32 Mev. In order that a $_3Li^6$ nucleus break up into individual nucleons, then, 32 Mev of energy must be supplied to it from an external source. The energy equivalent of the mass discrepancy in a nucleus is called its *binding energy* and is a measure of the stability of the nucleus. Binding energies arise from the action of the forces that hold nucleons together to form nuclei, just as ionization energies of atoms, which must be provided to remove electrons from them, arise from the action of electrostatic forces.

The binding energy *per nucleon,* arrived at by dividing the total binding energy of a nucleus by the number of nucleons it contains, is a most interesting quantity. The binding energy per nucleon is plotted as a function of A in Fig. 12-1. The curve rises steeply at first and then more gradually until it reaches a maximum in the vicinity of $A = 56$, corresponding to iron nuclei, and then drops slowly. Evidently nuclei of intermediate

FIGURE 12-1 Binding energy per nucleon.

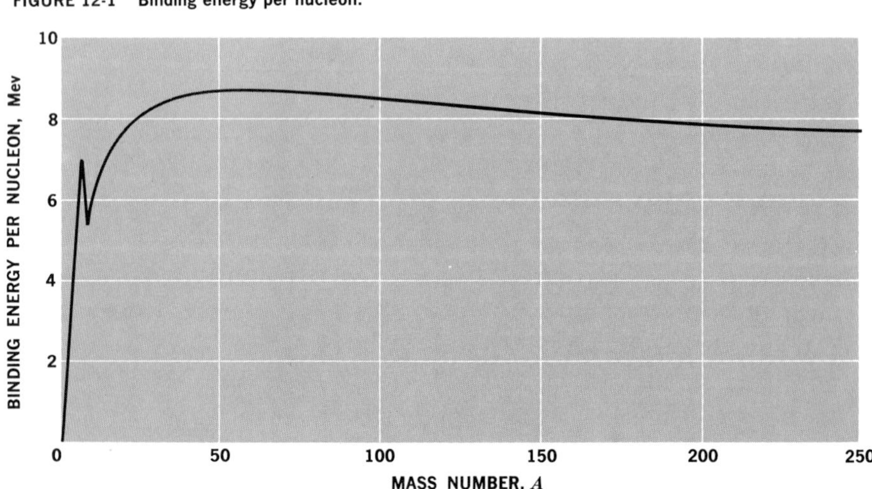

mass are the most stable, since the greatest amount of energy must be supplied to liberate their nucleons. This fact suggests that energy will be evolved if heavy nuclei can somehow be split into lighter ones or if light nuclei can somehow be joined to form heavier ones. The former process is known as nuclear *fission* and the latter as nuclear *fusion,* and both indeed occur under proper circumstances and do evolve energy as predicted.

12.6 The Deuteron

The unique short-range forces that bind nucleons so securely into nuclei constitute by far the strongest class of forces known. A number of detailed theories of nuclear forces have been proposed, and several of them have been fairly successful in accounting for various aspects of nuclear structure and behavior. Before looking into any of these theories, though, it is instructive to see what can be revealed by even a very general approach.

The simplest nucleus containing more than one nucleon is the *deuteron,* which consists of a proton and a neutron. The deuteron binding energy is 2.23 Mev, a figure that can be obtained either from the discrepancy in mass between m_{deuteron} and $m_p + m_n$ or from photodisintegration experiments which show that only gamma rays with $h\nu \geqslant 2.23$ Mev can disrupt deuterons into their constituent nucleons. In Chap. 7 we analyzed another two-body system, the hydrogen atom, with the help of quantum mechanics, but in that case the precise nature of the force between the proton and the electron was known. If a force law is known for an interaction, the corresponding potential energy function V can be found and substituted into Schrödinger's equation. Our understanding of nuclear forces is less complete than our understanding of Coulomb forces, however, and so it is not possible to discuss the deuteron in as much quantitative detail as the hydrogen atom.

The actual potential energy V of the deuteron, that is, the potential energy of either nucleon with respect to the other, depends upon the distance r between the centers of the neutron and proton more or less as shown by the black line in Fig. 12-2. (The repulsive "core" perhaps 0.4×10^{-15} m in radius reflects the inability of nucleons to mesh together more than a certain amount.) We shall approximate this $V(r)$ by the "square well" shown as a dashed line in the figure. This approximation means that we consider the nuclear force between neutron and proton to be zero when they are more than r_0 apart, and to have a constant magnitude, leading to the constant potential energy $-V_0$, when they are closer together than r_0. Thus the parameters V_0 and r_0 are representative of the strength and range, respectively, of the interaction holding the deuteron together, and the square-well potential itself is representative of the short-range character of the interaction.

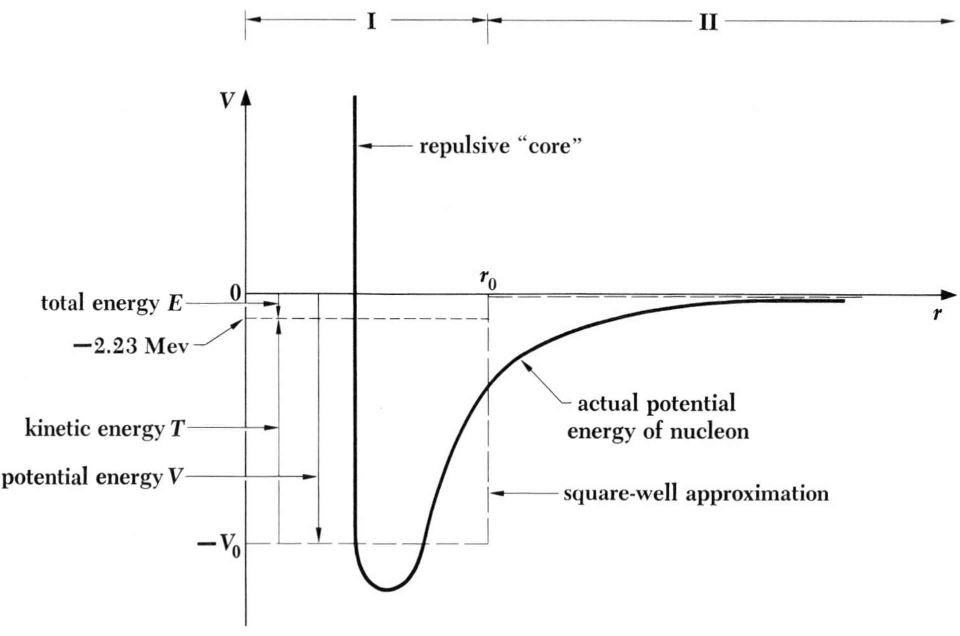

FIGURE 12-2 The actual potential energy of either proton or neutron in a deuteron and the square-well approximation to this potential energy as functions of the distance between proton and neutron.

A square-well potential means that V is a function of r alone, and therefore, as in the case of other central-force potentials, it is easiest to examine the problem in a spherical polar-coordinate system (see Fig. 7-1). In spherical polar coordinates Schrödinger's equation for a particle of mass m is, with $\hbar = h/2\pi$,

12.3
$$\frac{1}{r^2}\frac{\partial}{\partial r}\left(r^2\frac{\partial\psi}{\partial r}\right) + \frac{1}{r^2\sin\theta}\frac{\partial}{\partial\theta}\left(\sin\theta\frac{\partial\psi}{\partial\theta}\right) + \frac{1}{r^2\sin^2\theta}\frac{\partial^2\psi}{\partial\phi^2} + \frac{2m}{\hbar^2}(E - V)\psi = 0$$

Let us choose the particle in question to be the neutron, so that we imagine it moving in the force field of the proton. (The opposite choice would, of course, yield identical results.) We note from Fig. 12-2 that E, the total energy of the neutron, is negative and is the same as the binding energy of the deuteron.

In analyzing the hydrogen atom, where one particle is much heavier than the other, it is still necessary to consider the effects of nuclear motion, and we did this in Chap. 7 by replacing the electron mass m_e by its reduced mass m'. In this way the problem of a proton and an electron moving about a common center of mass is replaced by the problem of a single particle of mass m' moving about a fixed point. A similar procedure is even more appropriate here, since neutron and proton masses are almost the same. According to Eq. 5.30, the reduced mass of a neutron-proton system is

12.4 $$m' = \frac{m_n m_p}{m_n + m_p}$$

and so we replace the m of Eq. 12.3 with m' as given above.

We now assume that the solution of Eq. 12.3 can be written as the product of radial and angular functions,

12.5 $$\psi(r, \theta, \phi) = R(r)\Theta(\theta)\Phi(\phi)$$

As before, the function $R(r)$ describes how the wave function ψ varies along a radius vector from the nucleus, with θ and ϕ constant; the function $\Theta(\theta)$ describes how ψ varies with zenith angle θ along a meridian on a sphere centered at $r = 0$, with r and ϕ constant; and the function $\Phi(\phi)$ describes how ψ varies with azimuth angle ϕ along a parallel on a sphere centered at $r = 0$, with r and θ constant. Although angular motion can occur in a square-well potential, our interest is in radial motion, that is, in oscillations of the neutron and proton about their center of mass. If there is no angular motion, Θ and Φ are both constant and their derivatives are zero. With $\partial\psi/\partial\theta = \partial^2\psi/\partial\phi^2 = 0$, Eq. 12.3 becomes

12.6 $$\frac{1}{r^2}\frac{d}{dr}\left(r^2\frac{dR}{dr}\right) + \frac{2m'}{\hbar^2}(E - V)R = 0$$

A further simplification can be made by letting

12.7 $$u(r) = rR(r)$$

In terms of the new function u the wave equation becomes

12.8 $$\frac{d^2u}{dr^2} + \frac{2m'}{\hbar^2}(E - V)u = 0$$

Because V has two different values, $V = -V_0$ inside the well and $V = 0$ outside it, there are two different solutions to Eq. 12.8, u_{I} for $r \leqslant r_0$ and u_{II} for $r \geqslant r_0$. Inside the well the wave equation is

$$\frac{d^2u_{\mathrm{I}}}{dr^2} + \frac{2m'}{\hbar^2}(E + V_0)u_{\mathrm{I}} = 0$$

or, if we let

12.9 $$a^2 = \frac{2m'}{\hbar^2}(E + V_0)$$

it becomes simply

12.10 $$\frac{d^2u_{\rm I}}{dr^2} + a^2u_{\rm I} = 0$$

(We note from Fig. 12-2 that, since $|V_0| > |E|$, the quantities $E + V_0$ and hence a^2 are positive.) Equation 12.10 is the same as the wave equation for a particle in a box of Sec. 6.5, and again the solution is

12.11 $$u_{\rm I} = A \cos ar + B \sin ar$$

We recall that the radial wave function R is given by $R = u/r$, which means that the cosine solution must be discarded if R is not to be infinite at $r = 0$. Hence $A = 0$ and $u_{\rm I}$ inside the well is just

12.12 $$u_{\rm I} = B \sin ar$$

Outside the well $V = 0$ and so

12.13 $$\frac{d^2u_{\rm II}}{dr^2} + \frac{2m'}{\hbar^2} Eu_{\rm II} = 0$$

We note that the total energy E of the neutron is a negative quantity since it is bound to the proton. Therefore

12.14 $$b^2 = \frac{2m'}{\hbar^2}(-E)$$

is a positive quantity, and we can write

12.15 $$\frac{d^2u_{\rm II}}{dr^2} - b^2u_{\rm II} = 0$$

The solution of Eq. 12.15 is

12.16 $$u_{\rm II} = Ce^{-br} + De^{br}$$

Because it must be true that $u \rightarrow 0$ as $r \rightarrow \infty$, we conclude that $D = 0$. Hence outside the well

12.17 $$u_{\rm II} = Ce^{-br}$$

We now have expressions for u (and hence for ψ) both inside and outside the well, and it remains to match these expressions and their first derivatives at the well boundary; as discussed in Chap. 6, it is necessary that both u and du/dr be continuous everywhere in the region for which they are defined. At $r = r_0$, then,

$$u_{\rm I} = u_{\rm II}$$

12.18 $$B \sin ar_0 = Ce^{-br_0}$$

and

$$\frac{du_{\mathrm{I}}}{dr} = \frac{du_{\mathrm{II}}}{dr}$$

12.19 $$aB \cos ar_0 = -bCe^{-br_0}$$

By dividing Eq. 12.18 by Eq. 12.19 we eliminate the coefficients B and C and obtain the transcendental equation

12.20 $$\tan ar_0 = -\frac{a}{b}$$

Equation 12.20 cannot be solved analytically, but it can be solved either graphically or numerically to any desired degree of accuracy. We note that

12.21 $$\frac{a}{b} = \frac{\sqrt{2m'(E + V_0)}/\hbar}{\sqrt{2m'(-E)}/\hbar} = \sqrt{\frac{E + V_0}{-E}}$$

where $-E$ is the binding energy of the deuteron and V_0 is the depth of the potential well. Since $|V_0| > |E|$, in order to obtain a first crude approximation we might assume that a/b is so large that

$$\tan ar_0 \approx \infty$$

Since $\tan \theta$ becomes infinite at $\theta = \pi/2, \pi, 3\pi/2, \ldots, n\pi/2$, in this approximation the ground state of the deuteron, for which $n = 1$, corresponds to

$$ar_0 \approx \frac{\pi}{2}$$

(In fact, this is the only bound state of the deuteron.) Hence

$$\frac{\sqrt{2m'(E + V_0)}}{\hbar} r_0 \approx \frac{\sqrt{2m'V_0}}{\hbar} r_0 \approx \frac{\pi}{2}$$

since we are assuming that E is negligible compared with V_0, and

12.22 $$V_0 \approx \frac{\pi^2\hbar^2}{8m'{r_0}^2}$$

The above approximation is equivalent to assuming that the function u_{I} inside the well is at its maximum (corresponding to $ar = 90°$) at the boundary of the well. Actually, u_{I} must be somewhat past its maximum there in order to join smoothly with the function u_{II} outside the well, as shown in Fig. 12-3; a more detailed calculation shows that $ar \approx 105°$ at $r = r_0$. The difference between the two results is due to our neglect of the binding energy $-E$ relative to V_0 in obtaining Eq. 12.22, and when this neglect is remedied, the better approximation

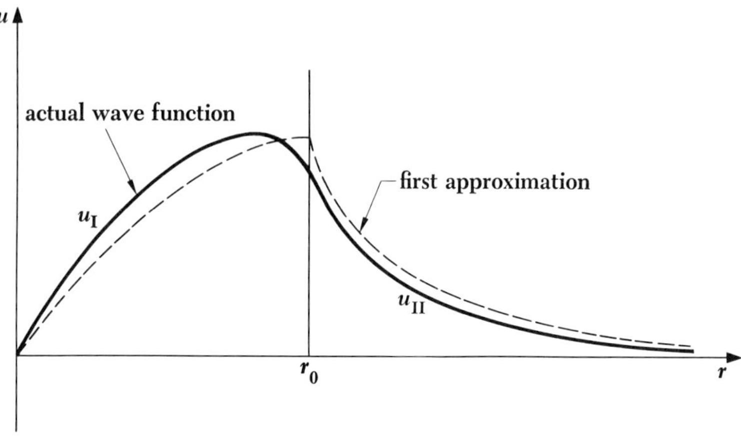

FIGURE 12-3 The wave function $u(r)$ of either proton or neutron in a deuteron.

12.23 $$V_0 \approx \frac{\pi^2\hbar^2}{8m'r_0^2} + \frac{2\hbar}{r_0}\sqrt{\frac{E}{2m'}}$$

is obtained.

Equation 12.23 is a relationship among r_0, the radius of the potential well and therefore representative of the range of nuclear forces, V_0, the depth of the well and therefore representative of the interaction potential energy between two nucleons due to nuclear forces, and $-E$, the binding energy of the deuteron. What we now ask is whether this relationship corresponds to reality in the sense that a reasonable choice of r_0 leads to a reasonable value for V_0. (There is nothing in our simple model that can point to a unique value for either r_0 or V_0.) Such a choice for r_0 might be 2×10^{-15} m. Substituting for the known quantities in Eq. 12.23 and expressing energies in million electron volts yields

$$V_0 \approx \frac{1.0 \times 10^{-28}\ \text{Mev-m}^2}{r_0^2} + \frac{1.9 \times 10^{-14}\ \text{Mev-m}}{r_0}$$

and so, for $r_0 = 2 \times 10^{-15}$ m,

$$V_0 \approx 35\ \text{Mev}$$

This is a plausible figure for V_0, from which we conclude that the basic features of our model—nucleons that maintain their identities in the nucleus instead of fusing together, strong nuclear forces with a short range and relatively minor angular dependence—are valid.

In anticipation of Sec. 12.9 we might remark that, in general, angular momentum *does* play a significant role in nuclear structure, although certain aspects of this structure are virtually independent of it. Even in the deuteron the proton and neutron interact in such a manner that a bound state exists only when their spins are parallel; the other possibility of opposite spins leads to a different potential well that does not give rise to a bound state.

12.7 Meson Theory of Nuclear Forces

In 1935 the Japanese physicist Hideki Yukawa proposed that nuclear forces arise from the constant exchange of particles (called *mesons*) back and forth between nearby nucleons. There were precedents for this kind of analysis: We saw in Chap. 9 how molecules are held together by the circulation of electrons among their component atoms, and the electric forces between charges can be formally interpreted as the result of the exchange of electromagnetic quanta between them.

According to the meson theory of nuclear forces, all nucleons consist of identical cores surrounded by a "cloud" of one or more mesons. Mesons may be neutral or carry either charge, and the sole difference between neutrons and protons is supposed to lie in the composition of their respective meson clouds. The forces that act between one neutron and another and between one proton and another are the result of the exchange of neutral mesons (designated π^0) between them. The force between a neutron and a proton is the result of the exchange of charged mesons (π^+ and π^-) between them. Thus a neutron emits a π^- meson and is converted into a proton:

$$n \rightarrow p + \pi^-$$

while the absorption of the π^- by the proton the neutron was interacting with converts it into a neutron:

$$p + \pi^- \rightarrow n$$

In the reverse process, a proton emits a π^+ meson whose absorption by a neutron converts it into a proton:

$$p \rightarrow n + \pi^+$$
$$n + \pi^+ \rightarrow p$$

While there is no simple mathematical way of demonstrating how the exchange of particles between two bodies can lead to attractive and repulsive forces, a rough analogy may make the process intuitively meaningful. Let us imagine two boys exchanging basketballs (Fig. 12-4). If they throw the balls at each other, they each move backward, and when they catch the balls thrown at them, their backward momentum increases. Thus this method of exchanging the basketballs yields the same effect as a repulsive force between the boys. If the boys snatch the basketballs from each other's hands, however, the result will be equivalent to an attractive force acting between them.

Although it is possible to prove, by more advanced mathematical techniques than we are using, that the exchange of mesons between nucleons can indeed lead to mutually attractive forces, a fundamental problem presents itself. If nucleons constantly emit and absorb mesons, why are neutrons or protons never found with other than their usual masses? The answer is based upon the uncertainty principle. The laws of physics refer exclusively to experimentally measurable quantities, and the uncertainty principle limits the accuracy with which certain combinations of measurements can be made. The emission of a meson by a nucleon which does not change in mass—a clear violation of the law of conservation of energy—can occur provided that the nucleon absorbs a meson emitted by the neighboring nucleon it is interacting with so soon afterward that *even in principle* it is impossible to determine whether or not any mass change actually has been involved.

repulsive force due to particle exchange

FIGURE 12-4 Attractive and repulsive forces can both arise from particle exchange.

attractive force due to particle exchange

Since the uncertainty principle may be written

$$\Delta E\ \Delta t \approx \hbar \tag{12.24}$$

an event in which an amount of energy ΔE is not conserved is not prohibited so long as the duration of the event does not exceed approximately $\hbar/\Delta E$.

We know that nuclear forces have a maximum range R of about 1.7×10^{-15} m, so that if we assume a meson travels between nuclei at approximately the speed of light c, the time interval Δt during which it is in flight is

$$\Delta t = \frac{R}{c} \tag{12.25}$$

The emission of a meson of mass m_π represents the nonconservation of

$$\Delta E = m_\pi c^2 \tag{12.26}$$

of energy. According to Eq. 12.24, this can occur provided that $\Delta E\ \Delta t \approx \hbar$; this means that

$$(m_\pi c^2)\left(\frac{R}{c}\right) \approx \hbar$$

Hence the meson mass must be

$$m_\pi \approx \frac{\hbar}{Rc}$$
$$\approx 1.9 \times 10^{-28}\ \text{kg}$$

which is about 200 m_e, that is, 200 electron masses.

Is the meson theory of nuclear forces correct? We are entitled to expect of any hypothesis that it make quantitative predictions in agreement with experiment; yet by and large the meson theory cannot account for nuclear properties in the detailed manner that quantum theory can account for atomic properties. There are exceptions to this disappointing picture, to be sure, for instance the problem of the magnetic moments of the proton and neutron. By analogy with the magnetic moment of the electron (Chap. 8), we would expect that of the proton to be $e\hbar/2m_p$ and that of the neutron, which is uncharged, to be 0. Actually, the magnetic moment of the proton is about 2.8 $e\hbar/2m_p$ and that of the neutron about 1.9 $e\hbar/2m_p$. A plausible explanation for these measurements is that both nucleons have structures that consist of "bare" cores of nuclear matter surrounded by clouds of charged mesons that contribute to the observed magnetic moments. Furthermore, particles *have* been discovered whose mass and behavior are in accord with meson theory, as we shall learn in Chap. 15. Hence there can be little doubt that Yukawa was on the right track, and it is possible, perhaps likely, that an elaboration of the meson theory will prove successful.

12.8 The Liquid-drop Model

While the attractive forces that nucleons exert upon one another are very strong, their range is so small that each particle in a nucleus interacts solely with its nearest neighbors. This situation is the same as that of atoms in a solid, which ideally vibrate about fixed positions in a crystal lattice, or that of molecules in a liquid, which ideally are free to move about while maintaining a fixed intermolecular distance. The analogy with a solid cannot be pursued because a calculation shows that the vibrations of the nucleons about their average positions would be too great for the nucleus to be stable. The analogy with a liquid, on the other hand, turns out to be extremely useful in understanding certain aspects of nuclear behavior.

Let us first see how the picture of a nucleus as a drop of liquid accounts for the observed variation of binding energy per nucleon with mass number. We start by assuming that the energy associated with each nucleon–nucleon bond has some value U; this energy is really negative, since attractive forces are involved, but is usually written as positive because binding energy is considered a positive quantity for convenience. (While it is plausible that the bonds arise from the interchange of mesons, we do not require a knowledge of their origin.) Because each bond energy U is shared by two nucleons, each has a binding energy of $\frac{1}{2}U$. When an assembly of spheres of the same size is packed together into the smallest volume, as we suppose is the case of nucleons within a nucleus, each interior sphere has 12 other spheres in contact with it (Fig. 12-5). Hence each interior nucleon in a nucleus has a binding energy of $12 \times \frac{1}{2}U$ or $6U$. If all A nucleons in a nucleus were in its interior, the total binding energy of the nucleus would be

$$E_v = 6AU \tag{12.27}$$

FIGURE 12-5 **In a tightly packed assembly of identical spheres, each interior sphere is in contact with twelve others.**

Equation 12.27 is often written simply as

12.28 $$E_v = a_1 A$$

The energy E_v is called the *volume energy* of a nucleus and is directly proportional to A.

Actually, of course, some nucleons are on the surface of every nucleus and therefore have fewer than 12 neighbors. The number of such nucleons depends upon the surface area of the nucleus in question. A nucleus of radius R has an area of

$$4\pi R^2 = 4\pi R_0^2 A^{2/3}$$

Hence the number of nucleons with fewer than the maximum number of bonds is proportional to $A^{2/3}$, reducing the total binding energy by

12.29 $$E_s = -a_2 A^{2/3}$$

The negative energy E_s is called the *surface energy* of a nucleus; it is most significant for the lighter nuclei since a greater fraction of their nucleons are on the surface. Because natural systems always tend to evolve toward configurations of minimum potential energy, nuclei tend toward configurations of maximum binding energy. (We recall that binding energy is the mass-energy difference between a nucleus and the same numbers of free neutrons and protons.) Hence a nucleus should exhibit the same surface-tension effects as a liquid drop, and in the absence of external forces it should be spherical since a sphere has the least surface area for a given volume.

The electrostatic repulsion between each pair of protons in a nucleus also contributes toward decreasing its binding energy. The *coulomb energy* E_c of a nucleus is the work that must be done to bring together Z protons from infinity into a volume equal to that of the nucleus. Hence E_c is proportional to $Z(Z - 1)/2$, the number of proton pairs in a nucleus containing Z protons, and inversely proportional to the nuclear radius $R = R_0 A^{1/3}$:

12.30 $$E_c = -a_3 \frac{Z(Z-1)}{A^{1/3}}$$

The coulomb energy is negative because it arises from a force that opposes nuclear stability.

The total *binding energy* E_b of a nucleus is the sum of its volume, surface, and coulomb energies:

12.31 $$\begin{aligned} E_b &= E_v + E_s + E_c \\ &= a_1 A - a_2 A^{2/3} - a_3 \frac{Z(Z-1)}{A^{1/3}} \end{aligned}$$

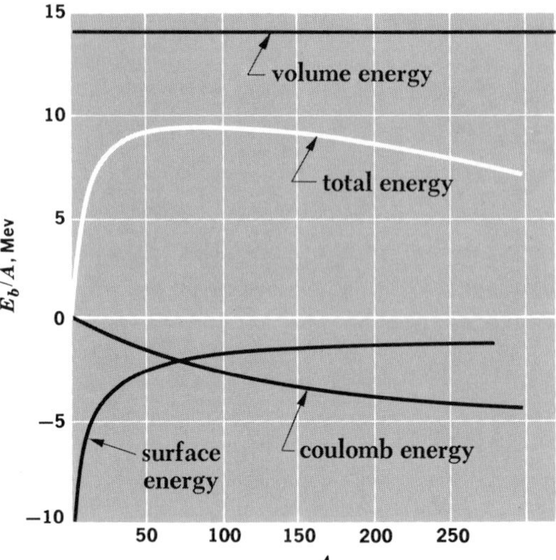

FIGURE 12-6 The binding energy per nucleon is the sum of the volume, surface, and Coulomb energies.

The binding energy *per nucleon* is therefore

12.32 $$\frac{E_b}{A} = a_1 - \frac{a_2}{A^{1/3}} - a_3\frac{Z(Z-1)}{A^{4/3}}$$

Each of the terms of Eq. 12.32 is plotted in Fig. 12-6 versus A, together with their sum, E_b/A. The latter is almost exactly the same as the empirical curve of E_b/A shown in Fig. 12-1. Hence the analogy of a nucleus with a liquid drop has some validity at least, and we may be encouraged to see what further aspects of nuclear behavior it can illuminate. We shall do this in Chap. 14 in connection with nuclear reactions.

Before leaving the subject of nuclear binding energy, it should be noted that effects other than those we have here considered also are involved. For instance, nuclei with equal numbers of protons and neutrons are especially stable, as are nuclei with even numbers of protons and neutrons. Thus such nuclei as $_2He^4$, $_6C^{12}$, and $_8O^{16}$ appear as peaks on the empirical binding energy per nucleon curve. These peaks imply that the energy states of neutrons and protons in a nucleus are almost identical and that each state can be occupied by two particles of opposite spin.

12.9 The Shell Model

The basic assumption of the liquid-drop model is that the constituents of a nucleus interact only with their nearest neighbors, like the molecules of a liquid. There is a good deal of empirical support for this assumption.

There is also, however, extensive experimental evidence for the contrary hypothesis that the nucleons in a nucleus interact primarily with a general force field rather than directly with one another. This latter situation resembles that of electrons in an atom, where only certain quantum states are permitted and no more than two electrons, which are Fermi particles, can occupy each state. Nucleons are also Fermi particles, and several nuclear properties vary periodically with Z and N in a manner reminiscent of the periodic variation of atomic properties with Z.

The electrons in an atom may be thought of as occupying positions in "shells" designated by the various principal quantum numbers, and the degree of occupancy of the outermost shell is what determines certain important aspects of an atom's behavior. For instance, atoms with 2, 10, 18, 36, 54, and 86 electrons have all their electron shells completely filled. Such electron structures are stable, thereby accounting for the chemical inertness of the rare gases. The same kind of situation is observed with respect to nuclei; nuclei having 2, 8, 20, 28, 50, 82, and 126 neutrons or protons are more abundant than other nuclei of similar mass numbers, suggesting that their structures are more stable. Since complex nuclei arose from reactions among lighter ones, the evolution of heavier and heavier nuclei became retarded when each relatively inert nucleus was formed; this accounts for their abundance.

Other evidence also points up the significance of the numbers 2, 8, 20, 28, 50, 82, and 126, which have become known as *magic numbers,* in nuclear structure. An example is the observed pattern of nuclear electric quadrupole moments, which are measures of the departures of nuclear charge distributions from sphericity. A spherical nucleus has no quadrupole moment, while one shaped like a football has a positive moment and one shaped like a pumpkin has a negative moment. Nuclei of magic N and Z are found to have zero quadrupole moments and hence are spherical, while other nuclei are distorted in shape, sometimes considerably so (the rare earth nuclei, for instance, have positive quadrupole moments that indicate major axes almost one-third greater than their minor axes).

The *shell model* of the nucleus is an attempt to account for the existence of magic numbers and certain other nuclear properties in terms of interactions between an individual nucleon and a force field produced by all the other nucleons. A potential energy function is used that corresponds to a square well about 50 Mev deep with rounded corners, so that there is a more realistic gradual change from $V = V_0$ to $V = 0$ than the sudden change of the pure square-well potential we used in treating the deuteron. Schrödinger's equation for a particle in a potential well of this kind is then solved, and it is found that stationary states of the system occur characterized by quantum numbers n, l, and m_l whose significance is the same as in the analogous case of stationary states of atomic electrons. Neutrons and protons occupy separate sets of states in a nucleus since the latter interact electrically as well as through the specifically nuclear charge.

In order to obtain a series of energy levels that leads to the observed magic numbers, it is merely necessary to assume a spin-orbit interaction whose magnitude is such that the consequent splitting of energy levels into sublevels is large for large l, that is, for large orbital angular momenta. It is assumed that LS coupling holds only for the very lightest nuclei, in which the l values are necessarily small in their normal configurations. In this scheme, as we saw in Sec. 8.5, the intrinsic spin angular momenta $\mathbf{S}_i$ of the particles concerned (the neutrons form one group and the protons another) are coupled together into a total spin momentum $\mathbf{S}$, and the orbital angular momenta $\mathbf{L}_i$ are separately coupled together into a total orbital momentum $\mathbf{L}$; $\mathbf{S}$ and $\mathbf{L}$ are then coupled to form a total angular momentum $\mathbf{J}$ of magnitude $\sqrt{J(J+1)}\,\hbar$. After a transition region in which an intermediate coupling scheme holds, the heavier nuclei exhibit jj coupling. In this case the $\mathbf{S}_i$ and $\mathbf{L}_i$ of each particle are first coupled to form a $\mathbf{J}_i$ for that particle of magnitude $\sqrt{j(j+1)}\,\hbar$, and the various $\mathbf{J}_i$ then couple together to form the total angular momentum $\mathbf{J}$. The jj coupling scheme holds for the great majority of nuclei.

When an appropriate strength is assumed for the spin-orbit interaction, the energy levels of either class of nucleon fall into the sequence shown in Fig. 12-7. The levels are designated by a prefix equal to the total quantum number n, a letter that indicates l for each particle in that level according to the usual pattern ($s, p, d, f, g, \ldots$ correspond respectively to $l = 0, 1, 2, 3, 4, \ldots$), and a subscript equal to j. The spin-orbit interaction splits each state of given j into $2j + 1$ substates, since there are $2j + 1$ allowed orientations of $\mathbf{J}_i$. Large energy gaps appear in the spacing of the levels at intervals that are consistent with the notion of separate shells. The number of available nuclear states in each nuclear shell is, in ascending order of energy, 2, 6, 12, 8, 22, 32, and 44; hence shells are filled when there are 2, 8, 20, 28, 50, 82, or 126 neutrons or protons in a nucleus.

The shell model is able to account for several nuclear phenomena in addition to magic numbers. Since each energy sublevel can contain two particles (spin up and spin down), only filled sublevels are present when there are even numbers of neutrons and protons in a nucleus ("even-even" nucleus). At the other extreme, a nucleus with odd numbers of neutrons and protons ("odd-odd" nucleus) contains unfilled sublevels for both kinds of particle. The stability we expect to be conferred by filled sublevels is borne out by the fact that 160 stable even-even nuclides are known, as against only four stable odd-odd nuclides. Even the latter are at the lower end of the periodic table: ${}_1H^2$, ${}_3Li^6$, ${}_5B^{10}$, and ${}_7N^{14}$.

Level $n(l)_j$	Nucleons per level $2j + 1$	Nucleons per shell	Total nucleons
$1j_{15/2}$	16		
$3d_{3/2}$	4		
$4s_{1/2}$	2		
$2g_{7/2}$	8		
$1i_{11/2}$	12		
$3d_{5/2}$	6		
$2g_{9/2}$	10		
$1i_{13/2}$	14	44	126
$3p_{1/2}$	2		
$3p_{3/2}$	4		
$2f_{5/2}$	6		
$2f_{7/2}$	8		
$1h_{9/2}$	10		
$1h_{11/2}$	12	32	82
$3s_{1/2}$	2		
$2d_{3/2}$	4		
$2d_{5/2}$	6		
$1g_{7/2}$	8		
$1g_{9/2}$	10	22	50
$2p_{1/2}$	2		
$1f_{5/2}$	6		
$2p_{3/2}$	4		
$1f_{7/2}$	8	8	28
$1d_{3/2}$	4	12	20
$2s_{1/2}$	2		
$1d_{5/2}$	6		
$1p_{1/2}$	2	6	8
$1p_{3/2}$	4		
$1s_{1/2}$	2	2	2

(Energy E increases upward.)

FIGURE 12-7 Sequence of nucleon energy levels according to the shell model (not to scale).

Further evidence in favor of the shell model is its ability to predict total nuclear angular momenta. In even-even nuclei, all of the protons and neutrons should pair off so as to cancel out one another's spin and orbital angular momenta. Thus even-even nuclei have zero nuclear angular momenta, as observed. In even-odd (even Z, odd N) and odd-even (odd Z, even N) nuclei, the half-integral spin of the single "extra" nucleon is combined with the integral angular momentum of the rest of the nucleus for a half-integral total angular momentum, and odd-odd nuclei each have an extra neutron and an extra proton whose half-integral spins yield integral total angular momenta, both of which are experimentally confirmed.

Both the liquid-drop and shell models of the nucleus are, in their very different ways, able to account for much that is known of nuclear behavior. Recently attempts have been made to devise theories that combine the best features of each of these models in a consistent scheme, and partial success has been achieved in the endeavor. The resulting *collective model* includes the possibility of a nucleus rotating as a whole, with excitation energies given by Eq. 9.23. The situation is complicated by the nonspherical shape of all but even-even nuclei and the centrifugal distortion experienced by a rotating nucleus; the detailed theory is consistent with the spacing of excited nuclear levels inferred from the gamma-ray spectra of nuclei and in other ways.

Problems

1. A beam of singly charged ions of $_3Li^6$ with energies of 400 ev enters a uniform magnetic field of flux density 0.08 weber/m^2. The ions move perpendicular to the field direction. Find the radius of their path in the magnetic field. (The $_3Li^6$ nuclear mass is 6.0134 amu.)

2. A beam of singly charged boron ions with energies of 1,000 ev enters a uniform magnetic field of flux density 0.2 weber/m^2. The ions move perpendicular to the field direction. Find the radii of the paths of the $_5B^{10}$ (10.013 amu) and $_5B^{11}$ (11.010 amu) isotopes in the magnetic field.

3. Ordinary boron is a mixture of the $_5B^{10}$ and $_5B^{11}$ isotopes and has a composite atomic weight of 10.82 amu. What percentage of each isotope is present in ordinary boron?

4. Show that the nuclear density of $_1H^1$ is 10^{14} times greater than its atomic density. (Assume the atom to have the radius of the first Bohr orbit.)

5. The binding energy of $_{17}Cl^{35}$ is 298 Mev. Find its mass in amu.

6. The mass of $_{10}Ne^{20}$ is 19.9924 amu. Find its binding energy in Mev.

7. Find the average binding energy per nucleon in $_8O^{16}$ (the mass of the neutral $_8O^{16}$ atom is 15.9949 amu).

8. How much energy is required to remove one proton from $_8O^{16}$? (The mass of the neutral $_7N^{15}$ atom is 15.0001 amu; that of the neutral $_8O^{15}$ atom is 15.0030 amu.)

9. How much energy is required to remove one neutron from $_8O^{16}$?

10. Compare the minimum energy a gamma-ray photon must possess if it is to disintegrate an alpha particle into a triton and a proton with that it must possess if it is to disintegrate an alpha particle into a $_2He^3$ nucleus and a neutron.

11. Show that the electrostatic potential energy of two protons 1.7×10^{-15} m apart is of the correct order of magnitude to account for the difference in binding energy between $_1H^3$ and $_2He^3$. How does this result bear upon the problem of the charge-independence of nuclear forces? (The masses of the neutral atoms are, respectively, 3.016049 and 3.016029 amu.)

12. Protons, neutrons, and electrons all have spins of ½. Why do $_2He^4$ atoms obey Bose-Einstein statistics while $_2He^3$ atoms obey Fermi-Dirac statistics?

13. Calculate the approximate value of a_3 in Eq. 12.30 using whatever assumptions seem appropriate.

13 NUCLEAR DECAY

Perhaps no single phenomenon has played so significant a role in the development of both atomic and nuclear physics as *radioactivity.* A nucleus undergoing radioactive decay spontaneously emits a $_2He^4$ nucleus (alpha particle), an electron (beta particle), or a photon (gamma ray), thereby ridding itself of nuclear excitation energy or achieving a configuration that is or will lead to one of greater stability. In this chapter we are concerned with the various physical processes involved in radioactive decay itself, and not with the many applications of radioactivity in the past and present.

13.1 Statistics of Radioactive Decay

The *activity* of a sample of any radioactive material is the rate at which the nuclei of its constituent atoms decay. If N is the number of nuclei present at a certain time in the sample, its activity R is given by

13.1 $$R = -\frac{dN}{dt}$$

The minus sign is inserted to make R a positive quantity, since dN/dt is, of course, intrinsically negative. While the natural units for activity are disintegrations per second, it is customary to express R in terms of the *curie* and its submultiples, the *millicurie* (mc) and *microcurie* (μc). By definition,

$$\begin{aligned} 1 \text{ curie} &= 3.70 \times 10^{10} \text{ disintegrations/sec} \\ 1 \text{ mc} = 10^{-3} \text{ curie} &= 3.70 \times 10^{7} \text{ disintegrations/sec} \\ 1\ \mu\text{c} = 10^{-6} \text{ curie} &= 3.70 \times 10^{4} \text{ disintegrations/sec} \end{aligned}$$

Experimental measurements on the activities of radioactive samples indicate that, in every case, they fall off exponentially with time. Figure 13-1 is a graph of R versus t for a typical radioisotope. We note that in every 5-hr period, regardless of when the period starts, the activity drops to half of what it was at the start of the period. Accordingly the *half life* $T_{1/2}$ of the isotope is 5 hr. Every radioisotope has a characteristic half life;

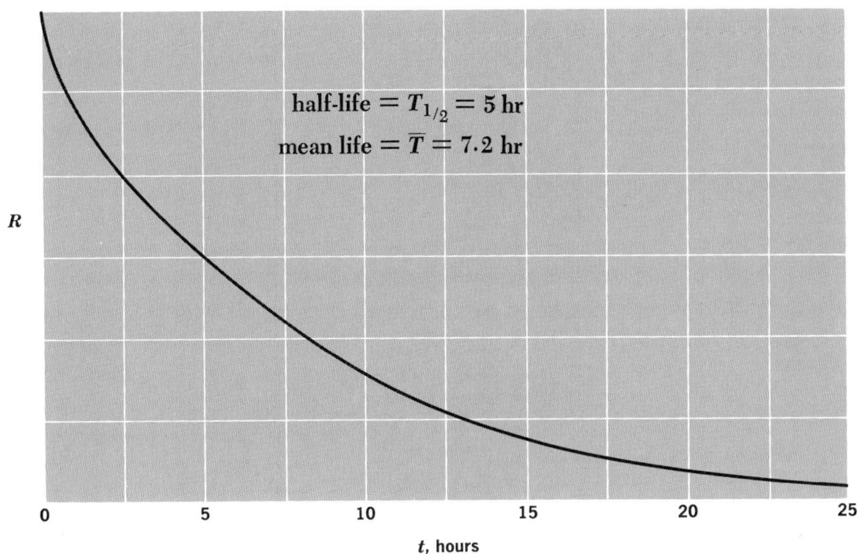

FIGURE 13-1 The activity of a radioisotope decreases exponentially with time.

some have half lives of a millionth of a second, others have half lives that range up to billions of years. When the observations plotted in Fig. 13-1 began, the activity of the sample was R_0. Five hr later it decreased to $0.5R_0$. After another 5 hr, R again decreased by a factor of 2 to $0.25R_0$. That is, the activity of the sample was only 0.25 its initial value after an interval of $2T_{1/2}$. With the lapse of another half life of 5 hr, corresponding to a total interval of $3T$, R became $\frac{1}{2}(0.25R_0)$, or $0.125R_0$.

The behavior illustrated in Fig. 13-1 indicates that we can express our empirical information about the time variation of activity in the form

13.2 $$R = R_0 e^{-\lambda t}$$

where λ, called the *decay constant*, has a different value for each radioisotope. The connection between decay constant λ and half life $T_{1/2}$ is easy to establish. After a half life has elapsed, that is, when $t = T_{1/2}$, the activity R drops to $\frac{1}{2}R_0$ by definition. Hence

$$R = R_0 e^{-\lambda t}$$
$$\tfrac{1}{2}R_0 = R_0 e^{-\lambda T_{1/2}}$$
$$e^{\lambda T_{1/2}} = 2$$

Taking natural logarithms of both sides of this equation,

$$\lambda T_{1/2} = \ln 2$$ **Half life**

13.3 $$T_{1/2} = \frac{\ln 2}{\lambda} = \frac{0.693}{\lambda}$$

The decay constant of the radioisotope whose half life is 5 hr is therefore

$$\lambda = \frac{0.693}{T_{1/2}}$$

$$= \frac{0.693}{5\ \text{hr} \times 3{,}600\ \text{sec/hr}}$$

$$= 3.85 \times 10^{-5}\ \text{sec}^{-1}$$

The fact that radioactive decay follows the exponential law of Eq. 13.2 is strong evidence that this phenomenon is statistical in nature: every nucleus in a sample of radioactive material has a certain probability of decaying, but there is no way of knowing in advance *which* nuclei will actually decay in a particular time span. If the sample is large enough—that is, if many nuclei are present—the actual fraction of it that decays in a certain time span will be very close to the a priori probability for any individual nucleus to decay. The statement that a certain radioisotope has a half life of 5 hr, then, signifies that every nucleus of this isotope has a 50 per cent chance of decaying in any 5-hr period. This does *not* mean a 100 per cent probability of decaying in 10 hr; a nucleus does not have a memory, and its decay probability per unit time is constant until it actually does decay. A half life of 5 hr implies a 75 per cent probability of decay in 10 hr, which increases to 87.5 per cent in 15 hr, to 93.75 per cent in 20 hr, and so on, because in every 5-hr interval the probability is 50 per cent.

The empirical activity law of Eq. 13.2 follows directly from the assumption of a constant probability λ per unit time for the decay of each nucleus of a given isotope. Since λ is the probability per unit time, $\lambda\, dt$ is the probability that any nucleus will undergo decay in a time interval dt. If a sample contains N undecayed nuclei, the number dN that decay in a time dt is the product of the number of nuclei N and the probability $\lambda\, dt$ that each will decay in dt. That is,

13.4 $$dN = -N\lambda\, dt$$

where the minus sign is required because N decreases with increasing t. Equation 13.4 can be rewritten

$$\frac{dN}{N} = -\lambda\, dt$$

and each side can now be integrated:

$$\int_{N_0}^{N} \frac{dN}{N} = -\lambda \int_0^t dt$$

$$\ln N - \ln N_0 = -\lambda t$$

13.5 $$N = N_0 e^{-\lambda t}$$

Equation 13.5 is a formula that gives the number N of undecayed nuclei at the time t in terms of the decay probability per unit time λ of the isotope involved and the number N_0 of undecayed nuclei at $t = 0$.

Since the activity of a radioactive sample is defined as

$$R = -\frac{dN}{dt}$$

we see that, from Eq. 13.5,

$$R = \lambda N_0 e^{-\lambda t}$$

This agrees with the empirical activity law if

$$R_0 = \lambda N_0$$

or, in general, if

13.6 $$R = \lambda N$$

Evidently the decay constant λ of a radioisotope is the same as the probability per unit time for the decay of a nucleus of that isotope.

Equation 13.6 permits us to calculate the activity of a radioisotope sample if we know its mass, atomic weight, and decay constant. As an example, let us determine the activity of a 1-gm sample of $_{38}Sr^{90}$, whose half life against beta decay is 28 years. The decay constant of $_{38}Sr^{90}$ is

$$\begin{aligned}\lambda &= \frac{0.693}{T_{1/2}} \\ &= \frac{0.693}{28 \text{ years} \times 3.16 \times 10^{7} \text{ sec/year}} \\ &= 7.83 \times 10^{-10} \text{ sec}^{-1}\end{aligned}$$

A kmole of an isotope has a mass very nearly equal to the mass number of that isotope expressed in kilograms. Hence 1 gm of $_{38}Sr^{90}$ contains

$$\frac{10^{-3} \text{ kg}}{90 \text{ kg/kmole}} = 1.11 \times 10^{-5} \text{ kmoles}$$

One kmole of any isotope contains Avogadro's number of atoms, and so 1 gm of $_{38}Sr^{90}$ contains

$$1.11 \times 10^{-5} \text{ kmole} \times 6.025 \times 10^{26} \text{ atoms/kmole} = 6.69 \times 10^{21} \text{ atoms}$$

Thus the activity of the sample is

$$\begin{aligned} R &= \lambda N \\ &= 7.83 \times 10^{-10} \times 6.69 \times 10^{21}\ \text{sec}^{-1} \\ &= 5.23 \times 10^{12}\ \text{sec}^{-1} \\ &= 141\ \text{curies} \end{aligned}$$

It is worth keeping in mind that the half life of a radioisotope is not the same as its *mean lifetime* $\overline{T}$. The mean lifetime of an isotope is the reciprocal of its decay probability per unit time:

13.7 $$\overline{T} = \frac{1}{\lambda}$$

Hence

13.8 $$\overline{T} = \frac{1}{\lambda} = \frac{T_{1/2}}{0.693} = 1.44T_{1/2}$$ **Mean lifetime**

$\overline{T}$ is nearly half again more than $T_{1/2}$. The mean lifetime of an isotope whose half life is 5 hr is 7.2 hr.

13.2 Radioactive Series

Most of the radioactive elements found in nature are members of four *radioactive series,* with each series consisting of a succession of daughter products all ultimately derived from a single parent nuclide. The reason that there are exactly four such series follows from the fact that alpha decay reduces the mass number of a nucleus by 4. Thus the nuclides whose mass numbers are all given by

13.9 $$A = 4n$$

where n is an integer, can decay into one another in descending order of mass number. Radioactive nuclides whose mass numbers obey Eq. 13.9 are said to be members of the $4n$ series. The members of the $4n + 1$ series have mass numbers specified by

13.10 $$A = 4n + 1$$

and members of the $4n + 2$ and $4n + 3$ series have mass numbers specified respectively by

13.11 $$A = 4n + 2$$

13.12 $$A = 4n + 3$$

The members of each of these series, too, can decay into one another in descending order of mass number.

Table 13.1

FOUR RADIOACTIVE SERIES

Mass numbers	Series	Parent	Half life, years	Stable end product
$4n$	Thorium	$_{90}Th^{232}$	1.39×10^{10}	$_{82}Pb^{208}$
$4n + 1$	Neptunium	$_{93}Np^{237}$	2.25×10^{6}	$_{83}Bi^{209}$
$4n + 2$	Uranium	$_{92}U^{238}$	4.51×10^{9}	$_{82}Pb^{206}$
$4n + 3$	Actinium	$_{92}U^{235}$	7.07×10^{8}	$_{82}Pb^{207}$

Table 13.1 is a list of the names of four important radioactive series, their parent nuclides and the half lives of these parents, and the stable daughters which are end products of the series. The half life of neptunium is so short compared with the estimated age ($\sim 10^{10}$ years) of the universe that the members of this series are not found in nature today. They have, however, been produced in the laboratory by the neutron bombardment of other heavy nuclei, as we shall describe in the next chapter. The sequences of alpha and beta decays that lead from parent to stable end product in each series are shown in Figs. 13-2 to 13-5. Some nuclides may decay either by beta or alpha emission, so that the decay chain *branches* at them. Thus $_{83}Bi^{212}$, a member of the thorium series, has a 66.3 per cent chance of beta decaying into $_{84}Po^{212}$ and a 33.7 per cent chance of alpha decaying into $_{81}Tl^{208}$. The beta decay is followed by a subsequent alpha decay and the alpha decay is followed by a subsequent beta decay, so that both branches lead to $_{82}Pb^{208}$.

Several alpha-radioactive nuclides whose atomic numbers are less than 82 are found in nature, though they are not very abundant. These nuclides are listed in Table 13.2.

13.3 Alpha Decay

Because the attractive forces between nucleons are of short range, the total binding energy in a nucleus is approximately proportional to its mass number A, the number of nucleons it contains. The repulsive electrostatic forces be-

Table 13.2

LIGHT ALPHA EMITTERS

Nuclide	Half life, years
$_{60}Nd^{144}$	1×10^{15}
$_{62}Sm^{147}$	1.4×10^{11}
$_{64}Gd^{152}$	1.1×10^{14}
$_{72}Hf^{174}$	2×10^{15}
$_{78}Pt^{190}$	6×10^{11}

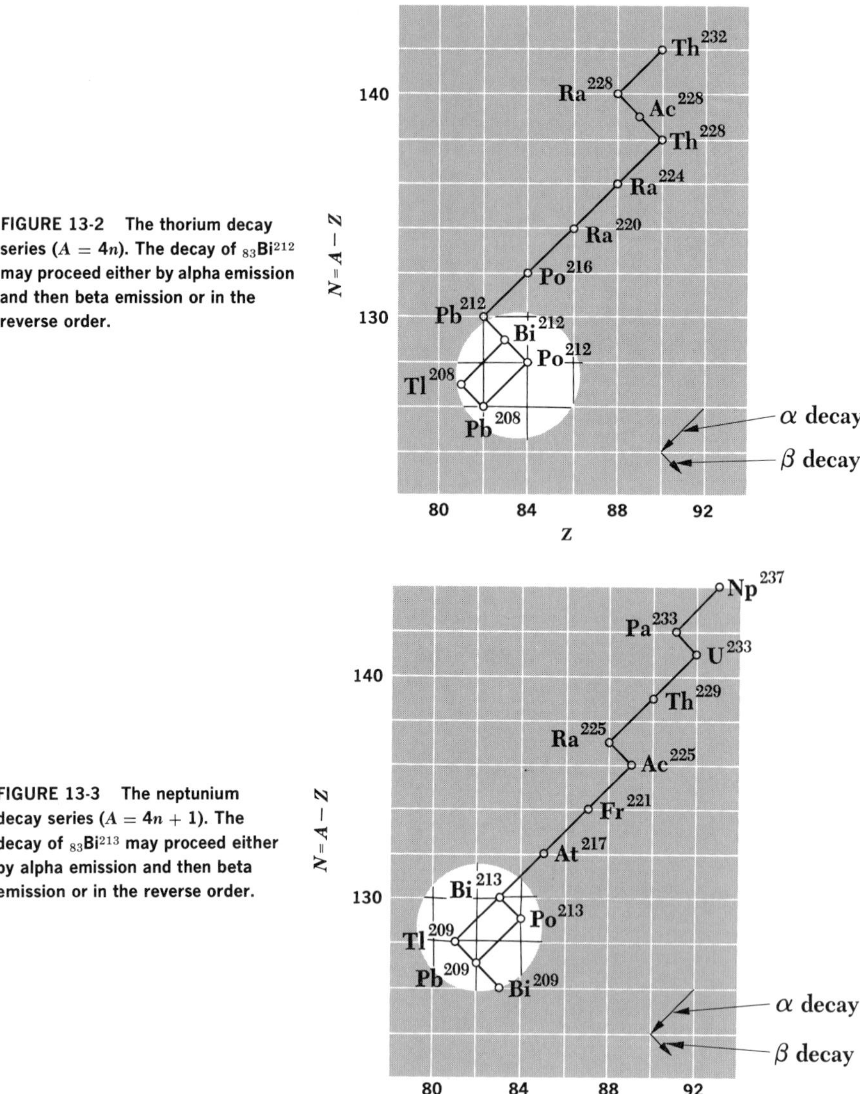

FIGURE 13-2 The thorium decay series ($A = 4n$). The decay of $_{83}Bi^{212}$ may proceed either by alpha emission and then beta emission or in the reverse order.

FIGURE 13-3 The neptunium decay series ($A = 4n + 1$). The decay of $_{83}Bi^{213}$ may proceed either by alpha emission and then beta emission or in the reverse order.

tween protons, however, are of unlimited range, and the total disruptive energy in a nucleus is approximately proportional to Z^2. Nuclei which contain 210 or more nucleons are so large that the short-range nuclear forces that hold them together are barely able to counterbalance the mutual repulsion of their protons. Alpha decay occurs in such nuclei as a means of increasing their stability by reducing their size.

Why are alpha particles invariably emitted rather than, say, individual protons or $_2He^3$ nuclei? The answer follows from the high binding energy of the alpha particle. To escape from a nucleus, a particle must have kinetic energy, and the alpha-particle mass is sufficiently smaller than that of its constituent nucleons for such energy to be available. To illustrate this point, we can compute, from the known masses of each particle and the parent and

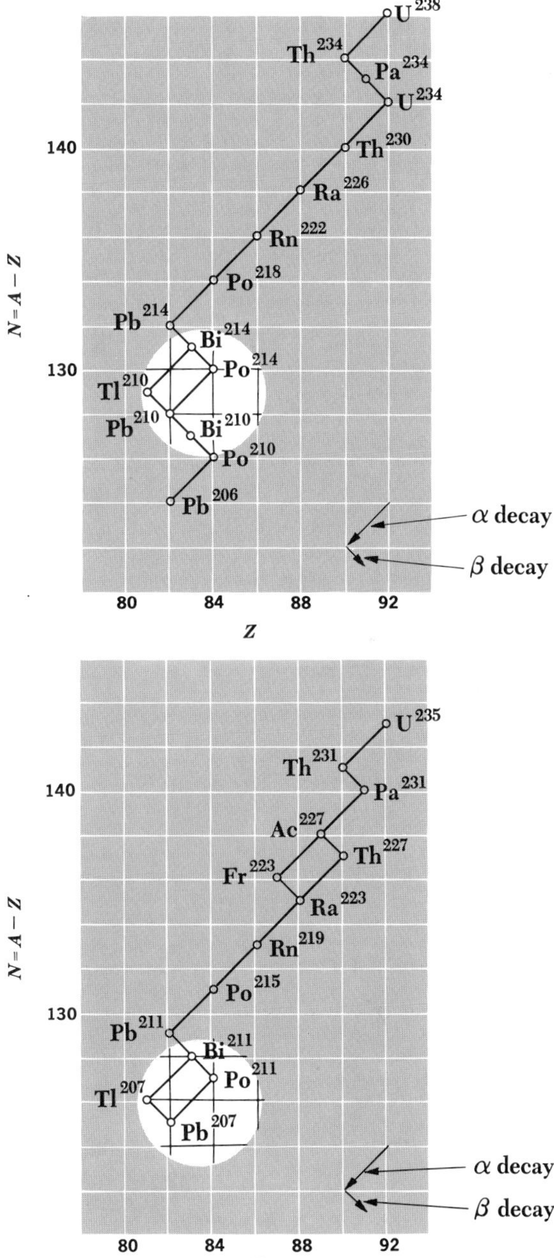

FIGURE 13-4 The uranium decay series ($A = 4n + 2$). The decay of $_{83}Bi^{214}$ may proceed either by alpha emission and then beta emission or in the reverse order.

FIGURE 13-5 The actinium decay series ($A = 4n + 3$). The decays of $_{89}Ac^{227}$ and $_{83}Bi^{211}$ may proceed either by alpha emission and then beta emission or in the reverse order.

daughter nuclei, the kinetic energy Q released when various particles are emitted by a heavy nucleus. This is given by

$$Q = (m_i - m_f - m_\alpha)c^2$$

where m_i is the mass of the initial nucleus, m_f the mass of the final nucleus, and m_α the alpha-particle mass. We find that *only* the emission of an alpha particle is energetically possible; other decay modes would require energy to be supplied from outside the nucleus. Thus alpha decay in ${}_{92}U^{232}$ is accompanied by the release of 5.4 Mev, while 6.1 Mev would somehow have to be furnished if a proton is to be emitted and 9.6 Mev if a ${}_2He^3$ nucleus is to be emitted. The observed disintegration energies in alpha decay agree with the corresponding predicted values based upon the nuclear masses involved.

The kinetic energy T_α of the emitted alpha particle is never quite equal to the disintegration energy Q because, since momentum must be conserved, the nucleus recoils with a small amount of kinetic energy when the alpha particle emerges. It is easy to show that, as a consequence of momentum and energy conservation, T_α is related to Q and the mass number A of the original nucleus by

$$T_\alpha \approx \frac{A-4}{A} Q$$

The mass numbers of nearly all alpha emitters exceed 210, and so most of the disintegration energy appears as the kinetic energy of the alpha particle. In the decay of ${}_{86}Rn^{222}$, $Q = 5.587$ Mev while $T_\alpha = 5.486$ Mev.

FIGURE 13-6 **The potential energy of an alpha particle as a function of its distance from the center of a nucleus.**

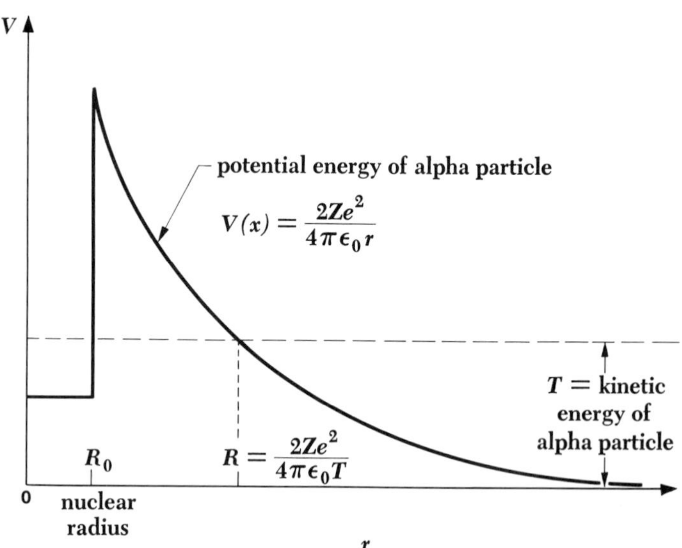

While a heavy nucleus can, in principle, spontaneously reduce its bulk by alpha decay, there remains the problem of *how* an alpha particle can actually escape from the nucleus. Figure 13-6 is a plot of the potential energy V of an alpha particle as a function of its distance r from the center of a certain heavy nucleus. The height of the potential barrier is about 25 Mev, which is equal to the work that must be done against the repulsive electrostatic force to bring an alpha particle from infinity to a position adjacent to the nucleus but just outside the range of its attractive forces. We may therefore regard an alpha particle in such a nucleus as being inside a box whose walls require an energy of 25 Mev to be surmounted. However, decay alpha particles have energies that range from 4 to 9 Mev, depending upon the particular nuclide involved—16 to 21 Mev short of the energy needed for escape!

13.4 Theory of Alpha Decay

While alpha decay is inexplicable on the basis of classical arguments, quantum mechanics provides a straightforward explanation. In fact, the theory of alpha decay developed independently in 1928 by Gamow and by Gurney and Condon was greeted as an especially striking confirmation of quantum mechanics. In this section we shall show how even a simplified treatment of the problem of the escape of an alpha particle from a nucleus gives results in agreement with experiment.

The basic notions of this theory are that an alpha particle may exist as an entity within a heavy nucleus, that such a particle is in constant motion and is contained in the nucleus by the surrounding potential barrier, and that there is a small—but definite—likelihood that the particle may pass through the barrier (despite its height) each time a collision with it occurs. Thus the decay probability per unit time λ can be expressed as

$$\lambda = \nu P$$

where ν is the number of times per second an alpha particle within a nucleus strikes the potential barrier around it and P is the probability that the particle will be transmitted through the barrier. If we suppose that at any moment only one alpha particle exists as such in a nucleus and that it moves back and forth along a nuclear diameter,

$$\nu = \frac{v}{2R}$$

where v is the alpha-particle velocity when it eventually leaves the nucleus and R is the nuclear radius. Typical values for v and R might be 2×10^7 m/sec and 10^{-14} m respectively, so that

$$\nu \simeq 10^{21} \text{ sec}^{-1}$$

The alpha particle knocks at its confining wall 10^{21} times per second and yet may have to wait an average of as much as 10^{10} years to escape from some nuclei!

Since $V > E$, classical physics predicts a transmission probability P of zero. In quantum mechanics a moving alpha particle is regarded as a wave, and the result is a small but definite value for P. The optical analog of this effect is well-known: a light wave undergoing reflection from even a perfect mirror nevertheless penetrates it with an exponentially decreasing amplitude before reversing direction.

Let us consider the case of a beam of particles of kinetic energy T incident from the left on a potential barrier of height V and width L, as in Fig. 13-7. On both sides of the barrier $V = 0$, which means that no forces act upon the particles there. In these regions Schrödinger's equation for the particles is

13.13 $$\frac{\partial^2 \psi_{\mathrm{I}}}{\partial x^2} + \frac{2m}{\hbar^2} E\psi_{\mathrm{I}} = 0$$

and

13.14 $$\frac{\partial^2 \psi_{\mathrm{III}}}{\partial x^2} + \frac{2m}{\hbar^2} E\psi_{\mathrm{III}} = 0$$

Let us assume that

13.15 $$\psi_{\mathrm{I}} = Ae^{iax} + Be^{-iax}$$

13.16 $$\psi_{\mathrm{III}} = Ee^{iax} + Fe^{-iax}$$

are solutions to Eqs. 13.13 and 13.14 respectively. The various terms in these solutions are not hard to interpret. As shown schematically in Fig. 13-8, Ae^{iax} is a wave of amplitude A incident from the left on the barrier. That is,

13.17 $$\psi_{\mathrm{I}+} = Ae^{iax}$$

This wave corresponds to the incident beam of particles in the sense that $\psi_{\mathrm{I}+}\psi_{\mathrm{I}+}^*$ is their probability density. If v is the group velocity of the wave, which equals the particle velocity,

13.18 $$\psi_{\mathrm{I}+}\psi_{\mathrm{I}+}^* \, v = AA^*v$$

is the flux of particles that arrive at the barrier. At $x = 0$ the incident wave strikes the barrier and is partially reflected, with

13.19 $$\psi_{\mathrm{I}-} = Be^{-iax}$$

representing the reflected wave. Hence

13.20 $$\psi_{\mathrm{I}} = \psi_{\mathrm{I}+} + \psi_{\mathrm{I}-}$$

On the far side of the barrier ($x > L$) there can be only a wave

$$\psi_{\mathrm{III}+} = Ee^{iax}$$

traveling in the $+x$ direction, since, by hypothesis, there is nothing in region III that could reflect the wave. Hence

13.21 $$F = 0$$

and

13.22 $$\begin{aligned}\psi_{\mathrm{III}} &= \psi_{\mathrm{III}+} \\ &= Ee^{iax}\end{aligned}$$

By substituting ψ_{I} and ψ_{III} back into their respective differential equations, we find that

$$a = \sqrt{\frac{2mT}{\hbar^2}}$$

It is evident that the transmission probability P for a particle to pass through the barrier is the ratio

13.23 $$P = \frac{\psi_{\mathrm{III}}\psi_{\mathrm{III}}^*}{\psi_{\mathrm{I}}\psi_{\mathrm{I}}^*} = \frac{EE^*}{AA^*}$$

between its probability density in region III and its probability density in region I. Classically $P = 0$ because the particle cannot exist inside the barrier; let us see what the quantum-mechanical result is.

In region II Schrödinger's equation for the particles is

13.24 $$\frac{\partial^2\psi_{\mathrm{II}}}{\partial x^2} + \frac{2m}{\hbar^2}(T - V)\,\psi_{\mathrm{II}} = 0$$

Its solution is

13.25 $$\psi_{\mathrm{II}} = Ce^{ibx} + De^{-ibx}$$

where

13.26 $$b = \sqrt{\frac{2m(T - V)}{\hbar^2}}$$

Since $V > T$, b is imaginary and we may define a new wave number b' by

13.27 $$\begin{aligned}b' &= -ib \\ &= \sqrt{\frac{2m(V - T)}{\hbar^2}}\end{aligned}$$

Hence

13.28 $$\psi_{II} = Ce^{-b'x} + De^{b'x}$$

The term

13.29 $$\psi_{II+} = Ce^{-b'x}$$

is an exponentially decreasing wave function that corresponds to a nonoscillatory disturbance moving to the right through the barrier. At the far end of the barrier (that is, at $x = L$) part of the disturbance is reflected, and

13.30 $$\psi_{II-} = De^{b'x}$$

is an exponentially decreasing wave function that corresponds to the reflected disturbance moving to the left.

Even though ψ_{II} does not oscillate, and therefore does not represent a moving particle of positive kinetic energy, the probability density $\psi_{II}\psi_{II}^*$ is not zero. There is a finite probability of finding a particle within the barrier. A particle at the far end of the barrier that is not reflected there will emerge into region III with the same kinetic energy T it originally had, and its wave function will be ψ_{III} as it continues moving unimpeded in the $+x$ direction. In the limit of an infinitely thick barrier, $\psi_{III} = 0$, which implies that all of the incident particles are reflected. The reflection process takes place *within* the barrier, however, not at its surface, and a barrier of finite width therefore permits a fraction P of the initial beam to pass through it.

In order to calculate P, we must apply certain boundary conditions to ψ_I, ψ_{II}, and ψ_{III}. Figure 13-7 is a schematic representation of the wave functions in regions I, II, and III which may help in visualizing the boundary

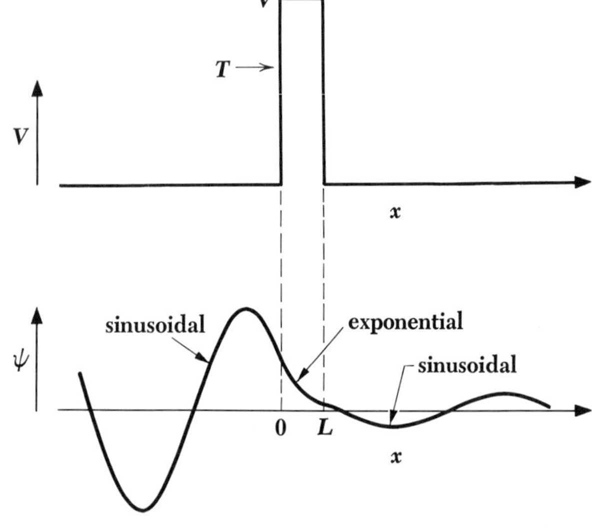

FIGURE 13-7 A beam of particles can "leak" through a finite barrier.

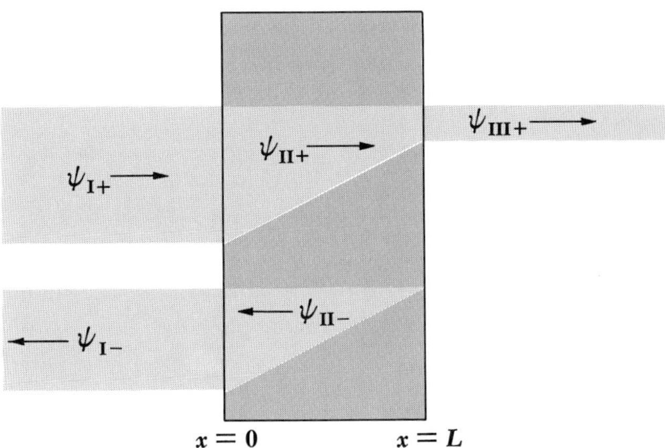

FIGURE 13-8 Schematic representation of barrier penetration.

conditions. As stated in Chap. 6, both ψ and its derivative $\partial\psi/\partial x$ must be continuous everywhere. With reference to Fig. 13-7, these conditions mean that, at each wall of the barrier, the wave functions inside and outside must not only have the same value but also the same slope, so that they match up perfectly. Hence at the left-hand wall of the barrier

13.31
$$\left.\begin{aligned} \psi_I &= \psi_{II} \\ \frac{\partial \psi_I}{\partial x} &= \frac{\partial \psi_{II}}{\partial x} \end{aligned}\right\} \qquad x = 0$$

and at the right-hand wall

13.32
$$\left.\begin{aligned} \psi_{II} &= \psi_{III} \\ \frac{\partial \psi_{II}}{\partial x} &= \frac{\partial \psi_{III}}{\partial x} \end{aligned}\right\} \qquad x = L$$

Substituting ψ_I, ψ_{II}, and ψ_{III} from Eqs. 13.15, 13.28, and 13.22 into the above equations yields

13.33
$$A + B = C + D$$

13.34
$$iaA - iaB = -b'C + b'D$$

13.35
$$Ce^{-b'L} + De^{b'L} = Ee^{iaL}$$

13.36
$$-aCe^{-b'L} + aDe^{b'L} = iaEe^{iaL}$$

Equations 13.33 to 13.36 may be readily solved to yield

13.37
$$\left(\frac{A}{E}\right) = \left[\frac{1}{2} + \frac{i}{4}\left(\frac{b'}{a} - \frac{a}{b'}\right)\right] e^{(ia+b')L} + \left[\frac{1}{2} - \frac{i}{4}\left(\frac{b'}{a} - \frac{a}{b'}\right)\right] e^{(ia-b')L}$$

The complex conjugate of A/E, which we require to compute the transmission probability P, is found by replacing i by $-i$ wherever it occurs in A/E:

13.38 $$\left(\frac{A}{E}\right)^* = \left[\frac{1}{2} - \frac{i}{4}\left(\frac{b'}{a} - \frac{a}{b'}\right)\right] e^{(-ia+b')L} + \left[\frac{1}{2} + \frac{i}{4}\left(\frac{b'}{a} - \frac{a}{b'}\right)\right] e^{(-ia-b')L}$$

Let us assume that the potential barrier is high relative to the kinetic energy of an incident particle; this means that $b' > a$ and

13.39 $$\left(\frac{b'}{a} - \frac{a}{b'}\right) \approx \frac{b'}{a}$$

Let us also assume that the barrier is wide enough for ψ_{II} to be severely attenuated between $x = 0$ and $x = L$; this means that $b'L \gg 1$ and

13.40 $$e^{b'L} \gg e^{-b'L}$$

Hence Eqs. 13.37 and 13.38 may be approximated by

13.41 $$\left(\frac{A}{E}\right) = \left(\frac{1}{2} + \frac{ib'}{4a}\right) e^{(ia+b')L}$$

and

13.42 $$\left(\frac{A}{E}\right)^* = \left(\frac{1}{2} - \frac{ib'}{4a}\right) e^{(-ia+b')L}$$

Multiplying (A/E) and $(A/E)^*$ yields

$$\left(\frac{A}{E}\right)\left(\frac{A}{E}\right)^* = \left(\frac{1}{4} + \frac{b'^2}{16a^2}\right) e^{2b'L}$$

and so the transmission probability P is

13.43 $$P = \frac{EE^*}{AA^*} = \left[\left(\frac{A}{E}\right)\left(\frac{A}{E}\right)^*\right]^{-1} = \left(\frac{16}{4 + (b'/a)^2}\right) e^{-2b'L}$$

Since

$$\left(\frac{b'}{a}\right)^2 = \frac{V}{T} - 1$$

the variation in the coefficient of the exponential of Eq. 13.43 with T and V is negligible compared with the variation in the exponential itself. The coefficient, furthermore, is never far from unity, and so

13.44 $$P \approx e^{-2b'L}$$

is a good approximation for the transmission probability. We shall find it convenient to write Eq. 13.44 as

13.45 $$\ln P = -2b'L$$

Equation 13.45 is derived for a rectangular potential barrier, while an alpha particle inside a nucleus is faced with a barrier of varying height, as in Fig. 13-6. We must therefore replace $\ln P = -2b'L$ by

13.46 $$\ln P = -2\int_0^L b'(x)\,dx = -2\int_{R_0}^{R} b'(x)\,dx$$

where R_0 is the radius of the nucleus and R the distance from its center at which $V = T$. Beyond R the kinetic energy of the alpha particle is positive, and it is able to move freely (Fig. 13-9). Since

$$V(x) = \frac{2Ze^2}{4\pi\varepsilon_0 x}$$

FIGURE 13-9 Alpha decay from the point of view of wave mechanics.

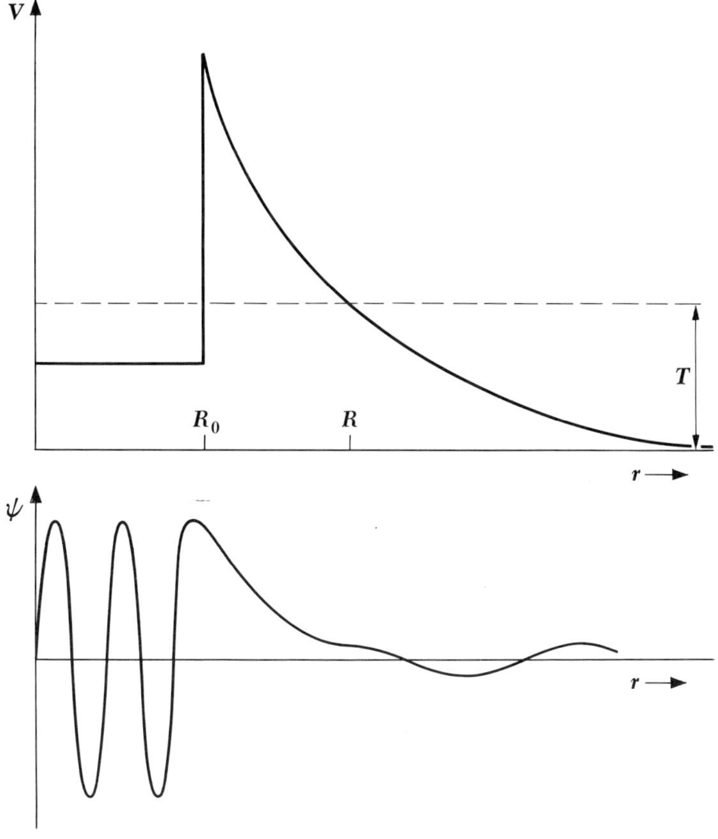

is the electrostatic potential energy of an alpha particle at a distance x from the center of a nucleus of charge Ze (that is, Ze is the nuclear charge *minus* the alpha particle charge of $2e$),

$$b' = \sqrt{\frac{2m(V-T)}{\hbar^2}}$$

$$= \left(\frac{2m}{\hbar^2}\right)^{1/2}\left(\frac{2Ze^2}{4\pi\varepsilon_0 x} - T\right)^{1/2}$$

and, since $T = V(R)$,

$$b' = \left(\frac{2mT}{\hbar^2}\right)^{1/2}\left(\frac{R}{x} - 1\right)^{1/2}$$

Hence

$$\ln P = -2\int_{R_0}^{R} b'(x)\,dx$$

$$= -2\left(\frac{2mT}{\hbar^2}\right)^{1/2}\int_{R_0}^{R}\left(\frac{R}{x} - 1\right)^{1/2} dx$$

13.47 $$= -2\left(\frac{2mT}{\hbar^2}\right)^{1/2} R\left[\cos^{-1}\left(\frac{R_0}{R}\right)^{1/2} - \left(\frac{R_0}{R}\right)^{1/2}\left(1 - \frac{R_0}{R}\right)^{1/2}\right]$$

Because the potential barrier is relatively wide, $R \gg R_0$, and

$$\cos^{-1}\frac{R_0}{R}^{1/2} \approx \frac{\pi}{2} - \left(\frac{R_0}{R}\right)^{1/2}$$

$$\left(1 - \frac{R_0}{R}^{1/2}\right) \approx 1$$

with the result that

$$\ln P = -2\left(\frac{2mT}{\hbar^2}\right)^{1/2} R\left[\frac{\pi}{2} - 2\left(\frac{R_0}{R}\right)^{1/2}\right]$$

Replacing R by

$$R = \frac{2Ze^2}{4\pi\varepsilon_0 T}$$

we obtain

$$\ln P = \frac{4e}{\hbar}\left(\frac{m}{\pi\varepsilon_0}\right)^{1/2} Z^{1/2} R_0^{1/2}$$

13.48 $$- \frac{e^2}{\hbar\varepsilon_0}\left(\frac{m}{2}\right)^{1/2} Z T^{-1/2}$$

The result of evaluating the various constants in Eq. 13.48 is

$$\ln P = 2.97Z^{1/2}R_0^{1/2} - 3.95ZT^{-1/2}$$

where T (the alpha particle kinetic energy) is expressed in Mev, R_0 (the nuclear radius) is expressed in units of 10^{-15} m, and Z is the atomic number of the nucleus minus the alpha particle. The decay constant λ, given by

$$\lambda = \nu P$$
$$= \frac{v}{2R} P$$

may therefore be written

13.49 $$\ln \lambda = \ln\left(\frac{v}{2R_0}\right) + 2.97Z^{1/2}R_0^{1/2} - 3.95ZT^{-1/2}$$ **Alpha decay**

Figure 13-10 is a plot of $\log_{10} \lambda$ versus $ZT^{-1/2}$ for a number of alpha-radioactive nuclides. The straight line fitted to the experimental data has the slope predicted by the theory throughout the entire range of decay constants. We can use the position of the line to determine R_0, the nuclear

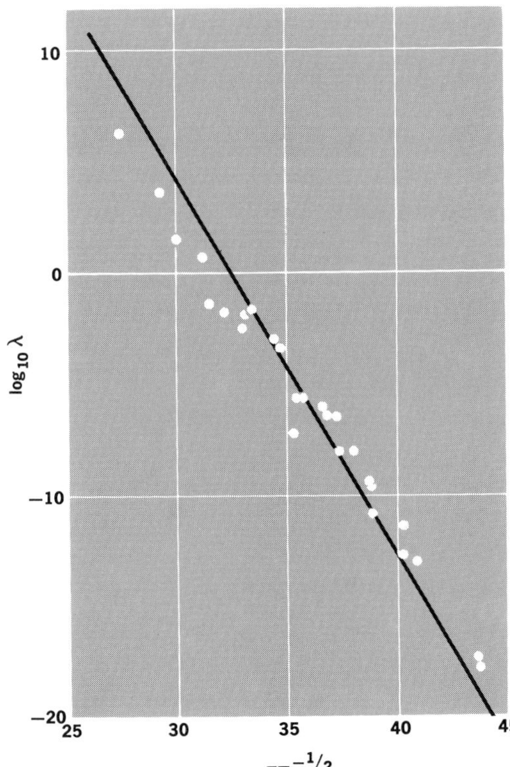

FIGURE 13-10 Experimental verification of the theory of alpha decay. Since $\log_{10} \lambda = \log_{10} e \times \ln \lambda = 0.4343 \ln \lambda$ the slope of the line in the figure ought to be 0.4343×-3.95 or -1.72, which it is.

radius. The result is just about what is obtained from nuclear scattering experiments like that of Rutherford, namely, $\sim 10^{-14}$ m, an independent means for determining nuclear sizes.

The quantum-mechanical analysis of alpha-particle emission, which is in complete accord with the observed data, is significant on two grounds. First, it makes understandable the enormous variation in half life with disintegration energy. The slowest decay is that of $_{90}Th^{232}$, whose half life is 1.3×10^{10} years, and the fastest decay is that of $_{84}Po^{212}$, whose half life is 3.0×10^{-7} sec. While its half life is 10^{24} greater, the distintegration energy of $_{90}Th^{232}$ (4.05 Mev) is only about half that of $_{84}Po^{212}$ (8.95 Mev)—behavior predicted by Eq. 13.49.

The second significant feature of the theory of alpha decay is its explanation of this phenomenon in terms of the penetration of a potential barrier by a particle which does not have enough energy to surmount the barrier. In classical physics such penetration cannot occur: a baseball thrown against the Great Wall of China has, classically, a 0 per cent probability of getting through. In quantum mechanics the probability is not much more than 0 per cent, but it is *not* identically equal to 0.

13.5 Beta Decay

Beta decay, like alpha decay, is a means whereby a nucleus can alter its Z/N ratio to achieve greater stability. Beta decay, however, presents a considerably greater problem to the physicist who seeks to understand natural phenomena. The most obvious difficulty is that in beta decay a nucleus emits an electron, while, as we have seen in the previous chapter, there are strong arguments against the presence of electrons in nuclei. Since beta decay is essentially the spontaneous conversion of a nuclear neutron into a proton and electron, this difficulty is disposed of if we simply assume that the electron leaves the nucleus immediately after its creation. A more serious problem is that observations of beta decay reveal that three conservation principles, those of energy, momentum, and angular momentum, are apparently being violated. We shall consider the latter aspect of beta decay first.

A device for determining the energies of the electrons emitted in beta decay is called a *beta-ray spectrometer.* A simple beta-ray spectrometer is shown schematically in Fig. 13-11. If r is the fixed radius of curvature and B the magnetic flux density, the electron momentum p is given by

$$p = eBr$$

where e is the electronic charge. In practice the magnetic field is varied with

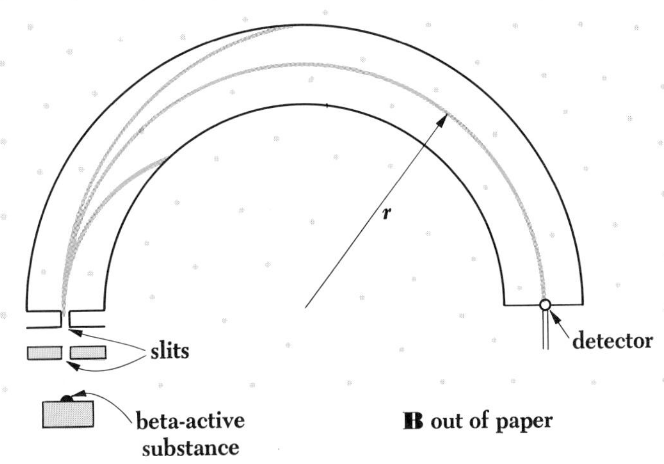

FIGURE 13-11 A beta-ray spectrometer.

a beta-radioactive nuclide as the electron source, and the relative momentum distribution determined. The electrons emitted in beta decay often have kinetic energies comparable with the rest energy of the electron, so the relativistic formula

$$T = \sqrt{m_0{}^2c^4 + p^2c^2} - m_0c^2$$

must be used to convert the observed momenta to kinetic energies.

The electron energies observed in the beta decay of a particular nuclide are found to vary *continuously* from 0 to a maximum value T_{max} characteristic of the nuclide. Figure 13-12 shows the energy spectrum of the electrons emitted in the beta decay of $_{83}Bi^{210}$; here $T_{max} = 1.17$ Mev. In every case the maximum energy

$$E_{max} = m_0c^2 + T_{max}$$

carried off by the decay electron is equal to the energy equivalent of the mass difference between the parent and daughter nuclei. Only seldom, however, is an emitted electron found with an energy of T_{max}.

It was suspected at one time that the "missing" energy was lost during collisions between the emitted electron and the atomic electrons surrounding the nucleus. An experiment first performed in 1927 showed that this hypothesis is not correct. In the experiment a sample of a beta-radioactive nuclide is placed in a calorimeter, and the heat evolved after a given number of decays is measured. The evolved heat divided by the number of decays gives the average energy per decay. In the case of $_{83}Bi^{210}$ the average evolved energy was found to be 0.35 Mev, which is very close to the 0.39-Mev average of

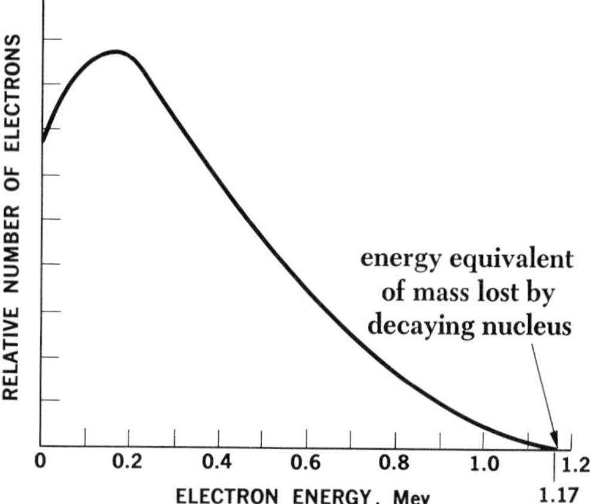

FIGURE 13-12 Energy spectrum of electrons from the beta decay of $_{83}Bi^{212}$.

the spectrum in Fig. 13-12 but far away indeed from the T_{max} value of 1.17 Mev. The conclusion is that the observed continuous spectra represent the actual energy distributions of the electrons emitted by beta-radioactive nuclei.

Linear and angular momenta are also found not to be conserved in beta decay. In the beta decay of certain nuclides the directions of the emitted electrons and of the recoiling nuclei can be observed; they are almost never exactly opposite as required for momentum conservation. The nonconservation of angular momentum follows from the known spins of ½ of the electron, proton, and neutron. Beta decay involves the conversion of a nuclear neutron into a proton:

$$n \rightarrow p + e^-$$

Since the spin of each of the particles involved is ½, this reaction cannot take place if spin (and hence angular momentum) is to be conserved.

13.6 The Neutrino

In 1930 Pauli proposed that if an uncharged particle of small or zero mass and spin ½ is emitted in beta decay together with the electron, the energy, momentum, and angular-momentum discrepancies discussed above would be removed. It was supposed that this particle, later christened the *neutrino*, carries off an energy equal to the difference between T_{max} and the actual electron kinetic energy (the recoil nucleus carries away negligible kinetic energy) and, in so doing, has a momentum exactly balancing those of the electron and the recoiling daughter nucleus. Subsequently it was found that there are *two* kinds of neutrino involved in beta decay, the neutrino itself (symbol ν) and the antineutrino (symbol $\bar{\nu}$). We shall discuss the distinction between them in Chap. 15. In ordinary beta decay it is an antineutrino that is emitted:

13.50 $$n \rightarrow p + e^- + \bar{\nu}$$ **Beta decay**

The neutrino hypothesis has turned out to be completely successful. The neutrino mass was not expected to be more than a small fraction of the electron mass because T_{max} is observed to be equal (within experimental error) to the value calculated from the parent-daughter mass difference; the neutrino mass is now believed to be zero. The reason neutrinos were not experimentally detected until recently is that their interaction with matter is extremely feeble. Lacking charge and mass, and not electromagnetic in nature as is the photon, the neutrino can pass unimpeded through vast amounts of matter. A neutrino would have to pass through 130 *light-years* of solid iron on the average before interacting! The only interaction with matter a neutrino can experience is through a process called inverse beta decay, which we shall consider shortly.

13.7 Positron Emission and Electron Capture

Positive electrons, usually called *positrons,* were discovered in 1932 and two years later were found to be spontaneously emitted by certain nuclei. The properties of the positron are identical with those of the electron except that it carries a charge of $+e$ instead of $-e$. Positron emission corresponds to the conversion of a nuclear proton into a neutron, a positron, and a neutrino:

13.51 $$p \rightarrow n + e^+ + \nu$$ **Positron emission**

While a neutron outside a nucleus can undergo negative beta decay into a proton because its mass is greater than that of the proton, the lighter proton cannot be transformed into a neutron except within a nucleus.

Positron emission leads to a daughter nucleus of lower atomic number Z while leaving the mass number A uncharged. Thus negative and positive beta decays may be represented as

13.52 $${}_ZX^A \rightarrow {}_{Z+1}X^A + e^- + \bar{\nu}$$

13.53 $${}_ZX^A \rightarrow {}_{Z-1}X^A + e^+ + \nu$$

Positron emission resembles electron emission in all respects except for the shapes of their respective energy spectra. The difference is most pronounced at the low-energy ends, as can be seen by comparing the spectrum of positrons from ${}_{29}Cu^{64}$ shown in Fig. 13-13 with the spectrum of electrons from ${}_{83}Bi^{210}$ shown in Fig. 13-12. There are many low-energy electrons, while

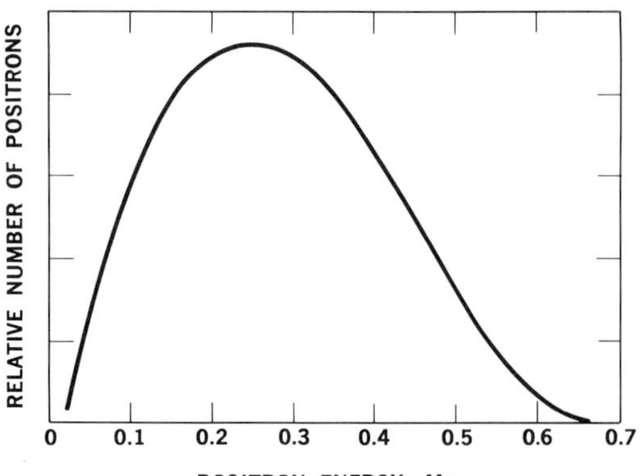

FIGURE 13-13 Energy spectrum of positrons from the beta decay of $_{29}Cu^{64}$.

there are very few low-energy positrons. The reason for this difference is not hard to find. An electron leaving a nucleus is attracted by the positive nuclear charge, which causes it to slow down somewhat. A positron leaving a nucleus, on the other hand, is repelled by the positive nuclear charge and is accordingly accelerated outward. Thus the average electron energy in beta decay is about $0.3T_{\text{max}}$, while the average positron energy is about $0.4T_{\text{max}}$.

Closely connected with positron emission is the phenomenon of *electron capture*. In electron capture a nucleus absorbs one of its inner orbital electrons, with the result that a nuclear proton becomes a neutron and a neutrino is emitted. Thus the fundamental reaction in electron capture is

13.54 $$p + e^- \rightarrow n + \nu$$ **Electron capture**

Electron capture is competitive with positron emission since both processes lead to the same nuclear transformation. Electron capture occurs more often than positron emission in heavy elements because the electron orbits in such elements have smaller radii; the closer proximity of the electrons promotes their interaction with the nucleus. Since almost the only unstable nuclei found in nature are of high Z, positron emission was not discovered until several decades after electron emission had been established.

13.8 Inverse Beta Decay

The beta decay of a proton within a nucleus follows the scheme

13.55 $$p \rightarrow n + e^+ + \nu$$

Because the absorption of an electron by a nucleus is equivalent to its emission of a positron, the electron capture reaction

13.56 $$p + e^- \rightarrow n + \nu$$

is essentially the same as the beta decay of Eq. 13.55. Similarly, the absorption of an antineutrino is equivalent to the emission of a neutrino, so that the reaction

$$p + \bar{\nu} \rightarrow n + e^+ \tag{13.57}$$

also involves the same physical process as that of Eq. 13.55. This latter reaction, called *inverse beta decay,* is interesting because it provides a method for establishing the actual existence of neutrinos.

Starting in 1953, a series of experiments were begun by F. Reines, C. L. Cowan, and others to detect the immense flux of neutrinos from the beta decays that occur in a nuclear reactor. A tank of water containing a cadmium compound in solution supplied the protons which were to interact with the incident neutrinos. Surrounding the tank were gamma-ray detectors. Immediately after a proton absorbed a neutrino to yield a positron and a neutron, the positron encountered an electron and both were annihilated. The gamma-ray detectors responded to the resulting pair of 0.51-Mev photons. Meanwhile the newly formed neutron migrated through the solution until, after a few microseconds, it was captured by a cadmium nucleus. The new, heavier cadmium nucleus then released about 8 Mev of excitation energy divided among three or four photons, which were picked up by the detectors several microseconds after those from the positron-electron annihilation. In principle, then, the arrival of the above sequence of photons at the detector is a sure sign that the reaction of Eq. 13.57 has occurred. To avoid any uncertainty, the experiment was performed with the reactor alternately on and off, and the expected variation in the frequency of neutrino-capture events was observed. Thus the existence of the neutrino may be regarded as experimentally established.

Inverse beta decay is the sole known means whereby neutrinos and antineutrinos interact with matter:

$$p + \bar{\nu} \rightarrow n + e^+$$
$$n + \nu \rightarrow p + e^-$$

The probability for these reactions is almost vanishingly small; this is why neutrinos are able to traverse freely such vast amounts of matter. Once liberated, neutrinos travel freely through space and matter indefinitely, constituting a kind of independent universe within the universe of other particles.

13.9 Gamma Decay

Nuclei can exist in states of definite energies, just as atoms can. An excited nucleus is denoted by an asterisk after its usual symbol; thus ${}_{38}Sr^{87*}$ refers to ${}_{38}Sr^{87}$ in an excited state. Excited nuclei return to their ground states by

X- and Gamma-ray Absorption

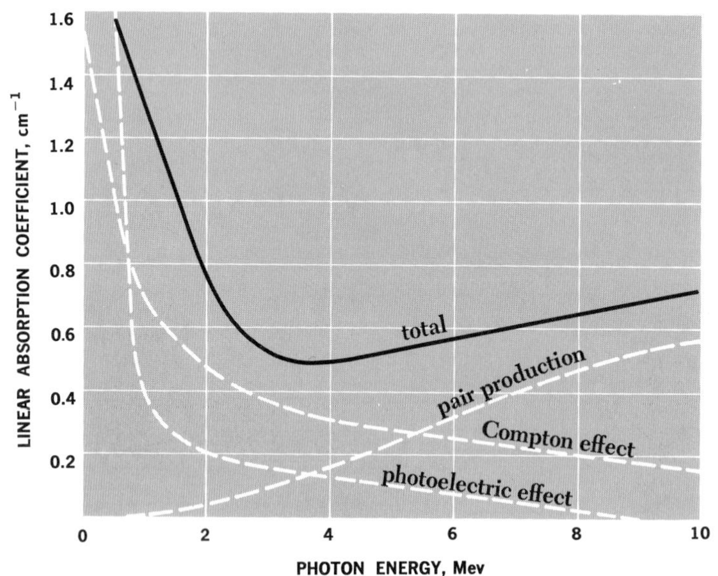

Three processes are responsible for the absorption of X and gamma rays in matter. At low photon energies Compton scattering is the sole mechanism, since there are definite thresholds for both the photoelectric effect (several ev) and electron pair production (1.02 ev). The likelihoods of both Compton scattering and the photoelectric effect decrease with increasing energy, as shown above in the case of a lead absorber, while the likelihood of pair production increases. At high photon energies the dominant mechanism of energy loss is pair production. The curve representing the total absorption in lead has its minimum at about 3.5 Mev. The ordinate of the graph is the linear absorption coefficient μ, which is equal to the ratio between the fractional decrease in radiation intensity $-dI/I$ and the absorber thickness dx. That is,

$$\frac{dI}{I} = -\mu\, dx$$

whose solution is

$$I = I_0 e^{-\mu x}$$

The intensity of the radiation decreases exponentially with the thickness of the absorber.

emitting photons whose energies correspond to the energy differences between the various initial and final states in the transitions involved. The photons emitted by nuclei range in energy up to several Mev, and are traditionally called *gamma rays.* Most excited nuclei have very short half lives against gamma decay, but a few remain excited for as long as several hours. A long-lived excited nucleus is called an *isomer* of the same nucleus in its ground state. The excited nucleus $_{38}Sr^{87*}$ has a half life of 2.8 hr and is accordingly an isomer of $_{38}Sr^{87}$.

As an alternate to gamma decay, an excited nucleus in some cases may return to its ground state by giving up its excitation energy to one of the orbital electrons around it. While we can think of this process, which is known as *internal conversion,* as a kind of photoelectric effect in which a nuclear photon is absorbed by an atomic electron, it is in better accord with experiment to regard internal conversion as representing a direct transfer of excitation energy from a nucleus to an electron. The emitted electron has a kinetic energy equal to the lost nuclear excitation energy minus the binding energy of the electron in the atom.

Problems

1. Tritium ($_1H^3$) has a half life of 12.5 years against beta decay. What fraction of a sample of pure tritium will remain undecayed after 25 years?

2. The half life of $_{11}Na^{24}$ is 15 hr. How long does it take for 93.75 per cent of a sample of this isotope to decay?

3. One g of radium has an activity of 1 curie. From this fact determine the half life of radium.

4. The mass of a millicurie of $_{82}Pb^{214}$ is 3×10^{-14} kg. From this fact find the decay constant of $_{82}Pb^{214}$. (Assume atomic mass in amu equal to mass number in Problems 4, 5, and 6.)

5. The half life of $_{92}U^{238}$ against alpha decay is 4.5×10^9 years. How many disintegrations per second occur in 1 g of $_{92}U^{238}$?

6. The potassium isotope $_{19}K^{40}$ undergoes beta decay with a half life of 1.83×10^9 years. Find the number of beta decays that occur per second in 1 g of pure $_{19}K^{40}$.

7. A 5.78-Mev alpha particle is emitted in the decay of radium. If the diameter of the radium nucleus is 2×10^{-14} m, how many alpha-particle de Broglie wavelengths fit inside the nucleus?

8. The isotope $_{89}Ac^{225}$ undergoes three alpha decays in succession. Why does it not simply emit a $_6C^{12}$ nucleus, as is energetically possible?

9. The isotope ${}_{6}C^{11}$ decays into ${}_{5}B^{11}$. What kind of particle is emitted?

10. The isotope ${}_{92}U^{239}$ undergoes two successive beta decays. What is the resulting isotope?

11. The isotope ${}_{92}U^{238}$ successively undergoes eight alpha decays and six beta decays. What is the resulting isotope?

12. Identify the nuclei that result from the negative beta decay of ${}_{79}Au^{198}$, ${}_{29}Cu^{66}$, and ${}_{15}P^{32}$.

13. Identify the nuclei that result from the positive beta decay of ${}_{48}Cd^{107}$, ${}_{19}K^{38}$, and ${}_{51}Sb^{120}$.

14. Calculate the maximum energy of the electrons emitted in the beta decay of ${}_{5}B^{12}$.

15. Why does ${}_{4}Be^{7}$ invariably decay by electron capture instead of by positron emission?

16. How much energy must a neutrino have if it is to react with a proton to yield a neutron and a positron?

17. The thickness of an absorbing slab that reduces the intensity of a gamma-ray beam by half is called the half-value thickness, $x_{1/2}$, of the material. Find $x_{1/2}$ in terms of μ, the linear absorption coefficient.

NUCLEAR REACTIONS 14

Nuclear reactions, like chemical reactions, provide both information and means for utilizing this information in a practical way. Our basic knowledge of nuclei has come from the study of their interactions with other nuclei and with such bombarding particles as electrons and neutrons, and fission and fusion reactions are the means whereby the vast reservoir of nuclear energy has been broached. We shall begin our examination of nuclear reactions with descriptions of the center-of-mass coordinate system and the idea of interaction cross section, concepts that are required in order to make sense of the experimental data, and then go on to examine some specific reactions.

14.1 Center-of-mass Coordinate System

Most of what we know about atomic nuclei and their interactions has come from experiments in which energetic bombarding particles collide with stationary target nuclei. The analysis of such collisions is greatly simplified by the use of a coordinate system that is moving with the center of mass of the colliding bodies, rather than the usual coordinate system that is fixed in the laboratory.

We shall restrict ourselves to the case of a particle of mass m_1 moving in the $+x$ direction with the velocity v (where $v \ll c$) which strikes another particle, initially at rest, whose mass is m_2 (Fig. 14-1). This corresponds to the situation usually found in nuclear physics. When particle 1 is at x_1 and particle 2 at x_2, the position X of the center of mass of the two particles is defined by the equation

14.1 $$(m_1 + m_2)X = m_1x_1 + m_2x_2$$ **Center-of-mass position**

Differentiating with respect to time,

14.2 $$(m_1 + m_2)\frac{dX}{dt} = m_1\frac{dx_1}{dt} + m_2\frac{dx_2}{dt}$$

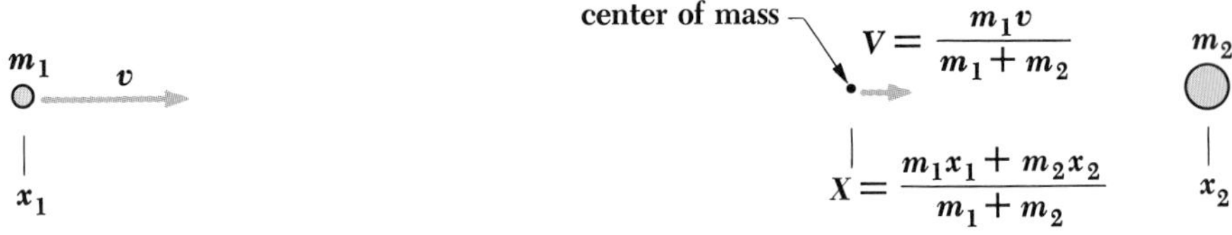

FIGURE 14-1 **Motion in the laboratory coordinate system before collision, where $m_2 = 2m_1$.**

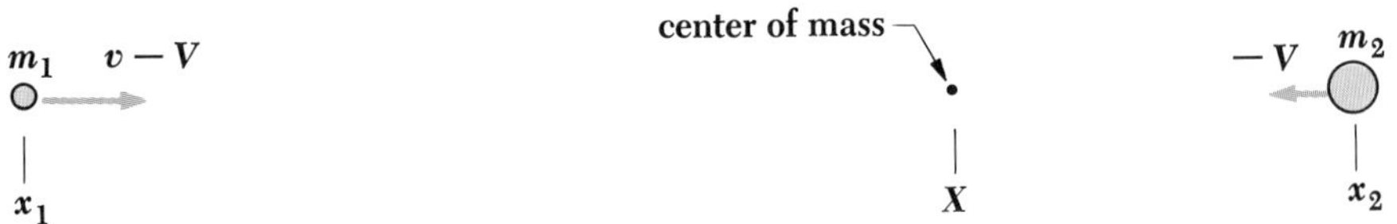

FIGURE 14-2 **Motion in the center-of-mass coordinate system before collision, where $m_2 = 2m_1$.**

No variation in mass is involved since only nonrelativistic velocities are being considered. In the present case

$$\frac{dx_1}{dt} = v$$

$$\frac{dx_2}{dt} = 0$$

and the velocity V of the center of mass is

$$V = \frac{dX}{dt}$$

Hence Eq. 14.2 becomes

$$(m_1 + m_2)V = m_1 v$$

and

14.3 $$V = \left(\frac{m_1}{m_1 + m_2}\right) v$$ **Center-of-mass velocity**

The center of mass moves in the same direction as v, but its speed is smaller by the factor $m_1/(m_1 + m_2)$.

Let us examine the two particles from the point of view of an observer located at the center of mass and moving with it (Fig. 14-2). This observer sees particle 1 approaching *from the left* at the velocity $v - V$ relative to him, while particle 2 approaches *from the right* at the velocity $-V$. In the center-of-mass coordinate system the total momentum of the two particles is therefore

$$m_1(v - V) - m_2 V = m_1 v - m_1\left(\frac{m_1}{m_1 + m_2}\right) v - m_2\left(\frac{m_1}{m_1 + m_2}\right) v$$
$$= 0$$

If the total momentum is 0 before the collision, it must be 0 afterward as well. Regardless of the nature of the interaction between the particles during the actual collision, they must move apart (in the center-of-mass system) with equal and opposite momenta. In the event of a glancing rather than head-on collision the particles will no longer move along the x axis, as in Fig. 14-3, but they must nevertheless have equal and opposite momenta for the total momentum to remain 0.

In the center-of-mass system the total kinetic energy T' also is unlike that in the laboratory system. In the center-of-mass system

$$\begin{aligned} T' &= \tfrac{1}{2}m_1(v - V)^2 + \tfrac{1}{2}m_2 V^2 \\ &= \tfrac{1}{2}m_1 v^2 - \tfrac{1}{2}(m_1 + m_2)V^2 \end{aligned}$$

while the total kinetic energy T in the laboratory system is

14.4 $$T = \tfrac{1}{2}m_1 v^2$$ **Laboratory system**

Therefore

14.5 $$T' = T - \tfrac{1}{2}(m_1 + m_2)V^2$$ **Center-of-mass system**

The total kinetic energy of the particles relative to the center of mass is their total kinetic energy in the laboratory system minus the kinetic energy $\tfrac{1}{2}(m_1 + m_2)V^2$ of the moving center of mass. It is appropriate to regard T' as the kinetic energy of the relative motion of the particles. The ratio between T' and T is

14.6 $$\frac{T'}{T} = \frac{m_2}{m_1 + m_2}$$

An elastic collision between two particles is one in which there is no loss of kinetic energy, although the distribution of kinetic energy between the par-

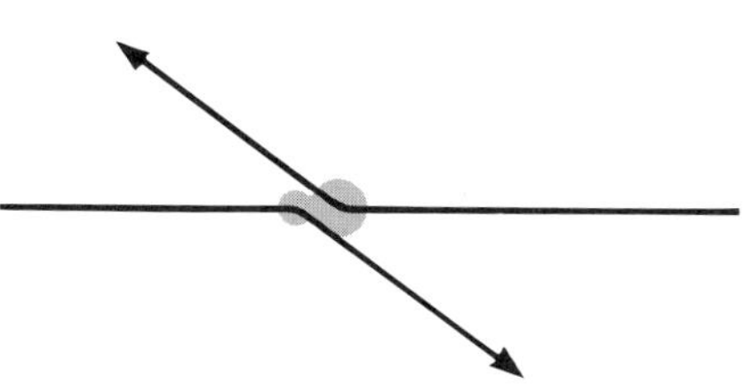

FIGURE 14-3 Head-on (top) and glancing (bottom) collisions as seen in center-of-mass coordinate system.

ticles may change. In an elastic collision as viewed from the center-of-mass system, the initial and final total momenta are 0, and the sum of the initial kinetic energies of the particles is the same as the sum of their final kinetic energies. To satisfy the first condition, as we noted above, the particles must recede from the center of mass after impact in exactly opposite directions along the same line of motion. Regardless of whether this line of motion is the same as the original line of motion, it must be true that

14.7 $$m_1v_1' = m_2v_2'$$

where v_1' and v_2' are the particle speeds after the collision. The conservation of kinetic energy requires that

14.8 $$\tfrac{1}{2}m_1(v - V)^2 + \tfrac{1}{2}m_2V^2 = \tfrac{1}{2}m_1v_1'^2 + \tfrac{1}{2}m_2v_2'^2$$

The only solution that satisfies both Eqs. 14.7 and 14.8 is

14.9 $$v_1' = v - V$$

14.10 $$v_2' = V$$

The particles recede from their center of mass after colliding with the *same speeds* with which they approached it before the collision.

In an inelastic collision, part of the initial kinetic energy is dissipated in some manner. In an inelastic collision of macroscopic bodies, the lost kinetic energy usually reappears as heat and sound. When a nucleus experiences an inelastic collision, the lost kinetic energy usually becomes excitation energy that is subsequently given off as gamma rays. If enough energy is involved, mesons may be emitted, as we shall learn in the next chapter. Analyzing a collision in center-of-mass coordinates makes it easy to determine the maximum amount of kinetic energy that can be dissipated while still conserving momentum. As shown in Fig. 14-4, a collision in which both particles lose all of their kinetic energy relative to the center of mass can still conserve momentum. Hence the maximum kinetic-energy decrease in a collision is T',

FIGURE 14-4 A completely inelastic collision as seen in laboratory and center-of-mass coordinate systems.

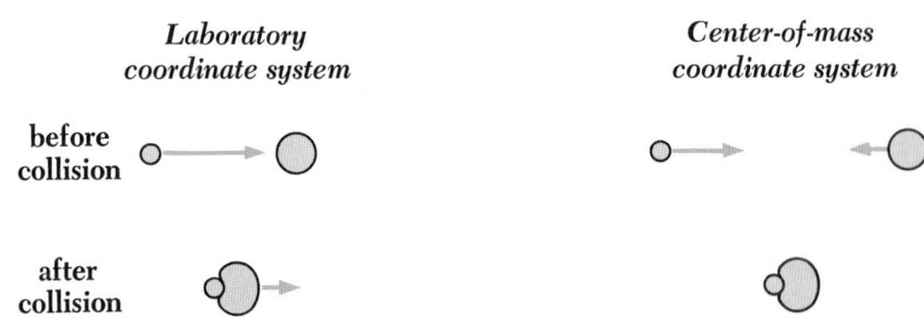

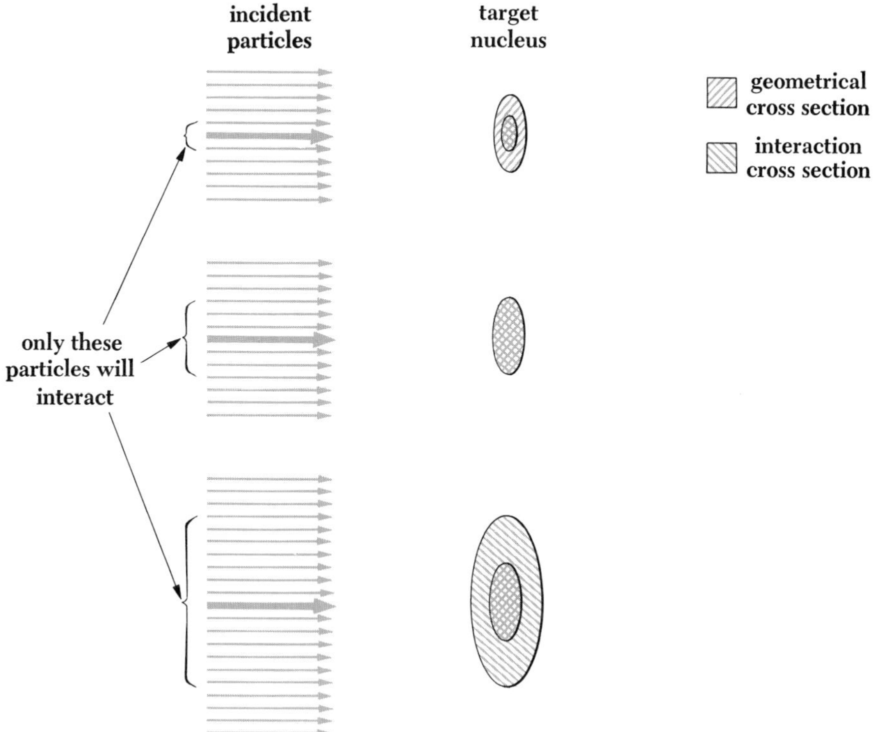

FIGURE 14-5 The concept of cross section.

the total initial kinetic energy in the center-of-mass system, which is related to the total initial kinetic energy T in the laboratory system by

$$T' = \left(\frac{m_2}{m_1 + m_2}\right) T \tag{14.11}$$

14.2 Cross Section

A very convenient way to express the probability that a bombarding particle will interact in a certain way with a target particle employs the idea of *cross section* that was introduced in Chap. 4 in connection with the Rutherford scattering experiment. What we do is visualize each target particle as presenting a certain area, called its cross section, to the incident particles, as in Fig. 14-5. Any incident particle that is directed at this area interacts with the target particle. Hence the greater the cross section, the greater the likelihood of an interaction. The interaction cross section of a target particle varies with the nature of the process involved and with the energy of the incident particle; it may be greater or less than the geometrical cross section of the particle.

Suppose we have a slab of some material whose area is A and whose thickness is dx (Fig. 14-6). If the material contains n atoms per unit volume, there are a total of $nAdx$ nuclei in the slab, since its volume is Adx. Each nucleus has a cross section of σ for some particular interaction, so that the aggregate cross section of all of the nuclei in the slab is $nA\sigma\, dx$. If there are N incident particles in a bombarding beam, the number dN that interact with nuclei in the slab is therefore specified by

$$\frac{\text{Interacting particles}}{\text{incident particles}} = \frac{\text{aggregate cross section}}{\text{target area}}$$

$$\frac{dN}{N} = \frac{nA\sigma\, dx}{A}$$

14.12 $$= n\sigma\, dx$$ **Cross section**

Equation 14.12 is valid only for a slab of infinitesimal thickness. To find the proportion of incident particles that interact with nuclei in a slab of finite thickness, we must integrate dN/N. If we assume that each incident particle is capable of only a single interaction, dN particles may be thought of as being removed from the beam in passing through the first dx of the slab. Hence we must introduce a minus sign in Eq. 14.12, which becomes

14.13 $$-\frac{dN}{N} = n\sigma\, dx$$

FIGURE 14-6 **The relationship between cross section and beam intensity.**

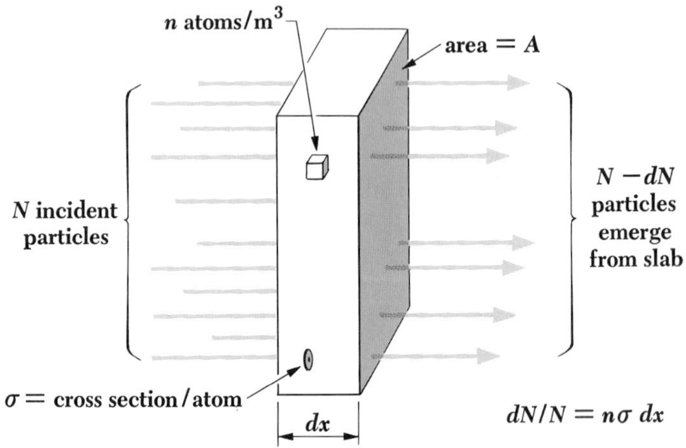

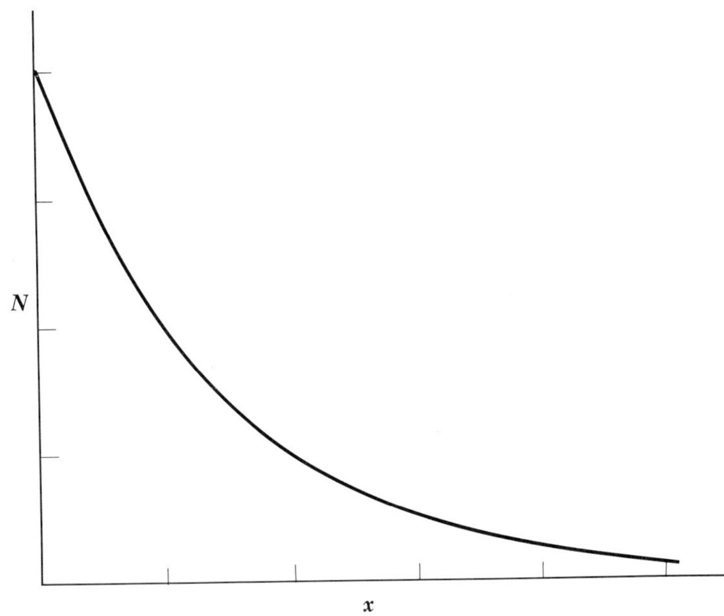

FIGURE 14-7 **The number of particles that penetrate a target decreases exponentially with target thickness.**

Denoting the initial number of incident particles by N_0, we have

14.14 $$\int_{N_0}^{N} \frac{dN}{N} = -n\sigma \int_0^x dx$$

$$\ln N - \ln N_0 = -n\sigma x$$

$$N = N_0 e^{-n\sigma x}$$

The number of surviving particles N decreases exponentially with increasing slab thickness x (Fig. 14-7). The number of particles that have undergone interactions in the slab is evidently

14.15 $$N_0 - N = N_0(1 - e^{-n\sigma x})$$

It is easy to show that Eq. 14.15 reduces to Eq. 14.12 as x decreases. If x is very small,

$$e^{-n\sigma x} \approx 1 - n\sigma x$$

$$N_0 - N \approx N_0 - N_0(1 - n\sigma x)$$

$$\approx N_0 n\sigma x$$

and

$$\frac{dN}{N} \approx \frac{N_0 - N}{N_0} \approx n\sigma x$$

The notion of cross section is often applied in situations where the interaction between incident and target particles is such that the former are absorbed. In such situations the quantity $n\sigma$ is called the *absorption coefficient* of the target material. The symbol for absorption coefficient is α:

14.16 $$\alpha = n\sigma$$

The greater the absorption coefficient, the more effective the material in stopping the incident particles.

The *mean free path* l of a particle in a material is the average distance it can travel in the material before interacting with a target nucleus. The probability f that an incident particle will undergo an interaction in a slab Δx thick of the material is

14.17 $$f = n\sigma\,\Delta x$$

The number of times H it must traverse the slab before interacting is therefore

14.18 $$H = \frac{1}{n\sigma\,\Delta x}$$

on the average. To verify this statement, we note that

$$Hf = 1$$

The product of the probability for interacting and the mean number of passages before interaction is 1, as it must be. The average distance the particle travels before interacting is therefore

$$H\,\Delta x = \frac{1}{n\sigma}$$

which is, by definition, the mean free path. Hence

14.19 $$l = \frac{1}{n\sigma}$$ **Mean free path**

a most useful formula.

While cross sections, which are areas, should be expressed in m^2, it is convenient and customary to express them in *barns*, where 1 barn $= 10^{-28}$ m^2. The barn is of the order of magnitude of the geometrical cross section of a nucleus. The cross sections for most nuclear reactions depend upon the energy of the incident particle. Figure 14-8 shows how the neutron-absorption cross section of $_{48}Cd^{113}$ varies with neutron energy; the narrow peak at 0.176 ev is associated with a specific energy level in the resulting $_{48}Cd^{114}$ nucleus.

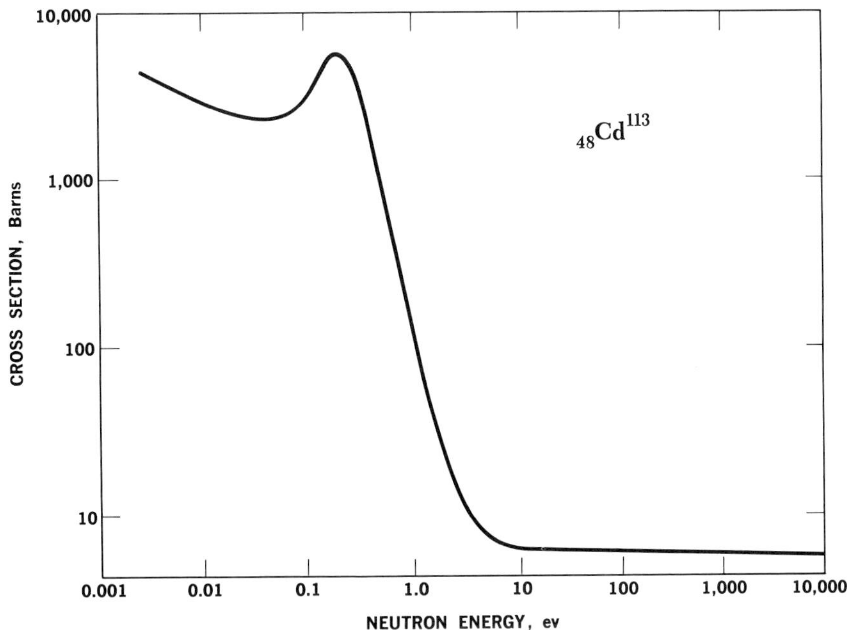

FIGURE 14-8 The cross section for neutron absorption in $_{48}Cd^{113}$ varies with neutron energy.

14.3 The Compound Nucleus

Many nuclear reactions actually involve two separate stages. In the first, an incident particle strikes a target nucleus and the two combine to form a new nucleus, called a *compound nucleus,* whose atomic and mass numbers are respectively the sum of the atomic numbers of the original particles and the sum of their mass numbers. The compound nucleus has no "memory" of how it was formed, since its nucleons are mixed together regardless of origin and the energy brought into it by the incident particle is shared among all of them. A given compound nucleus may therefore be formed in a variety of ways. To illustrate this, Table 14-1 shows six reactions whose product is the compound nucleus $_7N^{14*}$. (The asterisk signifies an excited state; compound nuclei are invariably excited by amounts equal to at least the binding energies of the incident particles in them.) While $_7N^{13}$ and $_6C^{11}$ are beta-radioactive with such short half lives as to preclude the detailed study of their reactions to form $_7N^{14*}$, there is no doubt that these reactions can occur.

Compound nuclei have lifetimes of the order of 10^{-16} sec or so, which, while so short as to prevent actually observing such nuclei directly, are nevertheless long relative to the 10^{-21} sec or so required for a nuclear particle with an energy of several Mev to pass through a nucleus. A given compound

Table 14.1

NUCLEAR REACTIONS WHOSE PRODUCT IS THE COMPOUND NUCLEUS $_7N^{14*}$

(The excitation energies given are calculated from the masses of the particles involved; the kinetic energy of an incident particle will add to the excitation energy of its reaction by an amount depending upon the dynamics of the reaction in the center-of-mass coordinate system.)

$$
\begin{aligned}
&{}_7\mathrm{N}^{13} + {}_0n^1 \rightarrow {}_7\mathrm{N}^{14*} \ (10.5\ \mathrm{Mev}) \\
&{}_6\mathrm{C}^{13} + {}_1\mathrm{H}^1 \rightarrow {}_7\mathrm{N}^{14*} \ (7.5\ \mathrm{Mev}) \\
&{}_6\mathrm{C}^{12} + {}_1\mathrm{H}^2 \rightarrow {}_7\mathrm{N}^{14*} \ (10.3\ \mathrm{Mev}) \\
&{}_6\mathrm{C}^{11} + {}_1\mathrm{H}^3 \rightarrow {}_7\mathrm{N}^{14*} \ (22.7\ \mathrm{Mev}) \\
&{}_5\mathrm{B}^{11} + {}_2\mathrm{He}^3 \rightarrow {}_7\mathrm{N}^{14*} \ (20.7\ \mathrm{Mev}) \\
&{}_5\mathrm{B}^{10} + {}_2\mathrm{He}^4 \rightarrow {}_7\mathrm{N}^{14*} \ (11.6\ \mathrm{Mev})
\end{aligned}
$$

nucleus may decay in one or more different ways, depending upon its excitation energy. Thus $_7N^{14*}$ with an excitation energy of, say, 12 Mev can decay via the reactions

$$
\begin{aligned}
&{}_7\mathrm{N}^{14*} \rightarrow {}_7\mathrm{N}^{13} + {}_0n^1 \\
&{}_7\mathrm{N}^{14*} \rightarrow {}_6\mathrm{C}^{13} + {}_1\mathrm{H}^1 \\
&{}_7\mathrm{N}^{14*} \rightarrow {}_6\mathrm{C}^{12} + {}_1\mathrm{H}^2 \\
&{}_7\mathrm{N}^{14*} \rightarrow {}_5\mathrm{B}^{10} + {}_2\mathrm{He}^4
\end{aligned}
$$

or simply emit one or more gamma rays whose energies total 12 Mev, but it *cannot* decay by the emission of a triton ($_1H^3$) or a helium-3 ($_2He^3$) particle since it does not have enough energy to liberate them. Usually a particular decay mode is favored by a compound nucleus in a specific excited state.

The formation and decay of a compound nucleus has an interesting interpretation on the basis of the liquid-drop nuclear model described in Chap. 12. In terms of this model, an excited nucleus is analogous to a drop of hot liquid, with the binding energy of the emitted particles corresponding to the heat of vaporization of the liquid molecules. Such a drop of liquid will eventually evaporate one or more molecules, thereby cooling down. The evaporation process occurs when statistical fluctuations in the energy distribution within the drop cause a particular molecule to have enough energy for escape. Similarly, a compound nucleus persists in its excited state until a particular nucleon or group of nucleons momentarily happens to have a sufficiently large fraction of the excitation energy to leave the nucleus. The time interval between the formation and decay of a compound nucleus fits in nicely with this picture.

14.4 The Coulomb Barrier

As we saw in the previous chapter, every nucleus is surrounded by an electrostatic potential barrier that opposes both the entry and escape of positively charged particles. Neutrons, which have no charge, are not faced with this *coulomb barrier* and accordingly are more readily absorbed and emitted by nuclei than are protons, deuterons, and alpha particles. Figure 14-9 is a plot of the potential energies of a proton and a neutron near a nucleus. The coulomb barrier is about 3 Mev in height for carbon nuclei, 13 Mev for silver nuclei, and 20 Mev for lead nuclei.

Evidently a neutron of any energy can enter a nucleus by "falling into" its potential well, just as a rolling ball falls into a hole in the ground, while an approaching proton is faced with a potential hill. Classically the proton would have to possess at least as much energy as the height of the barrier in order to enter the nucleus. Quantum-mechanically, of course, the proton *can* get in with less than this amount of energy, but the probability of its doing so is small. Hence the height of the coulomb barrier represents an effective threshold energy for nuclear reactions initiated by such particles as protons, deuterons, and alpha particles, while no such threshold is present for incoming neutrons. Figure 14-10 shows how the cross sections for comparable neutron- and proton-induced nuclear reactions vary with energy. The neutron cross section *decreases* with increasing energy, because the likelihood that a neutron be captured depends upon how much time it spends near a particular nucleus, which is inversely proportional to its speed. The proton cross section *increases* with increasing energy because of the presence of the coulomb barrier.

FIGURE 14-9 Proton and neutron potential energy near a nucleus.

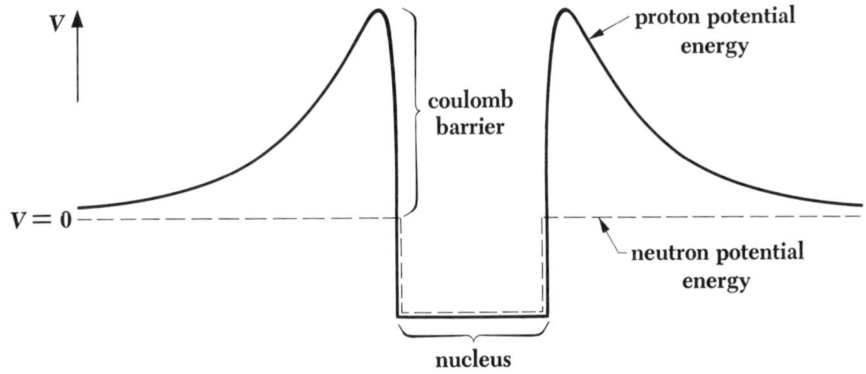

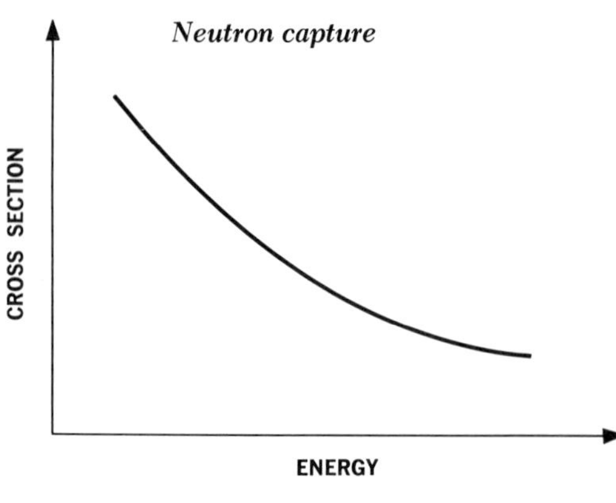

FIGURE 14-10 Neutron and proton capture cross sections vary differently with particle energy.

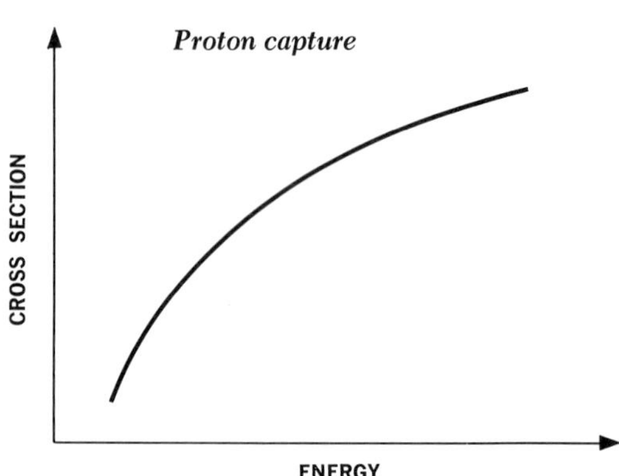

14.5 Nuclear Fission

Another type of nuclear-reaction phenomenon that can be analyzed with the help of the liquid-drop model is *fission,* in which a heavy nucleus ($A > \sim 230$) splits into two lighter ones. When a liquid drop is suitably excited, it may oscillate in a variety of ways. A simple one is shown in Fig. 14-11: the drop successively becomes a prolate spheroid, a sphere, an oblate spheroid, a sphere, a prolate spheroid again, and so on. The restoring force of its surface tension always returns the drop to spherical shape, but the inertia of the moving liquid molecules causes the drop to overshoot sphericity and go to the opposite extreme of distortion.

FIGURE 14-11 The oscillations of a liquid drop.

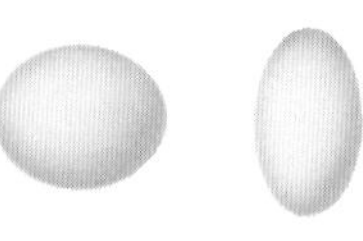
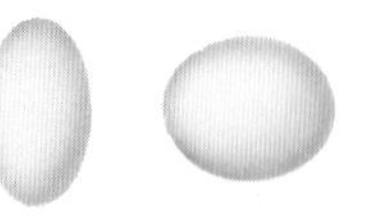
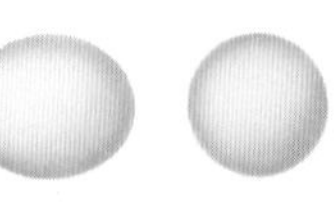

time ⟶

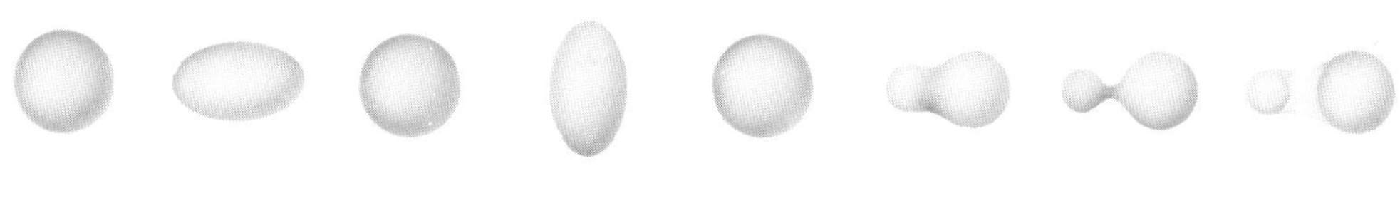

FIGURE 14-12 Nuclear fission according to the liquid drop model.

While nuclei may be regarded as exhibiting surface tension, and so can vibrate like a liquid drop when in an excited state, they also are subject to disruptive forces due to the mutual electrostatic repulsion of their protons. When a nucleus is distorted from a spherical shape, the short-range restoring force of surface tension must cope with the latter long-range repulsive force as well as with the inertia of the nuclear matter. If the degree of distortion is small, the surface tension is adequate to do both, and the nucleus vibrates back and forth until it eventually loses its excitation energy by gamma decay. If the degree of distortion is sufficiently great, however, the surface tension is not adequate to bring back together the now widely separated groups of protons, and the nucleus splits into two parts. This picture of fission is illustrated in Fig. 14-12.

The new nuclei that result from fission are called *fission fragments*. Usually fission fragments are of unequal size, and, because heavy nuclei have a greater neutron/proton ratio than lighter ones, they contain an excess of neutrons. To reduce this excess, two or three neutrons are emitted by the fragments as soon as they are formed, and subsequent beta decays bring their neutron/proton ratios to stable values.

A heavy nucleus undergoes fission when it acquires enough excitation energy (5 Mev or so) to oscillate violently. Certain nuclei, notably $_{92}U^{235}$, are sufficiently excited by the mere absorption of an additional neutron to split in two. Other nuclei, notably $_{92}U^{238}$ (which composes 99.3 per cent of natural uranium, with $_{92}U^{235}$ composing the remainder), require more excitation energy for fission than the binding energy released when another neutron is absorbed, and undergo fission only by reaction with fast neutrons whose kinetic energies exceed about 1 Mev. Fission can occur after excitation by other means besides neutron capture, for instance, by gamma-ray or proton bombardment. Some nuclides are so unstable as to be capable of spontaneous fission, but they are more likely to undergo alpha decay before this takes place.

The most striking aspect of nuclear fission is the magnitude of the energy evolved. This energy is readily computed. The heavy fissionable nuclides, whose mass numbers are about 240, have binding energies of ~7.6 Mev/nucleon, while fission fragments, whose mass numbers are about 120, have binding energies of ~8.5 Mev/nucleon. Hence 0.9 Mev/nucleon is released during fission—over 200 Mev for the 240 or so nucleons involved! Ordinary chemical reactions, such as those that participate in the combustion of coal and oil, liberate only a few electron volts per individual reaction, and even nuclear reactions (other than fission) liberate no more than several million electron volts. Most of the energy that is released during fission goes into the kinetic energy of the fission fragments: the emitted neutrons, beta and gamma rays, and neutrinos carry off perhaps 15 per cent of the total energy.

14.6 The Chain Reaction

Almost immediately after the discovery of nuclear fission in 1939 it was recognized that, because a neutron can induce fission in a suitable nucleus with the consequent evolution of additional neutrons, a self-sustaining sequence of fissions is, in principle, possible. The condition for such a chain reaction to occur in an assembly of fissionable material is simple: at least one neutron produced during each fission must, on the average, initiate another fission. If too few neutrons initiate fissions, the reaction will slow down and stop; if precisely one neutron per fission causes another fission, energy will be released at a constant rate; and if the frequency of fissions increases, the energy release will be so rapid that an explosion will occur. These situations are respectively called *subcritical*, *critical*, and *supercritical*.

Let us briefly examine the basic problems involved in the design of a *nuclear reactor*, which is a device for producing controlled power from nuclear fission. We shall use as an example a reactor fueled with natural uranium, similar to the one built in 1942 by Fermi in the first demonstration of the feasibility of chain reactions.

First we shall consider the loss of neutrons through the reactor surface and by non-fission-inducing absorption. The former problem can be met simply by increasing the reactor size, since a large object has less surface in proportion to its volume than a small one. The second problem is more difficult, since natural uranium contains only 0.7 per cent of the fissionable isotope ${}_{92}U^{235}$. The more abundant isotope ${}_{92}U^{238}$ readily captures fast neutrons, but it usually rids itself of the resulting excitation energy by merely emitting a gamma ray rather than by undergoing fission. However, ${}_{92}U^{238}$ has only a small cross section for the capture of *slow* neutrons, while the cross section of ${}_{92}U^{235}$ for slow-neutron-induced fission is 550 barns—many times larger than its geometrical cross section. Hence it is necessary to rapidly slow down the neutrons that are liberated in fission both to prevent their acquisition by ${}_{92}U^{238}$ and to promote further fissions in ${}_{92}U^{235}$.

To accomplish the slowing down of fission neutrons, the uranium in a reactor is dispersed in a matrix of a *moderator*, a substance whose nuclei absorb energy from incident fast neutrons that collide with them without capture occurring. While the exact amount of energy lost by a moving particle that elastically collides with another depends upon the details of the interaction, in general the energy transfer is greatest when the participants are of equal mass. Hydrogen, however, whose proton nuclei have masses almost identical with those of neutrons, readily captures neutrons to form deuterons:

$$_1H^1 + {}_0n^1 \rightarrow {}_1H^2 + \gamma \tag{14.20}$$

Deuterons are less likely to react with neutrons, and so *heavy water*, whose molecules contain deuterium atoms instead of ordinary hydrogen atoms, makes a suitable moderating material. The first reactor employed carbon in the form of graphite as a moderator, because graphite was more readily available than heavy water in the necessary quantities and $_6C^{12}$ also has only a small cross section for neutron capture.

The cross section of $_{92}U^{235}$ for fission by slow neutrons is 550 barns so that, since this isotope composes 0.7 per cent of natural uranium, the "average" uranium nucleus has a fission cross section of 0.007×550, or 3.9 barns. The cross section of $_{92}U^{235}$ for radiative neutron capture,

$$_{92}U^{235} + {}_0n^1 \rightarrow {}_{92}U^{236*} \rightarrow {}_{92}U^{236} + \gamma \tag{14.21}$$

is 101 barns, and the average uranium nucleus accordingly has for this process a cross section of 0.007×101, or 0.7 barns. The 99.3 per cent abundant $_{92}U^{238}$ has a radiative slow-neutron-capture cross section of 2.8 barns, for an average uranium cross section of 0.993×2.8, or 2.8 barns. Hence the average fission cross section is 3.9 barns while the average absorption cross section is $0.7 + 2.8 = 3.5$ barns, and so only about half the slow neutrons captured in a block of uranium induce fissions. Because each fission in $_{92}U^{235}$ releases an average of 2.5 neutrons, no more than 0.5 neutron per fission can be lost if a self-sustaining chain reaction is to occur.

The actual operation of a reactor begins when a sufficiently large amount of fissionable material is brought together in the presence of a moderator. A single stray neutron—from the cosmic radiation, perhaps, or from a spontaneous fission—strikes a $_{92}U^{235}$ nucleus and causes it to split, releasing two or three additional neutrons. These neutrons are slowed down from energies of several Mev to energies of less than 1 ev by collisions with moderator nuclei, and then proceed to induce further fissions. The period of time between the release of a fission neutron and its later absorption is under a millisecond (0.001 sec). It is necessary to provide a means for controlling the chain-reaction rate. This is accomplished by means of rods made of a material, such as cadmium or boron, which readily absorbs slow neutrons; as these rods are inserted into the reactor, the reaction rate is progressively damped. The

The Cyclotron

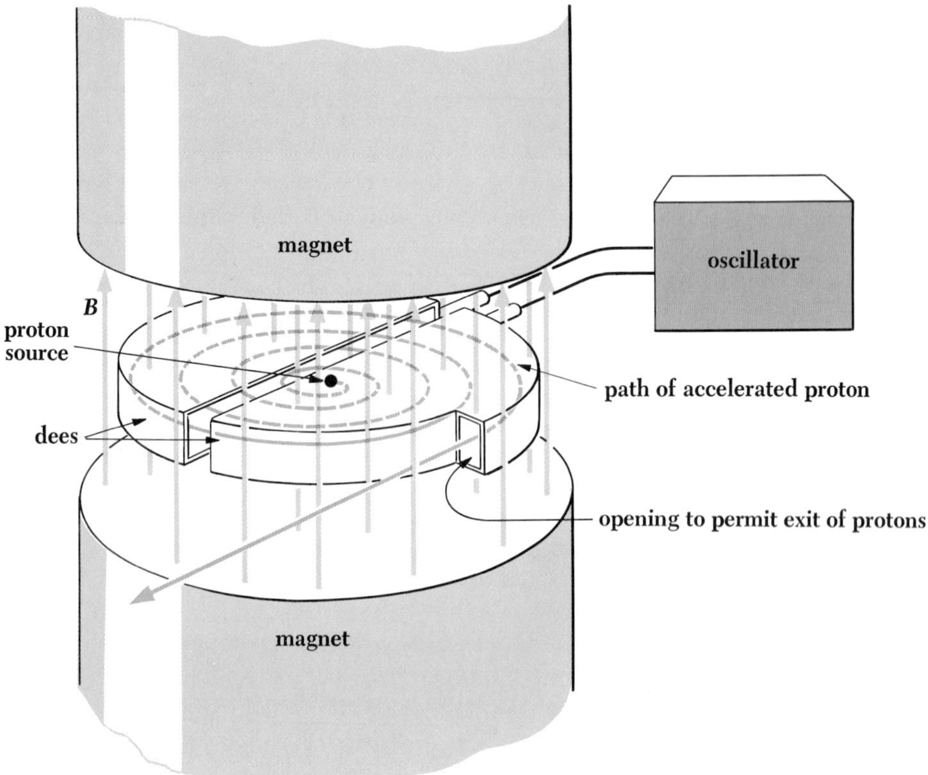

The cyclotron is a device for accelerating charged particles to high energies by subjecting them to repeated voltage impulses. Devised in 1932 by E. O. Lawrence, it consists of a pair of hollow copper electrodes called "dees" (because they are shaped like the letter D) that are located in a uniform magnetic field. A particle of charge e, mass m, and speed v which is directed perpendicular to a magnetic field B moves in a circular path of radius

$$R = \frac{mv}{eB}$$

The period of time T required by the particle to complete each orbit is the circumference $2\pi R$ divided by the speed v, namely,

$$T = \frac{2\pi R}{v} = \frac{2\pi m}{eB}$$

Hence T is independent of the particle's speed provided that v is much less than c. If the dees of a cyclotron are connected to a source of alternating potential of frequency $1/T$ so that their polarities reverse sign with a period of $\frac{1}{2}T$, the particle will be accelerated by the electric field between the dees each time it passes through the gap. (There is no electric field within the dees.) The particle goes faster and faster, but T is independent of v and the accelerations continue as it spirals outward in the magnetic field. A typical cyclotron might have a flux density of 2 webers/m^2 and a dee diameter of a meter; when used with an oscillator of appropriate frequency (several megacycles) and appropriate voltage (200 kvolts or so), such a cyclotron can produce a 1-ma current of 25-Mev deuterons or 12-Mev protons.

At higher energies the relativistic increase in their mass causes the particles to fall out of synchronism with the applied alternating potential, arriving "too late" at the gap between the dees to be accelerated. The synchrocyclotron, developed in 1945, surmounts this limitation by decreasing the oscillator frequency to compensate for the increasing mass. A synchrocyclotron therefore generates bursts of energetic particles, usually 60 per second, instead of the essentially continuous output of a cyclotron. Energies of up to about 1 bev can be produced in a synchrocyclotron; higher energies require machines of impractical dimensions. Other instruments, which employ series of small magnets along the particle orbits in place of a single large magnet, have been used to achieve energies of scores of bev per particle.

energy generated by a nuclear reactor is manifested as heat, and it can be extracted by circulating a suitable liquid or gaseous coolant through the reactor interior.

14.7 Transuranic Elements

Elements of atomic number greater than 92, which is that of uranium, have such short half lives that, had they been formed when the universe came into being, they would have disappeared long ago. Such *transuranic elements* may be produced in the laboratory by the bombardment of certain heavy nuclides with neutrons. Thus ${}_{92}\mathrm{U}^{238}$ may absorb a neutron to become ${}_{92}\mathrm{U}^{239}$, which beta-decays ($T_{1/2} = 23$ min) into ${}_{93}\mathrm{Np}^{239}$, an isotope of the transuranic element *neptunium:*

$$
\begin{aligned}
&{}_{92}\mathrm{U}^{238} + {}_0n^1 \rightarrow {}_{92}\mathrm{U}^{239} \\
&{}_{92}\mathrm{U}^{239} \rightarrow {}_{93}\mathrm{Np}^{239} + e^-
\end{aligned}
$$

This neptunium isotope is itself radioactive, undergoing beta decay with a half life of 2.3 days into an isotope of the transuranic element *plutonium:*

$${}_{93}\mathrm{Np}^{239} \rightarrow {}_{94}\mathrm{Pu}^{239} + e^-$$

Plutonium alpha-decays into ${}_{92}U^{235}$ with a half life of 24,000 years:

$$ {}_{94}Pu^{239} \rightarrow {}_{92}U^{235} + {}_{2}He^{4} $$

It is interesting to note that ${}_{94}Pu^{239}$, like ${}_{92}U^{235}$, is fissionable and can be used in nuclear reactors and weapons. Plutonium is chemically different from uranium; its separation from the remaining ${}_{92}U^{238}$ after neutron irradiation is more easily accomplished than the separation of ${}_{92}U^{235}$ from the much more abundant ${}_{92}U^{238}$ in natural uranium.

Transuranic elements past einsteinium ($Z = 99$) have half-lives too short for their isolation in weighable quantities, though they can be identified by chemical means. The transuranic element of the highest atomic number yet discovered is lawrencium, $Z = 103$.

14.8 Thermonuclear Energy

The basic exothermic reaction in stars—and hence the source of nearly all of the energy in the universe—is the fusion of hydrogen nuclei into helium nuclei. This can take place under stellar conditions in two different series of processes. In one of them, the *proton-proton cycle,* direct collisions of protons result in the formation of heavier nuclei whose collisions in turn yield helium nuclei. The other, the *carbon cycle,* is a series of steps in which carbon nuclei absorb a succession of protons until they ultimately disgorge alpha particles to become carbon nuclei once more.

The initial reaction in the proton-proton cycle is

$$ {}_{1}H^{1} + {}_{1}H^{1} \rightarrow {}_{1}H^{2} + e^{+} + \nu $$

the formation of deuterons by the direct combination of two protons accompanied by the emission of a positron. A deuteron may then join with a proton to form a ${}_{2}He^{3}$ nucleus:

$$ {}_{1}H^{1} + {}_{1}H^{2} \rightarrow {}_{2}He^{3} + \gamma $$

Finally two ${}_{2}He^{3}$ nuclei react to produce a ${}_{2}He^{4}$ nucleus plus two protons:

$$ {}_{2}He^{3} + {}_{2}He^{3} \rightarrow {}_{2}He^{4} + {}_{1}H^{1} + {}_{1}H^{1} $$

The total evolved energy is $(\Delta m)c^2$, where Δm is the difference between the mass of four protons and the mass of an alpha particle plus two positrons; it turns out to be 24.7 Mev.

The carbon cycle proceeds in the following way:

$$
\begin{aligned}
{}_1H^1 + {}_6C^{12} &\rightarrow {}_7N^{13} \\
{}_7N^{13} &\rightarrow {}_6C^{13} + e^+ + \nu \\
{}_1H^1 + {}_6C^{13} &\rightarrow {}_7N^{14} + \gamma \\
{}_1H^1 + {}_7N^{14} &\rightarrow {}_8O^{15} + \gamma \\
{}_8O^{15} &\rightarrow {}_7N^{15} + e^+ + \nu \\
{}_1H^1 + {}_7N^{15} &\rightarrow {}_6C^{12} + {}_2He^4
\end{aligned}
$$

The net result again is the formation of an alpha particle and two positrons from four protons, with the evolution of 24.7 Mev; the initial ${}_6C^{12}$ acts as a kind of catalyst for the process, since it reappears at its end.

Self-sustaining fusion reactions can occur only under conditions of extreme temperature and pressure, to ensure that the participating nuclei have enough energy to react despite their mutual electrostatic repulsion and that reactions occur frequently enough to counterbalance losses of energy to the surroundings. Stellar interiors meet these specifications. In the sun, whose interior temperature is estimated to be 2×10^6 °K, the carbon and proton-proton cycles have about equal probabilities for occurrence. In general, the carbon cycle is more efficient at high temperatures, while the proton-proton cycle is more efficient at low temperatures. Hence stars hotter than the sun obtain their energy largely from the former cycle, while those cooler than the sun obtain the greater part of their energy from the latter cycle.

The energy liberated in the fusion of light nuclei into heavier ones is often called *thermonuclear energy,* particularly when the fusion takes place under man's control. On the earth neither the proton-proton nor carbon cycle offers any hope of practical application, since their several steps require a great deal of time. The fusion reactions that seem most promising as terrestrial energy sources are the direct combination of two deuterons in either of the following ways:

$$
\begin{aligned}
{}_1H^2 + {}_1H^2 &\rightarrow {}_2He^3 + {}_0n^1 + 3.3 \text{ Mev} \\
{}_1H^2 + {}_1H^2 &\rightarrow {}_1H^3 + {}_1H^1 + 4.0 \text{ Mev}
\end{aligned}
$$

and the direct combination of a deuteron and a triton to form an alpha particle,

$$
{}_1H^3 + {}_1H^2 \rightarrow {}_2He^4 + {}_0n^1 + 17.6 \text{ Mev}
$$

Capitalizing upon the above reactions requires an abundant, cheap source of deuterium. Such a source is the oceans and seas of the world, which contain about 0.015 per cent deuterium—a total of perhaps 10^{15} tons! In addition, a more efficient means of promoting fusion reactions than merely bombarding a target with fast particles from an accelerator is required, since the operation of an accelerator consumes far more power than can be evolved by the relatively few reactions that occur in the target. Current approaches to this problem all involve very hot fully ionized gases containing deuterium or deuterium-tritium mixtures which are contained by strong magnetic fields. The purpose of high temperature is to ensure that the individual $_1H^2$ and $_1H^3$ nuclei have enough energy to come together and react despite their electrostatic repulsion. A magnetic field is used as a container to keep the reactive gas from contacting any other material which might cool it down or contaminate it; there is little likelihood that the wall will melt since the gas, though at a temperature of several million degrees K, actually does not have a high energy density. While nuclear-fusion reactors present more severe practical difficulties than fission reactors, there is little doubt that they will eventually become a reality.

Problems

(The masses in amu of neutral atoms of nuclides mentioned below are: $_1H^1$, 1.007825; $_1H^3$, 3.016049; $_2He^3$, 3.016029; $_2He^4$, 4.002603; $_5B^{10}$, 10.0129; $_6C^{13}$, 13.0034; $_7N^{14}$, 14.0031; $_8O^{16}$, 15.9949; $_8O^{17}$, 16.9994. The neutron mass is 1.008665 amu, and the atomic weight of iron is 55.85.)

1. Find the minimum energy in the laboratory system that an alpha particle must have in order to initiate the reaction

$$_2He^4 + {}_7N^{14} + 1.18 \text{ Mev} \rightarrow {}_8O^{17} + {}_1H^1$$

2. Find the minimum energy in the laboratory system that a neutron must have in order to initiate the reaction

$$_0n^1 + {}_8O^{16} + 2.20 \text{ Mev} \rightarrow {}_6C^{13} + {}_2He^4$$

3. A particle of mass m_1 and velocity v_1 strikes a particle of mass m_2 which is at rest. If the latter particle acquires a velocity v_2 in the same direction as v_1 as the result of the collision, and if the collision is perfectly elastic, find the ratio between v_1 and v_2.

4. A particle of mass m_1 collides with a particle of mass m_2 which is at rest. Show that, if $m_1 > m_2$, the maximum angle in the laboratory system through which m_1 can be scattered is given by $\theta_{max} = \sin^{-1} m_2/m_1$.

5. There are approximately 6×10^{28} atoms/m^3 in solid aluminum. A beam of 0.5-Mev neutrons is directed at an aluminum foil 0.1 mm thick. If the capture cross section for neutrons of this energy in aluminum is 2×10^{-31} m^2, find the fraction of incident neutrons that are captured.

6. The density of $_5B^{10}$ is 2.5×10^3 kg/m^3. The capture cross section of $_5B^{10}$ is about 4,000 barns for "thermal" neutrons, that is, neutrons in thermal equilibrium with matter at room temperature. How thick a layer of $_5B^{10}$ is required to absorb 99 per cent of an incident beam of thermal neutrons?

7. The density of iron is about 8×10^3 kg/m^3. The neutron-capture cross section of iron is about 2.5 barns. What fraction of an incident beam of neutrons is absorbed by a sheet of iron 1 cm thick?

8. The cross section of iron for neutron capture is 2.5 barns. What is the mean free path of neutrons in iron?

9. The cross section of iron for neutrino capture is 10^{-20} barn. What is the mean free path of neutrinos in iron?

10. How much energy must a photon have if it is to split an alpha particle into a $_2He^3$ nucleus and a neutron?

11. How much energy must a photon have if it is to split an alpha particle into a triton ($_1H^3$) and a proton?

12. A $_7N^{15}$ nucleus reacts with an incident proton to form a compound nucleus which then undergoes alpha decay. What is the remaining nucleus?

13. Complete the following nuclear reactions:

$$_{17}Cl^{35} + ? \rightarrow {}_{16}S^{32} + {}_2He^4$$
$$_5B^{10} + ? \rightarrow {}_3Li^7 + {}_2He^4$$
$$_3Li^6 + ? \rightarrow {}_4Be^7 + {}_0n^1$$

14. Complete the following nuclear reactions:

$$_{11}Na^{23} + {}_1H^1 \rightarrow {}_{10}Ne^{20} + ?$$
$$_{11}Na^{23} + {}_1H^2 \rightarrow {}_{12}Mg^{24} + ?$$
$$_{11}Na^{23} + {}_1H^2 \rightarrow {}_{11}Na^{24} + ?$$

15. The fission of $_{92}U^{235}$ releases approximately 200 Mev. What percentage of the original mass of $_{92}U^{235} + n$ disappears?

16. When a neutron is absorbed by a target nucleus, the resulting compound nucleus is usually more likely to emit a gamma ray than a proton, deuteron, or alpha particle. Why?

17. Certain stars obtain part of their energy by the fusion of three alpha particles to form a $_6C^{12}$ nucleus. How much energy does each such reaction evolve?

15 ELEMENTARY PARTICLES

While nuclei are apparently composed solely of protons and neutrons, several score other elementary particles have been observed to be emitted by nuclei under appropriate circumstances. These particles, christened "strange particles" soon after their discovery somewhat over a decade ago, bring the total number merely of relatively stable elementary particles to over 30. To discern order in this multiplicity of particles has not proved to be an easy task. While certain regularities in elementary-particle properties have been established, and while such particles as the electron, the neutrino, and the π meson are relatively well-understood, no comprehensive theory of elementary particles has yet found wide acceptance. It is fitting to conclude our survey of modern physics with this topic, then, as a reminder that there remains much to be learned about the natural world.

15.1 The Theory of the Electron

The electron is the only elementary particle for which a satisfactory theory is known. This theory was developed in 1928 by P. A. M. Dirac, who obtained a wave equation for a charged particle in an electromagnetic field that incorporated the results of special relativity. When the observed mass and charge of the electron are inserted in the appropriate solutions of this equation, the intrinsic angular momentum of the electron is found to be ½ $\hbar$ (that is, spin ½) and its magnetic moment is found to be $e\hbar/2m$, one Bohr magneton. These predictions agree with experiment, and the agreement is strong evidence for the correctness of the Dirac theory.

Perhaps the most striking result of the Dirac theory is its prediction that electrons can exist in *negative* energy states. Dirac found that *both* the positive and negative roots of the relativistic energy equation

15.1 $$E = \pm\sqrt{m_0{}^2c^2 + p^2c^2}$$

are equally permissible. The positive root permits the total energy E of an electron to have any value from m_0c^2, its rest energy, to ∞. The negative root permits E to have any value from $-m_0c^2$ to $-\infty$. There is nothing in the theory to prevent an electron in a positive energy state from undergoing a transition to a negative energy state by radiating a photon of appropriate energy, or to prevent an electron in a negative energy state from falling to a still more negative state in the same way. Since systems always tend to evolve toward configurations of minimum energy, all the electrons in the universe should ultimately have energies of $E = -\infty$! There being no evidence that this peculiar destiny is indeed forthcoming, Dirac proposed that all negative-electron energy states are normally occupied. The exclusion principle therefore prevents positive-energy electrons from undergoing transitions into any of these states. Dirac further proposed that the "sea" of negative-energy electrons is not observable directly.

While positive-energy electrons are prevented from falling into negative energy states, since the latter are all occupied, the reverse process can occur. Given enough energy in some way, an electron of negative total energy can become one of positive total energy and thus become observable. This process is shown schematically in Fig. 15-1. The removal of an electron from the negative-energy electron sea leaves behind a "hole." Such a hole has interesting properties. Since it represents the absence of a particle of negative mass and kinetic energy, it behaves as though it is a particle of positive mass and kinetic energy. Moreover, a hole responds to electric and magnetic fields precisely as if it has a *positive* charge, as in the case of a hole in the electron structure of a crystal. Thus a hole has all the characteristics of an ordinary electron except that its charge is $+e$.

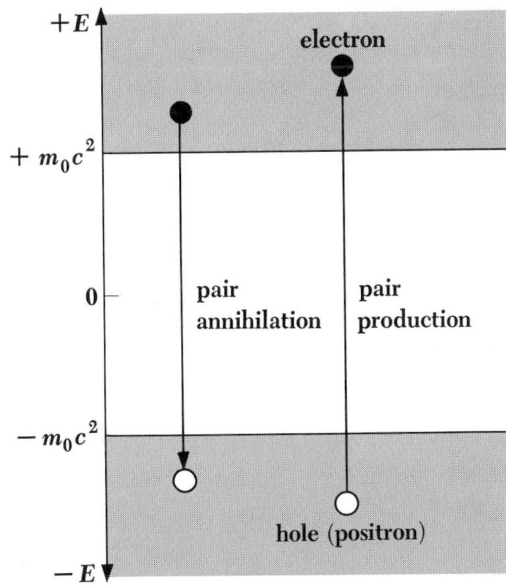

FIGURE 15-1 Pair production and annihilation according to the Dirac theory.

While Dirac's reasoning had no obvious flaws, his notion of holes in an infinite sea of negative-energy electrons was too strange to find acceptance when proposed. In 1932, however, positive electrons were actually detected in the flux of cosmic radiation at the earth's surface, completely verifying the hole hypothesis. Positive electrons are usually called *positrons.*

The formation of a positron requires a minimum energy of $2m_0c^2$ (1.02 Mev), since this amount of energy is needed to bring an electron in a state whose energy is $-m_0c^2$ to a state whose energy is $+m_0c^2$. When a positron is formed, an electron simultaneously appears, since it is the absence of this electron from the unobservable negative-energy sea which constitutes the positron. If we wish, we can regard the process of electron-positron pair creation as one involving the *materialization* of matter from energy, since $2m_0c^2$ of energy disappears whenever such a pair comes into being. Any available energy in excess of $2m_0c^2$ goes into the kinetic energy of the electron and positron. Experimentally electron-positron pairs are found to be produced when gamma rays of $h\nu > 1.02$ Mev pass near nuclei (Fig. 15-2); the presence of the relatively heavy nuclei is required in order that momentum as well as energy be conserved in the process.

The reverse of pair creation occurs when an electron of positive energy falls into a vacant negative energy state (Fig. 15-1). Since the latter is observed as a positron, we may regard the process as one in which an electron and a positron *annihilate* each other. The simultaneous disappearance of an electron and a positron liberates $2m_0c^2$ of energy. No nucleus or other particle is needed for annihilation to take place, since the energy evolved appears as two gamma rays whose directions are such as to conserve both momentum and energy.

15.2 Antiparticles

The positron is often spoken of as the *antiparticle* of the electron, since it is able to undergo mutual annihilation with an electron. All other known elementary particles except for the photon and the π^0 meson also have antiparticle counterparts; the photon and π^0 meson are their own antiparticles. The antiparticle of a particle has the same mass, spin, and lifetime if unstable, but its charge (if any) has the opposite sign and the alignment or antialignment between its spin and magnetic moment is also opposite to that of the particle.

The annihilation of a particle-antiparticle pair need not always result in a pair of photons, as it does in the case of electron-positron annihilation. When an antiproton is annihilated with a proton or neutron, for example, or an antineutron with a neutron or proton, several neutral and charged π

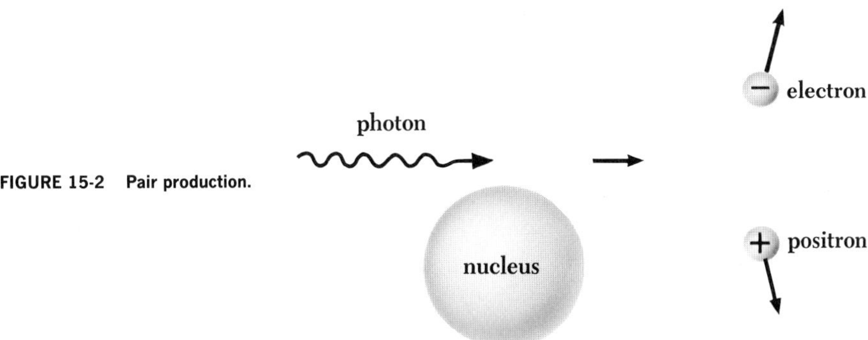

FIGURE 15-2 Pair production.

mesons are usually produced. This is further evidence that π mesons may be regarded as quanta of the nuclear force field in the same sense that photons are quanta of the electromagnetic field.

The distinction between the neutrino and the antineutrino is a particularly interesting one. The spin of the neutrino is opposite in direction to the direction of its motion; viewed from behind, as in Fig. 15-3, the neutrino spins counterclockwise. The spin of the antineutrino, on the other hand, is in the same direction as its direction of motion; viewed from behind, it spins clockwise. Thus the neutrino moves through space in the manner of a left-handed screw, while the antineutrino does so in the manner of a right-handed screw.

Prior to 1956 it had been universally assumed that neutrinos could be either left-handed or right-handed, implying that, since no difference was possible

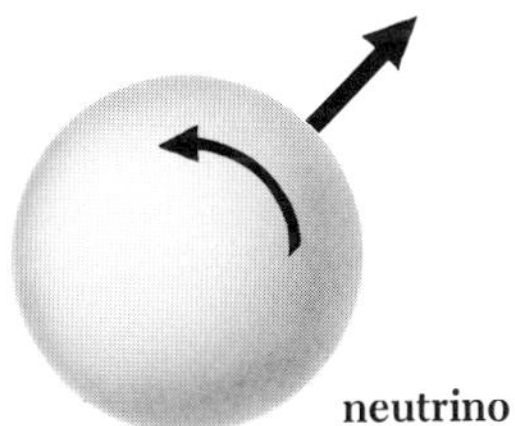

FIGURE 15-3 Neutrinos and antineutrinos have opposite directions of spin.

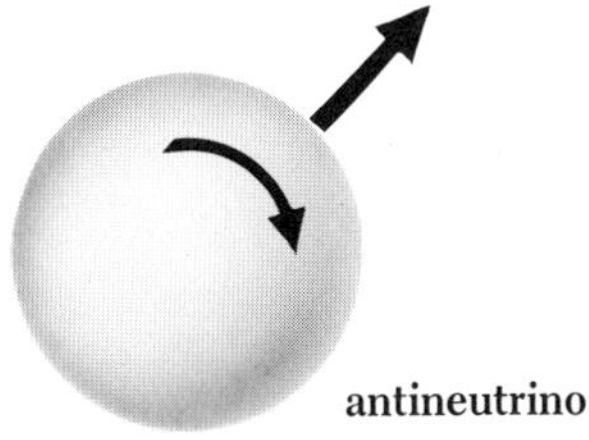

Radiation Counters

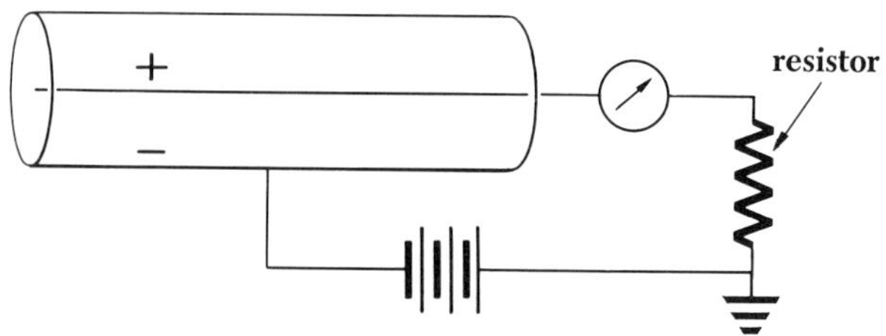

Gas-filled detector

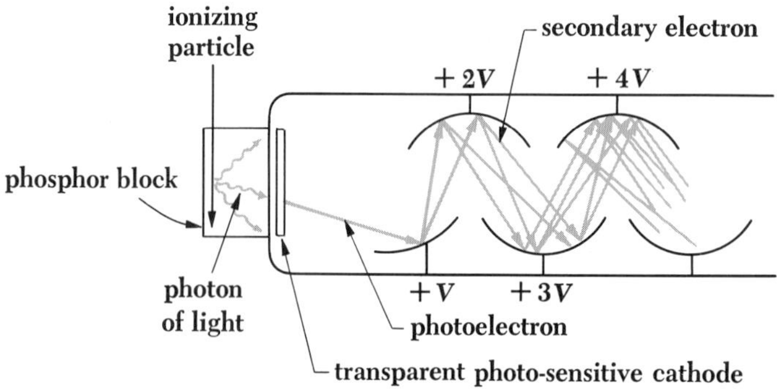

Scintillation counter

Two basic types of instruments are widely used to detect ionizing radiation. One type is sketched at top above. It consists of a tube filled with a gas and containing two electrodes. The central wire is maintained at a positive potential relative to the metal cylinder around it, and there is a strong electric field in its vicinity. A charged particle or a gamma-ray photon entering the gas creates ions in its path; the heavy positive ions migrate more or less slowly to the cathode, while the electrons are accelerated to the anode. The resulting current pulse in the resistor can be detected and thus indicates the passage of the particle through the tube. When the applied voltage is relatively low, the current is directly proportional to the amount of ionization produced in the tube. Such a device is called an ionization chamber. When the number of particles or photons passing through the tube is required rather than the ionization they produce, a high potential difference is established across the electrodes. Now the electrons that go to the central wire acquire enough energy to create further ions along the way, and an "avalanche" of secondary electrons is formed that arrives at the wire to produce a pulse whose amplitude is independent of the initiating event. This device is a Geiger counter, and, as its name suggests, its function is to count particles rather than to measure ionization.

The second class of particle detectors is based upon the flashes of light emitted by certain materials when struck by radiation. Such materials, called phosphors, are familiarly used in thin layers in television picture tubes, where they emit light when electron beams fall upon them. In a scintillation counter use is made of an especially sensitive phosphor crystal that is transparent to the light produced. The light flashes are detected by means of a photomultiplier tube, which amplifies the initial output of photoelectrons from a suitable cathode by the secondary emission of further electrons at each succeeding anode.

between them except one of spin direction, the neutrino and antineutrino are identical. This assumption had roots going all the way back to Leibniz, Newton's contemporary and an independent inventor of calculus. The argument may be stated as follows: if we observe an object or a physical process of some kind both directly and in a mirror, we cannot ideally distinguish which object or process is being viewed directly and which by reflection. By definition, distinctions in physical reality must be capable of discernment or they are meaningless. Now the only difference between something seen directly and the same thing seen in a mirror is the interchange of right and left, and so *all* objects and processes must occur with equal probability with right and left interchanged. This plausible doctrine is indeed experimentally valid for nuclear and electromagnetic interactions, but until 1956 its applicability to neutrinos had never been actually tested. In that year T. D. Lee and C. N. Yang suggested that several serious theoretical discrepancies would be removed if neutrinos and antineutrinos have different handedness, even though it meant that neither particle could therefore be reflected in a mirror. Experiments performed soon after their proposal showed unequivocally that neutrinos and antineutrinos are distinguishable, having left-handed and right-handed spins respectively. We might note that the absence of right-left symmetry in neutrinos can occur only if the neutrino mass is exactly zero, thereby resolving what had been the very difficult experimental problem of measuring the neutrino mass.

15.3 π Mesons

As we saw in Chap. 12, the meson theory of nuclear forces is based upon the presumed exchange of particles called π mesons between interacting nucleons. Such exchanges violate the law of conservation of energy, so that a limit is set by the uncertainty principle to the time available for them to take place. This limit leads to a predicted π-meson mass of about 200 m_e. Twelve years after the meson theory was formulated, particles with the required properties (though somewhat higher mass) were actually found to exist outside nuclei.

Two factors contributed to the belated discovery of the free π meson. First, enough energy must be supplied to a nucleon so that its emission of a π meson conserves energy. Thus at least $m_\pi c^2$ of energy, about 140 Mev, is required. To furnish a nucleon with this much energy in a collision, the incident particle must have considerably more kinetic energy than $m_\pi c^2$ in order that momentum as well as energy be conserved. Let us suppose that we wish to create a π meson by the impact of a fast proton on a stationary proton. In this case the amount of energy available for dissipation is the kinetic energy T' in the center-of-mass system, which is related to the kinetic energy T in the laboratory system by the formula

$$T' = \left(\frac{m_t}{m_i + m_t}\right)T$$

Here $T' = 140$ Mev and the masses of the incident and target particles are the same, so that

$$T = 2T' = 280 \text{ Mev}$$

Particles with kinetic energies of several hundred Mev are therefore required to produce free π mesons, and such particles are found in nature only in the diffuse stream of cosmic radiation that bombards the earth. Hence the discovery of the π meson had to await the development of sufficiently sensitive and precise methods of investigating cosmic-ray interactions. More recently high-energy accelerators were placed in operation; they yielded the necessary particle energies, and the profusion of mesons that were created with their help could be studied readily.

The second reason for the lag between the prediction and experimental discovery of the π meson is its instability: the mean lifetime of the charged π meson is only 2.56×10^{-8} sec, and that of the neutral π meson is 2.3×10^{-16} sec. The lifetime of the π^0 meson is so short, in fact, that its existence was not established until 1950. Charged π mesons almost invariably decay into lighter mesons called μ *mesons* and neutrinos:

$$\pi^+ \rightarrow \mu^+ + \nu_\mu$$
$$\pi^- \rightarrow \mu^- + \bar{\nu}_\mu$$

These neutrinos are not the same as those involved in beta decay, which is why their symbols are ν_μ and $\bar{\nu}_\mu$. The existence of two classes of neutrino was established in 1962. A metal target was bombarded with high-energy protons, and π mesons were created in profusion. Inverse reactions traceable to the neutrinos from the decay of these mesons produced μ mesons only, and no electrons. Hence these neutrinos must be somehow different from those associated with beta decay.

The neutral π^0 meson decays into a pair of gamma rays:

$$\pi^0 \rightarrow \gamma + \gamma$$

The π^+ and π^- mesons have rest masses of $273m_e$, while that of the π^0 mesons is slightly less, $264m_e$. The π^- meson is the antiparticle of the π^+ meson, and the π^0 meson is its own antiparticle, a distinction it shares only with the photon.

15.4 μ Mesons

While the existence of π mesons is so readily understandable that they were predicted many years before their actual discovery, μ mesons even today represent something of a puzzle. Their physical properties are known quite accurately. Positive and negative μ mesons have the same rest mass, $207m_e$, and the same spin, ½. Both decay with a mean life of 2.26×10^{-6} sec into electrons and neutrino-antineutrino pairs:

$$\mu^+ \rightarrow e^+ + \nu + \overline{\nu}$$
$$\mu^- \rightarrow e^- + \nu + \overline{\nu}$$

As with electrons, the positive-charge state of the μ meson represents the antiparticle. There is no neutral μ meson.

Unlike the case of π mesons, which, as we would expect, interact strongly with nuclei, the only interaction between μ mesons and matter is an electrostatic one. Accordingly μ mesons readily penetrate considerable amounts of matter before being absorbed. The majority of cosmic-ray particles at sea level are μ mesons from the decay of π mesons created in nuclear collisions caused by fast primary cosmic-ray atomic nuclei, since nearly all of the other particles in the cosmic-ray stream either decay or lose energy rapidly and are absorbed far above the earth's surface.

The mysterious aspect of the μ meson is its function—or, rather, its apparent lack of any function. Only in its mass and instability does the μ meson differ significantly from the electron, leading to the hypothesis that the μ meson is merely a kind of "heavy electron" rather than a unique entity. Other evidence, which we shall examine later in this chapter, is less unflattering to the μ meson, although it is still not wholly clear why, for instance, π mesons should preferentially decay into μ mesons rather than directly into electrons; only about 0.01 per cent of π mesons decay directly into electrons and neutrinos.

Cloud and Bubble Chambers

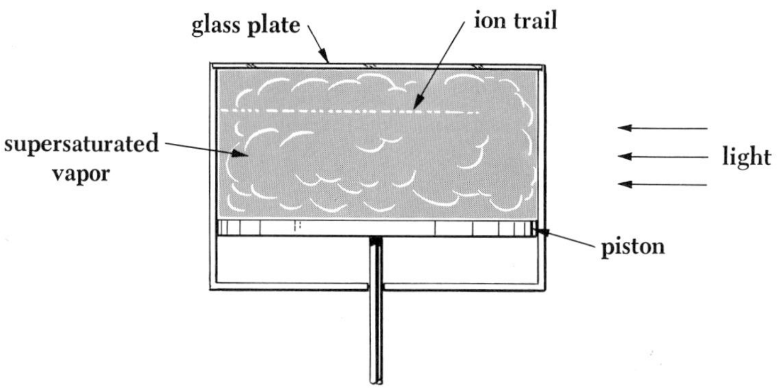

A cloud chamber is a device for rendering visible the path of a moving charged particle. Invented by C. T. R. Wilson in 1907, the cloud chamber makes use of the fact that a supersaturated vapor condenses into liquid droplets around any ions that may be present within it. In its simplest version, a cloud chamber consists of a glass-fronted cylinder containing a mixture of air and water vapor. When the piston is moved back rapidly, the vapor expands and cools to a supersaturated state. If a charged particle passes through the chamber at precisely this time, the ions it leaves behind serve as the nuclei of water droplets that condense from the vapor. This trail can be observed and photographed by illuminating the chamber from the side. The identity and initial energy of a particle that stops in the chamber can be determined from the length of the track (the longer the track, the greater the energy), from the density of the track (the greater the density, the heavier the particle and the slower it is), and from the character of the track (the straighter the track, the heavier the particle). Cloud chambers are especially useful in studying the interactions between an incoming particle and the nuclei of gas atoms it may strike.

Ordinary cloud chambers are not ideal because they are sensitive only for a fraction of a second after the expansion and because the gas density is so low that high-energy particles usually neither stop within the chamber nor experience collisions there. Another type of cloud chamber eliminates the first objection by using a container kept cold at its bottom. If a heavy gas fills the chamber and a lighter one is allowed to diffuse downward, the latter becomes supersaturated as it nears the cooler bottom and thus establishes a layer of continuous sensitivity. The density objection is overcome in the bubble chamber, a device that is essentially a cloud chamber operating in reverse. A bubble chamber contains a superheated liquid (a liquid heated beyond its normal boiling point), and bubbles of vapor form around any ions created within it. Because a liquid rather than a gas is involved, the likelihood of finding interesting events along the path of an incident particle is increased; modern high-energy accelerators make extensive use of bubble chambers in investigating nuclear and elementary-particle phenomena.

The photograph above shows an event that occurred in a bubble chamber at the University of California at Berkeley. The chamber was filled with liquid hydrogen and placed in a magnetic field to permit the charges and momenta of the various particles to be determined. An antiproton enters at the upper left and collides with a proton to form a neutron and an antineutron, neither of whose tracks is visible, since they are uncharged. After a short distance the antineutron is annihilated, and several π mesons are created and move off in random directions.

15.5 *K* Mesons

The π and μ mesons do not exhaust the list of known particles with masses intermediate between those of the electron and proton. A third class of mesons, called *K mesons*, has been discovered; its members, all unstable, may decay in a variety of ways. Charged *K* mesons have rest masses of $966m_e$, spins of 0, and mean lifetimes of 1.22×10^{-8} sec. The following decay schemes are possible for K^+ mesons:

$$\begin{aligned} K^+ &\rightarrow \pi^+ + \pi^+ + \pi^- \\ &\rightarrow \pi^+ + \pi^0 + \pi^0 \\ &\rightarrow \pi^+ + \pi^0 \\ &\rightarrow \mu^+ + \pi^0 + \nu \\ &\rightarrow \mu^+ + \nu \\ &\rightarrow e^+ + \pi^0 + \nu \end{aligned}$$

The possible decays of K^- mesons, which are classed as antiparticles, follow a similar pattern.

There are apparently two distinct varieties of neutral *K* mesons, the ${K_1}^0$ and ${K_2}^0$. Both have rest masses of $974m_e$ and spins of 0, but the former has a mean lifetime of about 1×10^{-10} sec, while that of the latter is about 6×10^{-8} sec. The following decay modes are known for neutral *K* mesons:

$$\begin{aligned} {K_1}^0 &\rightarrow \pi^+ + \pi^- \\ &\rightarrow \pi^0 + \pi^0 \\ {K_2}^0 &\rightarrow \pi^+ + \pi^- + \pi^0 \\ &\rightarrow \pi^0 + \pi^0 + \pi^0 \\ &\rightarrow \pi^- + \mu^+ + \nu \\ &\rightarrow \pi^+ + \mu^- + \overline{\nu} \\ &\rightarrow \pi^- + e^+ + \nu \\ &\rightarrow \pi^+ + e^- + \overline{\nu} \end{aligned}$$

In addition to their electromagnetic interaction with matter through which they pass, *K* mesons exhibit varying degrees of specifically nuclear interactions. The K^+ and K^0 mesons interact only weakly with nuclei, while their antiparticle counterparts are readily scattered and absorbed by nuclei in their paths.

15.6 Hyperons

Elementary particles heavier than protons are called *hyperons*. The known hyperons fall into four classes, Λ, Σ, Ξ, and Ω hyperons, in order of increasing mass. (Λ, Σ, Ξ, and Ω are, respectively, the Greek capital letters *lambda, sigma, xi,* and *omega.*) All are unstable with extremely brief mean lifetimes. The spin of all hyperons is ½ except that of the Ω hyperon, which is $\frac{3}{2}$. The masses, lifetimes, and decay schemes of various hyperons are given in Table 15.1.

Hyperons exhibit definite interactions with nuclei. The Λ^0 hyperon is even able to act as a nuclear constituent. A nucleus containing a bound Λ^0 hyperon is called a *hyperfragment;* eventually the Λ^0 decays, of course, with the resulting nucleon and π meson either reacting with the parent nucleus or emerging from it entirely.

15.7 Systematics of Elementary Particles

Despite the multiplicity of elementary particles and the diversity of their properties, it is possible to discern an underlying order in their behavior. The fact of this order does not constitute a theory of elementary particles, however, any more than the order found in atomic spectra constitutes a theory of the atom, but it does provide hope that there may indeed be a single theoretical picture that can encompass elementary-particle phenomena in the manner that the quantum theory encompasses atomic phenomena. Thus far no such picture has emerged, although the striking success of Dirac's theory of the electron suggests that it will be a part of some more general theory. In the remainder of this chapter we shall examine the regularities observed in elementary particles and their apparent significance.

Table 15.2 is a listing in order of rest mass of the relatively stable elementary particles we have thus far mentioned plus the η meson, which we shall dis-

Table 15.1

HYPERON PROPERTIES

Particle	Mass, m_e	Half life, sec	Decay
Λ^0	2,182	1.7×10^{-10}	$\Lambda^0 \rightarrow p + \pi^-$
			$\rightarrow n + \pi^0$
			$\rightarrow p + e^- + \tilde{\nu}$
Σ^+	2,328	0.6×10^{-10}	$\Sigma^+ \rightarrow p + \pi^0$
			$\rightarrow n + \pi^+$
Σ^-	2,341	1.2×10^{-10}	$\Sigma^- \rightarrow n + \pi^-$
Σ^0	2,332	$<10^{-12}$	$\Sigma^0 \rightarrow \Lambda + \gamma$
Ξ^-	2,583	0.9×10^{-10}	$\Xi^- \rightarrow \Lambda + \pi^-$
Ξ^0	2,571	1.0×10^{-10}	$\Xi^0 \rightarrow \Lambda + \pi^0$
Ω^-	3,290	$\sim 10^{-10}$	$\Omega^- \rightarrow \Lambda + K^-$
			$\rightarrow \Xi^0 + \pi^-$
			$\rightarrow \Xi^- + \pi^0$

Table 15.2

ELEMENTARY PARTICLES STABLE AGAINST DECAY BY THE STRONG NUCLEAR INTERACTION

Class	Name	Particle $+e$	Particle 0	Particle $-e$	Antiparticle $+e$	Antiparticle 0	Antiparticle $-e$	Spin	Rest mass, m_e	Rest mass, Mev	Half life, sec	L	M	B	S	Y	I
												(Antiparticles have opposite signs)					
PHOTON	photon		γ			(γ)		1	0	0	stable	0	0	0			
LEPTON	e-neutrino		ν_e			$\overline{\nu_e}$		½	0	0	stable	+1	0	0			
	μ-neutrino		ν_μ			$\overline{\nu_\mu}$		½	0	0	stable	0	+1	0			
	electron			e^-	e^+			½	1	0.51	stable	+1	0	0			
	μ meson			μ^-	μ^+			½	207	106	1.5×10^{-6}	0	+1	0			
MESON	π meson		π^0			(π^0)		0	264	135	7×10^{-17}	0	0	0	0	0	1
		π^+					π^-		273	140	1.8×10^{-8}						
	K meson	K^+					K^-	0	966	494	8×10^{-9}	0	0	0	+1	+1	½
			K^0			$\overline{K^0}$			974	498	7×10^{-11}; 4×10^{-8}						
	η meson		η^0			(η^0)		0	1,073	548	$\sim 10^{-18}$	0	0	0	0	0	0
BARYON	nucleon proton	p					$\overline{p}$	½	1,836	938	stable	0	0	+1	0	+1	½
	nucleon neutron		n			$\overline{n}$			1,839	940	7×10^2						
	Λ hyperon		Λ^0			$\overline{\Lambda^0}$		½	2,182	1,115	1.7×10^{-10}	0	0	+1	−1	0	0
	Σ hyperon	Σ^+					$\overline{\Sigma^-}$	½	2,328	1,192	0.6×10^{-10}	0	0	+1	−1	0	1
			Σ^0			$\overline{\Sigma^0}$			2,332	1,194	$<10^{-12}$						
				Σ^-	$\overline{\Sigma^+}$				2,341	1,197	1.2×10^{-10}						
	Ξ hyperon		Ξ^0			$\overline{\Xi^0}$		½	2,571	1,310	1.0×10^{-10}	0	0	+1	−2	−1	½
				Ξ^-	$\overline{\Xi^+}$				2,583	1,320	0.9×10^{-10}						
	Ω hyperon			Ω^-	$\overline{\Omega^+}$			3/2	3,290	1,676	$\sim 10^{-10}$	0	0	+1	−3	−2	0

cuss shortly. By relatively stable is meant that the half lives of the particles all greatly exceed the time required for light to travel a distance equal to the "diameter" of an elementary particle. This diameter is probably a little over 10^{-15} m, and the characteristic time required to traverse it at the speed of light is therefore of the order of magnitude of 10^{-23} sec. Thus the particles in Table 15.2 are almost all capable of traveling through space as distinct entities along paths of measurable length in such devices as bubble chambers.

A considerable body of experimental evidence also points to the existence of many different "particles" whose lifetimes against decay are only about 10^{-23} sec. What can be meant by a particle which exists for so brief an interval? Indeed, how can a time of 10^{-23} sec even be measured? Such particles cannot be detected by observing their formation and subsequent decay in a bubble chamber or other instrument, but instead appear as resonant states in the interaction of more stable (and hence more readily observable) particles. Resonant states occur in atoms as energy levels; in Chap. 5 we reviewed the Franck-Hertz experiment, which showed the existence of atomic energy levels through the occurrence of inelastic electron scattering from atoms at certain energies only. An atom in a specific excited state is not the same as that atom in its ground state or in another excited state, but we do not usually speak of such an excited atom as though it were a member of a special species only because the interaction that gives rise to the excited state—the electromagnetic interaction—is well understood. A rather different situation holds in the case of elementary particles, where the various interactions involved are, except for the electromagnetic one, only partially understood, and much of our information comes from the properties of the resonances.

Let us see what is involved in a resonance in the case of elementary particles. An experiment is performed, for instance the bombardment of protons by energetic π^+ mesons, and a certain reaction is studied, for instance

$$\pi^+ + p \rightarrow \pi^+ + p + \pi^+ + \pi^- + \pi^0$$

The effect of the interaction of the π^+ meson and the proton is the creation of three new π mesons. In each such reaction the new mesons have a certain total energy that consists of their rest energies plus their kinetic energies relative to their center of mass. If we plot the number of events observed versus the total energy of the new mesons in each event, we obtain a graph like that of Fig. 15-4. Evidently there is a strong tendency for the total meson energy to be 785 Mev and a somewhat weaker tendency for it to be 548 Mev. We can say that the reaction exhibits resonances at 548 and 785 Mev or, equivalently, we can say that this reaction proceeds via the creation of an intermediate particle which can be either one whose mass is 548 Mev or one whose mass is 785 Mev. From the graph we can even esti-

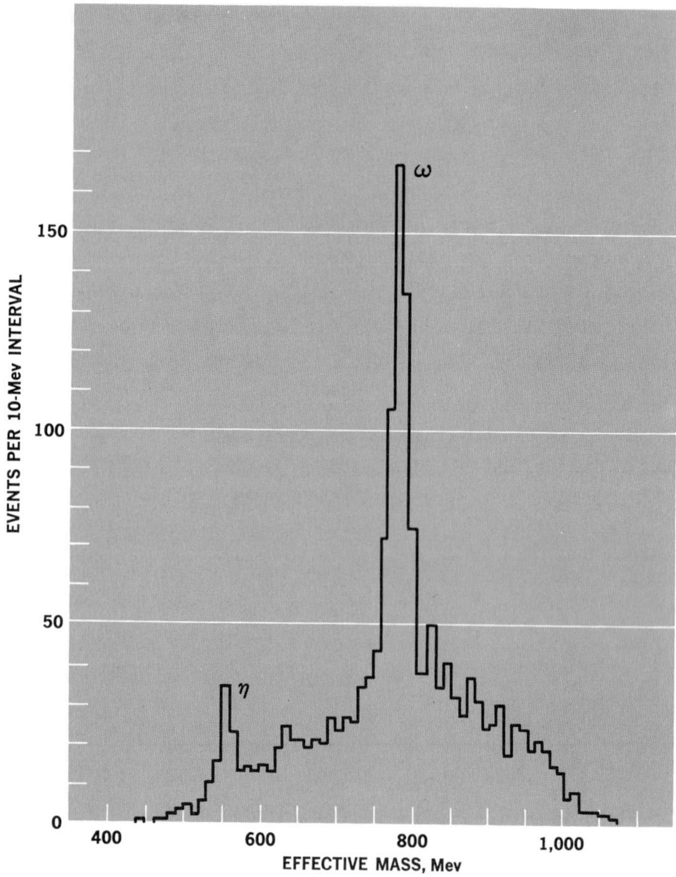

FIGURE 15-4 Resonant states in the reaction $\pi^+ + p \rightarrow \pi^+ + p + \pi^+ + \pi^- + \pi^0$ occur at effective masses of 548 and 785 Mev. By effective mass is meant the total energy, including mass energy, of the three new mesons relative to their center of mass.

mate the mean lifetimes of these intermediate particles, which are known as the η and ω mesons, respectively. According to the uncertainty principle, the uncertainty in decay time of an unstable particle—which is its mean lifetime τ—will give rise to an uncertainty in the determination of its energy—which is the width ΔE at half maximum of the corresponding peak in Fig. 15-4—whose relationship is

$$\tau \, \Delta E \approx \hbar$$

Hence the lifetimes of the resonances, or, just as well, the lifetimes of the η and ω mesons, can be established. The η lifetime is sufficiently long for it to be regarded as a relatively stable particle and it is included in Table 15.2, while the ω lifetime is too short by many orders of magnitude. We shall return to the resonance particles later in this chapter.

The particles in Table 15.2 seem to fall naturally into four general categories. In a class by itself is the photon, a stable particle with zero rest mass and unit spin. If there is a *graviton,* a particle that is the quantum of the gravitational field in the same sense that the photon is the quantum of the electromagnetic field or the π meson the quantum of the nuclear force field, it would be another member of this class. The graviton, as yet undetected, should be massless and stable, and should have a spin of 2. Its interaction with matter would be extremely weak, and it is unlikely that present techniques are capable of verifying its existence. (The zero mass of the graviton can be inferred from the unlimited range of gravitational forces. As we saw in Sec. 12.7, the mutual forces between two bodies can be regarded as transmitted by the exchange of particles between them. If energy conservation is to be preserved, the uncertainty principle requires that the range of the force be inversely proportional to the mass of the exchanging particles, so gravitational forces can have an infinite range only if the graviton mass is zero. A similar argument holds for the photon mass.)

After the photon in Table 15.2 come the e-neutrino and the μ-neutrino, the electron, and the μ meson, all with spins of ½. These particles are jointly called *leptons.* The π, K, and η mesons, all with spins of 0, are classed as *mesons.* (Despite its name, the μ meson has more in common with the other leptons than with the π, K, and η mesons.) The heaviest particles, namely the nucleons and hyperons, comprise the *baryons.*

While this grouping is reasonable on the basis of mass and spin alone, there is further evidence in its favor. Let us introduce three new quantum numbers, L, M, and B as follows. We assign the number $L = 1$ to the electron and the e-neutrino, and $L = -1$ to their antiparticles; all other particles have $L = 0$. We assign the number $M = 1$ to the μ meson and its neutrino, and $M = -1$ to their antiparticles; all other particles have $M = 0$. Finally, we assign $B = 1$ to all baryons, and $B = -1$ to all antibaryons; all other particles have $B = 0$. The significance of these numbers is that, in every process of whatever kind that involves elementary particles, the total values of L, M, and B remain constant. The classical conservation laws of energy, momentum, angular momentum, and electric charge plus the new conservation laws of L, M, and B are what determine whether any given process is capable of taking place or not. An example is the decay of the neutron,

$$n^0 \rightarrow p^+ + e^- + \bar{\nu}$$

While $L = 0$ for the neutron, $L = 1$ for the electron and -1 for the antineutrino, so the total value of L before and after the decay is 0. Similarly, $B = 1$ for both neutron and proton, so that the total value of B before and after the decay is 1. The stability of the proton is a consequence of energy and baryon-number conservation: There are no baryons of smaller mass than the proton, and so the proton cannot decay.

15.8 Strangeness Number

Despite the introduction of the quantum numbers L, M, and B, certain aspects of elementary-particle behavior still defied explanation. For instance, it was hard to see why certain heavy particles decay into lighter ones together with the emission of a gamma ray while others do not undergo apparently equally permissible decays. Thus the Σ^0 baryon decays into a Λ^0 baryon and a gamma ray,

$$\Sigma^0 \rightarrow \Lambda^0 + \gamma$$

while the Σ^+ baryon is never observed to decay into a proton and a gamma ray,

$$\Sigma^+ \rightarrow p^+ + \gamma$$

Another peculiarity is based upon the general observation that physical processes in nature that release large amounts of energy take place more rapidly than processes that release small amounts. However, many strange particles whose decay releases considerable energy have relatively long lifetimes, well over a billion times longer than theoretical calculations predict. A third odd feature is that strange particles are never created singly, but always two or more at a time. These and still other considerations led to the introduction of a quantity called *strangeness number* S. Table 15.2 shows the values of S that are assigned to the various elementary particles. We note that L, B, and S are 0 for the photon and π^0 meson. Since these particles are also uncharged, there is no conceivable way of distinguishing between them and their antiparticles. For this reason the photon and π^0 meson are regarded as their own antiparticles. Before we consider the interpretation of the strangeness number, we shall have to examine the various kinds of particle interaction.

There are apparently four types of interaction between elementary particles that, in principle, give rise to all of the physical processes in the universe. The feeblest of these is the gravitational interaction. Next is the so-called weak interaction that is present between leptons and other leptons, mesons, or baryons in addition to any electromagnetic forces that may exist. The weak interaction is responsible for particle decays in which neutrinos are involved, notably beta decays. Stronger than gravitational and weak interactions are the electromagnetic interactions between all charged particles and also those with electric or magnetic moments. Finally, strongest of all are the nuclear forces (usually called simply strong forces when elementary particles are being discussed) that are found between mesons, baryons, and mesons and baryons.

The relative strengths of the strong, electromagnetic, weak, and gravitational interactions are in the ratios $1:1.5 \times 10^{-4}:3 \times 10^{-15}:10^{-40}$. Of course, the distances through which the corresponding forces act are very different. While the strong force between nearby nucleons is many powers of 10 greater than the gravitational force between them, when they are a meter apart the proportion is the other way. The structure of nuclei is determined by the properties of the strong interaction, while the structure of atoms is determined by those of the electromagnetic interaction. Matter in bulk is electrically neutral, and the strong and weak forces are severely limited in their range. Hence the gravitational interaction, utterly insignificant on a small scale, becomes the dominant one on a large scale. The role of the weak force in the structure of matter is apparently that of a minor perturbation that sees to it that nuclei with inappropriate neutron/proton ratios undergo corrective beta decays.

Let us now return to the strangeness number S. It is found that in all processes involving strong and electromagnetic interactions the strangeness number is conserved. The decay

$$\begin{array}{rccc} & \Sigma^0 & \rightarrow \Lambda^0 & + \gamma \\ S = & -1 & -1 & 0 \end{array}$$

conserves S and is observed to occur, while the superficially similar decay

$$\begin{array}{rccc} & \Sigma^+ & \not\rightarrow p^+ & + \gamma \\ S = & -1 & 0 & 0 \end{array}$$

does not conserve S and has never been observed. Strange particles are created in high-energy nuclear collisions which involve strong interactions, and their multiple appearance results from the necessity of conserving S. The relative slowness with which all unstable elementary particles save the π^0 meson and Λ^0 hyperon decay is accounted for if we assume that weak interactions are also characteristic of mesons and baryons as well as leptons, though normally dominated by strong or electromagnetic interactions. With strong or electromagnetic processes impossible except in the above cases owing to the lack of conservation of S, only the weak interaction is available for processes in which the total value of S changes. Events governed by weak interactions take place slowly, as borne out by experiment. Even the weak interaction, however, is unable to permit S to change by more than $+1$ or -1 in a decay. Thus the Ξ^- hyperon does not decay directly into a neutron since

$$\begin{array}{rccc} & \Xi^- & \not\rightarrow n^0 & + \pi^- \\ S = & -2 & 0 & 0 \end{array}$$

but instead via the two steps

$$\begin{array}{rlll} & \Xi^- \rightarrow & \Lambda^0 & + \pi^- \\ S = & -2 & -1 & 0 \end{array}$$

$$\begin{array}{rlll} & \Lambda^0 \rightarrow & n^0 & + \pi^0 \\ S = & -1 & 0 & 0 \end{array}$$

A quantity called *hypercharge,* Y, has also been found useful in characterizing particle families; it is conserved in strong interactions. Hypercharge is equal to the sum of the strangeness and baryon numbers of the particle families:

15.2 $$Y = S + B$$

For mesons the hypercharge is equal to the strangeness. The various hypercharge assignments are listed in Table 15.2.

15.9 Isotopic Spin

It is obvious from Table 15.2 that there are a number of particle families each of whose members has essentially the same mass and interaction properties but different charge. These families are called *multiplets,* and it is natural to think of the members of a multiplet as representing different charge states of a single fundamental entity. It has proved useful to categorize each multiplet according to the number of charge states it exhibits by a number I such that the multiplicity of the state is given by $2I + 1$. Thus the nucleon multiplet is assigned $I = \frac{1}{2}$, and its $2 \cdot \frac{1}{2} + 1 = 2$ states are the neutron and the proton. The π meson multiplet has $I = 1$, and its $2 \cdot 1 + 1 = 3$ states are the π^+, π^-, and π^0 mesons. The η meson has $I = 0$ since it occurs in only a single state and $2 \cdot 0 + 1 = 1$. There is evidently an analogy here with the splitting of an angular-momentum state of quantum number l into $2l + 1$ substates, and this has led to the somewhat misleading name of *isotopic spin quantum number* for I.

Pursuing the analogy with angular momentum, isotopic spin can be represented by a vector $\mathbf{I}$ in "isotopic spin space" whose component in a certain direction is governed by a quantum number customarily denoted I_3. The possible values of I_3 are restricted to $I, I - 1, \ldots, 0, \ldots, -(I - 1), -I$, so that I_3 is half-integral if I is half-integral and integral or zero if I is integral. The isotopic spin of the nucleon is $I = \frac{1}{2}$, which means that I_3 can be either $\frac{1}{2}$ or $-\frac{1}{2}$; the former is taken to represent the proton and the latter the neutron. In the case of the π meson, $I = 1$ and $I_3 = 1$ corresponds to the π^+ meson, $I_3 = 0$ to the π^0 meson, and $I_3 = -1$ to the π^- meson. The values of I_3 for the other mesons and baryons are assigned in a similar way.

The charge of a meson or baryon is related to its baryon number B, its strangeness number S, and the component I_3 of its isotopic spin by the formula

15.3 $$q = e\left(I_3 + \frac{B}{2} + \frac{S}{2}\right)$$

Each allowed orientation of the isotopic spin vector **I** hence is directly connected to the charge of the particle thus represented. In the case of the nucleon multiplet, the proton has $I_3 = ½$, $B = 1$, and $S = 0$, so that $q = e$, while the neutron has $I_3 = -½$, $B = 1$, and $S = 0$, so that $q = 0$. In the case of the π meson multiplet, $B = S = 0$ and the three values of I_3 of 1, 0, and -1 respectively yield $q = e$, 0, and $-e$. Charge and baryon number B are conserved in all interactions. Thus I_3 must be conserved whenever S is conserved, namely in strong and electromagnetic interactions. Only in weak interactions does the total I_3 change.

An additional conservation law is suggested by the observed charge independence of nuclear forces, which result from the strong interaction. Such properties of a nucleus as its binding energy and pattern of energy levels change when a neutron is substituted for a proton or vice versa only by amounts that follow from purely electromagnetic considerations, implying that the strong interaction itself does not depend upon electric charge. Now the difference between a proton and a neutron in isotopic spin space lies only in the orientation of their isotopic spin vectors, so we can say that the charge independence of the strong interaction means that this interaction is independent of orientation in isotopic spin space. Angular momentum is likewise independent of orientation in real space and it is conserved in all interactions, which might lead us to surmise that isotopic spin is conserved in strong interactions. This surmise happens to be correct, and as a result the isotopic spin quantum number I is conserved in strong, but not in weak or in electromagnetic, interactions. We shall return to the relation between conservation principles and invariance with respect to symmetry operations in Sec. 15.10.

We note that, although I_3 is conserved in electromagnetic interactions, I itself need not be. An example of a process in which I changes while I_3 does not is the decay of the π^0 meson into two photons:

$$\pi^0 \rightarrow \gamma + \gamma$$

A π^0 meson has $I = 1$ and $I_3 = 0$, while I is not defined for photons; there is no component I_3 of isotopic spin on either side of the equation, which is consistent with its conservation, although I has changed.

15.10 Symmetries and Conservation Principles

In Sec. 15.9 the charge independence of the strong interaction was expressed in terms of the isotropy of isotopic spin space, so that isotopic spin is independent of orientation in this space. By analogy with angular momentum, this symmetry was said to imply the conservation of isotopic spin in strong interactions. It is a remarkable fact that *all* known symmetries in the physical world lead directly to conservation laws, so the relationship between symmetry under rotations of **I** and the conservation of **I** is wholly plausible. Let us survey some of these symmetry-conservation relationships before continuing our discussion of elementary particles.

What is meant by a "symmetry"? Formally, if rather vaguely, we might say that a symmetry of a particular kind exists when a certain operation leaves something unchanged. A candle is symmetric about a vertical axis because it can be rotated about that axis without changing in appearance or any other feature; it is also symmetric with respect to reflection in a mirror. Table 15.3 lists the principal symmetry operations which leave the laws of physics unchanged under some or all circumstances. The simplest symmetry operation is translation in space, which means that the laws of physics do not depend upon where we choose the origin of our coordinate system to be. By more advanced methods than we are employing in this book, it is possible to show that the invariance of the description of nature to translations in space has as a consequence the conservation of linear momentum! Another

Table 15.3

SOME SYMMETRY OPERATIONS AND THEIR ASSOCIATED CONSERVATION PRINCIPLES

Symmetry operation	Conserved quantity
All interactions are independent of:	
Translation in space	Linear momentum **p**
Translation in time	Energy E
Rotation in space	Angular momentum **L**
Electromagnetic gauge transformation	Electric charge q
Lorentz transformation	Velocity of center of mass **V**
Interchange of identical particles	Type of statistical behavior
Inversion of space, time, and charge	Product of charge parity, space parity, and time parity CPT
?	Baryon number B
?	Lepton number L
?	Lepton number M
The strong and electromagnetic interactions only are independent of:	
Inversion of space	Parity P
Reflection of charge	Charge parity C, I_3, and strangeness S
The strong interaction only is independent of:	
Charge	Isotopic spin I

simple symmetry operation is translation in time, which means that the laws of physics do not depend upon when we choose $t = 0$ to be, and this invariance has as a consequence the conservation of energy. Invariance under rotations in space, which means that the laws of physics do not depend upon the orientation of the coordinate system in which they are expressed, has as a consequence the conservation of angular momentum.

Conservation of electric charge is related to gauge transformations, which are shifts in the zeros of the scalar and vector electromagnetic potentials V and $\mathbf{A}$. (As elaborated in electromagnetic theory, the electromagnetic field can be described in terms of the potentials V and $\mathbf{A}$ instead of in terms of $\mathbf{E}$ and $\mathbf{B}$, where the two descriptions are related by the vector calculus formulas $\mathbf{E} = -\nabla V$ and $\mathbf{B} = \nabla \times \mathbf{A}$.) Gauge transformations leave $\mathbf{E}$ and $\mathbf{B}$ unaffected since they are obtained by differentiating the potentials, and this invariance leads to charge conservation.

Invariance under a Lorentz transformation signifies the isotropy of space-time and leads to the conservation of the velocity of the center of mass of an isolated system. In Chap. 1 we saw an indication of this relationship when $E = mc^2$ was derived starting from the latter conservation principle.

The interchange of identical particles in a system is a type of symmetry operation which leads to the preservation of the character of the wave function of a system. The wave function may be symmetric under such an interchange (Sec. 8.2), in which case the particles do not obey the exclusion principle and the system follows Bose-Einstein statistics, or it may be antisymmetric, in which case the particles obey the exclusion principle and the system follows Fermi-Dirac statistics. Conservation of statistics (or, equivalently, of wave-function symmetry or antisymmetry) signifies that no process occurring within an isolated system can change the statistical behavior of that system. A system exhibiting Bose-Einstein behavior cannot spontaneously alter itself to exhibit Fermi-Dirac statistical behavior or vice versa. This conservation principle has applications in nuclear physics, where it is found that nuclei that contain an odd number of nucleons (odd mass number A) obey Fermi-Dirac statistics while those with even A obey Bose-Einstein statistics; conservation of statistics is thus a further condition a nuclear reaction must observe.

The conservations of the baryon number B and the lepton numbers L and M are alone among the principal conservation principles in having no known symmetries associated with them.

Apart from the charge independence of the strong interaction and its associated conservation of isotopic spin, which we have already mentioned, the remaining symmetry operations in Table 15.3 all involve *parities* of one kind or another. The term parity with no qualification refers to the behavior of a wave function under an inversion in space. By inversion in space is meant the reflection of spatial coordinates through the origin, with $-x$ replacing x, $-y$ replacing y, and $-z$ replacing z. If the sign of the wave function ψ does not change under such an inversion,

$$\psi(x, y, z) = \psi(-x, -y, -z)$$

and ψ is said to have *even* parity. If the sign of ψ changes,

$$\psi(x, y, z) = -\psi(-x, -y, -z)$$

and ψ is said to have *odd* parity. Thus the function $\cos x$ has even parity since $\cos x = \cos(-x)$, while $\sin x$ has odd parity since $\sin x = -\sin(-x)$.

If we write

$$\psi(x, y, z) = P\psi(-x, -y, -z)$$

we can regard P as a quantum number characterizing ψ whose possible values are $+1$ (even parity) and -1 (odd parity). Every elementary particle has a certain parity associated with it, and the parity of a system such as an atom or a nucleus is the product of the parity of the wave function that describes the coordinates of its constituent particles and the intrinsic parities of the particles themselves. Since ψ^2 is independent of P, the parity of a system is not a quantity that has an obvious physical consequence. However, it *is* found that the initial parity of an isolated system does not change during whatever events occur within it, which can be ascertained by comparing the parities of known final states of a reaction or transformation with the parities of equally plausible final states that are not observed to occur. A system of even parity retains even parity, a system of odd parity retains odd parity; this principle is known as conservation of parity.

The conservation of parity is an expression of the inversion symmetry of space, that is, of the lack of dependence of the laws of physics upon whether a left-handed or a right-handed coordinate system is used to describe phenomena. In Sec. 15.2 it was noted that the neutrino has a left-handed spin and the antineutrino a right-handed spin, so that there is a profound difference between the mirror image of either particle and the particle itself. This asymmetry implies that interactions in which neutrinos and antineutrinos participate—the weak interactions—need not conserve parity, and indeed parity conservation is found to hold true only in the strong and electromagnetic interactions. Historically the fact that spatial inversion is not invariably a symmetry operation was suggested by the failure of parity conservation in the decay of the K^+ meson, and was later confirmed by experiments showing the specific handedness of ν and $\bar{\nu}$.

Two other parities occur in Table 15.3, time parity T and charge parity C, which respectively describe the behavior of a wave function when t is replaced by $-t$ and when q is replaced by $-q$. The symmetry operation that corresponds to the conservation of time parity is *time reversal.* Time reversal symmetry implies that the direction of increasing time is not significant, so that the reverse of any process that can occur is also a process that can occur. In other words, if symmetry under time reversal holds, it is impossible to establish by viewing it whether a motion picture of an event is being run forwards or backwards. Although time parity T was long considered to be conserved in every interaction, it was discovered in 1964 that the $K_2{}^0$ meson can decay into a π^+ and a π^- meson, which violates the conservation of T. The symmetry of phenomena under time reversal thus has an ambiguous status at present. The symmetry operation that corresponds to the conservation of charge parity C is *charge conjugation,* which is the replacement of every particle in a system by its antiparticle. Charge parity C, like space parity P, is not conserved in weak interactions. However, despite the limited validities of the conservation of C, P, and T, there are good theoretical reasons for believing that the product CPT of the charge, space, and time parities of a system is invariably conserved. The conservation of CPT means that for every process there is an antimatter mirror-image counterpart that takes place in reverse, and this particular symmetry seems to hold even though its component symmetries sometimes fail individually.

15.11 Theories of Elementary Particles

In addition to the particles listed in Table 15.2 there are, as mentioned earlier, a great many "particles" of extremely brief lifetimes whose existence is revealed by resonances in interactions involving their longer-lived brethren. These resonant states are characterized by definite values of mass, charge, angular momentum, isotopic spin, parity, strangeness, and so on, and it is no more logical to disqualify them as legitimate particles because their existences are so transient than it is to consider the neutron, say, as merely an unstable state of the proton. Of course, it is possible to make out an excellent case for supposing the latter, and then to go on to generalize that *all* the various "elementary" particles are actually excited states of a very few truly elementary particles, as yet unidentified; this constitutes one line of attack toward a comprehensive theory of elementary particles. On the other hand, if we accept the particles of Table 15.2 as legitimate, then it is consistent to include the resonant states as well, and to seek a theoretical framework that embraces the entire collection of several score entities.

The latter point of view seems especially applicable to the strongly interacting particles, each of which may be regarded in a sense as a composite of the others. To see what this notion means, let us consider the strange case of the π^0 meson, which cannot participate in electromagnetic interactions since it has neither charge nor magnetic moment and yet which is observed to decay into a pair of photons, which are electromagnetic quanta. Apparently what happens is that the π^0 first undergoes a virtual transition to a nucleon-antinucleon pair, which the uncertainty principle permits to exist for a very short time even though energy is not conserved, and this nucleon-antinucleon pair can interact electromagnetically to produce two photons whose energies are consistent with the π^0 mass. This situation can be described by saying that the π^0 meson can communicate with otherwise inaccessible particle states through the intermediacy of the nucleon-antinucleon pair, with which it communicates via the strong interaction. Since the π^0 cannot be regarded as being itself (so to speak) at all times in view of its anomalous decay, it must be regarded as a composite of all the particle states with which it can communicate via the strong interaction. Similar arguments can be given for all the other strongly interacting particles, and this approach to a theory of elementary particles is under active study. To be sure, these incestuous relationships are confined to particles linked by the strong interaction, namely the mesons and baryons; the ground state of an atom, for instance, can be understood without reference to the properties of its excited states, although the proton, the "ground state" of the baryon family in the sense that it alone is stable in space, *cannot* be understood without bringing the other baryons into the description as well.

Several interesting and suggestive classification schemes have been devised for the strongly interacting particles based upon the abstract theory of groups. One of these schemes, the so-called *eightfold way*, collects isotopic spin multiplets into supermultiplets whose members have the same spin and parity

FIGURE 15-5 Supermultiplets of spin-½ baryons and spin-0 mesons stable against decay by the strong nuclear interaction. Arrows indicate possible transformations according to the eightfold way.

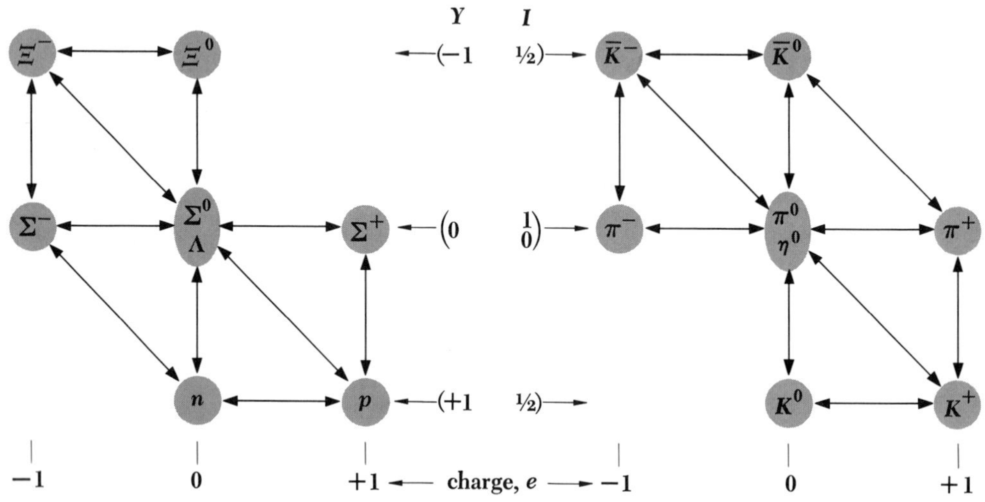

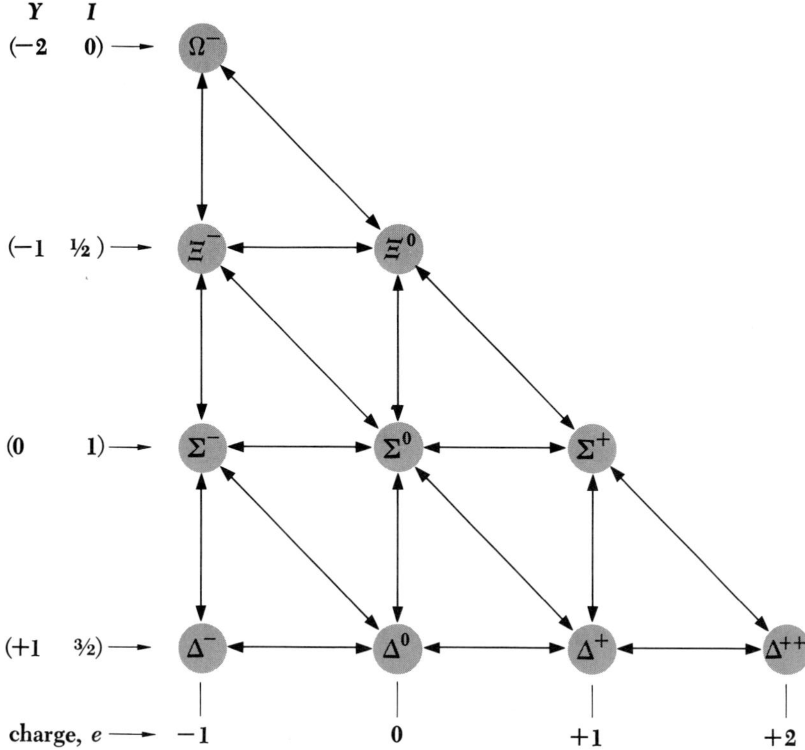

FIGURE 15-6 Baryon supermultiplet whose members have spin 3/2 and (except Ω^-) are short-lived resonance particles; the Ξ and Σ particles here are heavier and have different spins from the ones in Table 15.2. Arrows indicate possible transformations according to the eightfold way. The Ω^- particle was predicted from this scheme.

but differ in charge and hypercharge (Figs. 15-5 and 15-6). The scheme prescribes the number of members each particular supermultiplet should have and also relates mass differences among these members. The great triumph of the eightfold way was its prediction of a previously unknown particle, the Ω^- hyperon, which was subsequently searched for and finally discovered in 1964. Other group-theoretic approaches have related the supermultiplets of the eightfold way to one another and have attempted to incorporate relativistic considerations into the comprehensive picture that is emerging.

The success of the eightfold way in organizing our knowledge of the strongly interacting particles implies that the symmetry of its mathematical structure has a counterpart in a symmetry in nature. The further we probe into nature, the more hints we receive of a profound order that underlies the complications and confusions of experience. But for all the elegance of the symmetries that have been revealed, there still remains the problem of the fundamental interactions themselves, what they signify, and how they are related to one another and to the properties of the particles through which they are manifested.

Problems

1. Find the maximum kinetic energy of the electron emitted in the beta decay of the free neutron.

2. How much energy must an antineutrino have if it is to react with a proton to produce a neutron and a positron?

3. What is the energy of each of the gamma rays produced when a π^0 meson decays?

4. How much energy must a gamma-ray photon have if it is to materialize into a neutron-antineutron pair?

5. A positron collides head-on with an electron and both are annihilated. Each particle had a kinetic energy of 1 Mev. Find the wavelength of each of the resulting photons.

6. A π^0 meson whose kinetic energy is equal to its rest energy decays in flight. Find the angle between the two gamma-ray photons that are produced.

7. Why does a free neutron not decay into an electron and a positron?

8. Why are photons and π mesons the only particles multiply produced in collisions between elementary particles?

9. Why does a Λ hyperon not decay into a π^+ and a π^- meson?

10. Trace the decay of the Ξ^0 particle into stable particles.

11. A μ^- meson collides with a proton, and a neutron plus another particle are created. What is the other particle?

12. Does the reaction $\pi^- + p \rightarrow \pi^0 + n$ violate any conservation principles?

13. Under what circumstances can a proton collide with another proton to produce a π^0 meson while conserving both isotopic spin I and its third component I_3?

14. How long does the uncertainty principle permit a π^0 meson to exist as a nucleon-antinucleon pair?

15. A proton of kinetic energy T_0 collides with a stationary proton, and a proton-antiproton pair is produced. If the momentum of the bombarding proton is shared equally by the four particles that emerge from the collision, find the minimum value of T_0.

FOR FURTHER READING

The books in each category are listed in approximate order of increasing difficulty.

Similar coverage to CONCEPTS OF MODERN PHYSICS:

R. W. Christy and A. Pytte: *The Structure of Matter,* W. A. Benjamin, Inc., New York, 1965.

F. K. Richtmyer, E. H. Kennard, and T. Lauritsen: *Introduction to Modern Physics,* McGraw-Hill Book Co., Inc., New York, 1955.

R. M. Eisberg: *Fundamentals of Modern Physics,* John Wiley and Sons, Inc., New York, 1961.

R. B. Leighton: *Principles of Modern Physics,* McGraw-Hill Book Co., Inc., New York, 1959.

Quantum Mechanics:

P. Fong: *Elementary Quantum Mechanics,* Addison-Wesley Publishing Co., Inc., Reading, Mass., 1962.

D. Park: *Introduction to the Quantum Theory,* McGraw-Hill Book Co., Inc., New York, 1964.

D. Bohm: *Quantum Theory,* Prentice-Hall, Inc., Englewood Cliffs, N. J., 1951.

Chemical Binding

J. E. Spice: *Chemical Binding and Structure,* Pergamon Press, Ltd., London, 1964.

F. O. Rice and E. Teller: *The Structure of Matter,* John Wiley and Sons, Inc., New York, 1949, and Interscience Publishers, Inc., New York, 1961 (paperback).

H. B. Gray: *Electrons and Chemical Bonding,* W. A. Benjamin, Inc., New York, 1965.

J. C. Slater: *Quantum Theory of Molecules and Solids,* Vol. 1, McGraw-Hill Book Co., Inc., New York, 1963.

The Solid State

L. V. Azaroff: *Introduction to Solids,* McGraw-Hill Book Co., Inc., New York, 1960.

C. A. Wert and R. M. Thomson: *Physics of Solids,* McGraw-Hill Book Co., Inc., New York, 1964.

C. Kittel: *Elementary Solid State Physics,* John Wiley and Sons, Inc., New York, 1962.

Nuclear Physics

S. Glasstone: *Sourcebook on Atomic Energy,* D. Van Nostrand Co., Inc., Princeton, N. J., 1958.

I. Kaplan: *Nuclear Physics,* Addison-Wesley Publishing Co., Inc., Reading, Mass., 1963.

W. E. Burcham: *Nuclear Physics,* McGraw-Hill Book Co., Inc., New York, 1963.

ANSWERS TO ODD-NUMBERED PROBLEMS

Chapter 1

1. If the speed of light in free space were not the same to all observers, certain laws of physics would assume different forms in different reference frames in uniform relative motion.
5. 4.2×10^7 m/sec
7. 6.3×10^4 years
9. 213 m
11. 4.2×10^7 m/sec
13. 1.87×10^8 m/sec; 1.64×10^8 m/sec
15. 39°
17. 200 Mev
19. 8.9×10^{-28} kg
21. 3.6×10^{17} joules
23. 2.7×10^{11} kg
25. $1{,}960m_e$
27. 0.294 Mev
29. $F = m_0\,(dv/dt)/(1 - v^2/c^2)^{3/2}$
31. 1,630 Mev/c
33. $0.987c$

Chapter 2

1. 1,800 A
3. 5,400 A; 3.9 ev
5. 2.83×10^{-19} joule
7. 3,970 A
9. 1.24×10^5 volts
11. 3.14 A
13. 5×10^{18} sec^{-1}
15. 0.015 A
17. 2.4×10^{19} sec^{-1}
19. 0.565 A

Chapter 3

3. 0.10 A
5. 6.62×10^{-34} m
7. $\lambda = \dfrac{12.27}{V^{1/2}}\left(\dfrac{eV}{2m_0c^2} + 1\right)^{-1/2}$
9. $u = \frac{1}{2}w$
11. 6.2%
13. 7.3×10^5 m/sec; 4.0×10^2 m/sec

Chapter 4

1. 10°
3. 0.876
7. 1.14×10^{-13} m
9. $m/m_0 = 1.002$
11. $f = \dfrac{1}{2\pi}\sqrt{\dfrac{Qe}{4\pi\varepsilon_0 mR^3}}$. For the hydrogen atom, $f = 6.6 \times 10^{15}$ sec^{-1}, which is comparable with the highest frequencies in the hydrogen spectrum.

Chapter 5

1. 18,750 A
3. 920 A
5. 12 volts
7. 1.04 Mev
9. 8.2×10^6 rev
11. 1.05×10^5 °K
13. 2.4 A
15. They are the same.

Chapter 6

1. 2.1 Mev

Chapter 7

5. 68%

Chapter 8

1. 182
5. 1.39×10^{-4} ev
7. 0.0283 A

9. 2.46×10^{15}
11. 20 webers/m^2
13. $5\mu_B$; 3

Chapter 9

1. Sodium ions have closed shells, while sodium atoms each have a single outer electron.
3. Both have one electron outside a closed shell.
5. 0.5 ev; 2.9×10^{-9} m
7. 1.9 ev; 7.6×10^{-10} m
9. 1.27 A
11. 2.23 A
13. No.
15. (*a*) For rotation, $T = 173°$K; for vibration, $T = 6350°$K. (*b*) 3.5×10^5 °K. (*c*) 26

Chapter 10

1. $e^{13\times 10^6}$
7. $(2/\sqrt{\pi})v_m$
9. 1.6×10^9 neutrons/m^3
11. 2.7; 1.4×10^{-4}

Chapter 11

1. 4.26 ev
3. There would be many electrons per state.
5. $\frac{3}{5}u_F$
7. *p*-type
9. Owing to the exclusion principle their energies far exceed kT.

Chapter 12

1. 8.83 cm
3. 19% B^{10}, 81% B^{11}
5. 34.97 amu
7. 7.98 Mev
9. 15.6 Mev
11. Nuclear forces cannot be strongly charge-dependent.
13. ~0.6 Mev

Chapter 13

1. $\frac{1}{4}$
3. 1,620 years
5. 1.23×10^4 sec^{-1}
7. 3.37
9. A positron
11. $_{82}Pb^{206}$
13. $_{47}Hg^{107}$, $_{18}A^{38}$, $_{50}Sn^{120}$
15. The mass of $_4Be^7$ is not sufficiently larger than that of $_3Li^7$ to permit the creation of a positron.
17. $x_{1/2} = (\ln 2)/\mu$

Chapter 14

1. 1.52 Mev
3. $v_1/v_2 = (m_1 + m_2)/2m_1$
5. 1.2×10^{-6}
7. 0.21
9. 1.2×10^{16} km
11. 18.8 Mev
13. $_1H^1$, $_0n^1$, $_1H^2$
15. 0.1%
17. 7.28 Mev

Chapter 15

1. 0.78 Mev
3. 68 Mev
5. 8.6×10^{-13} m
7. This decay does not conserve baryon number or spin.
9. This decay does not conserve baryon number or spin.
11. A neutrino
13. This can occur if initially and finally the isotopic spins of the protons are parallel while the isotopic spin of the π^0 is opposite to the final spin of the protons.
15. 5630 Mev

APPENDIX

SYMBOLS

Å	Angstrom unit (10^{-10} m)
A	Mass number
b	Impact parameter
B	Baryon number
$\mathbf{B}$	Magnetic flux density
E	Total energy
$\mathbf{E}$	Electric field intensity
E_n	Energy of quantum state of quantum number n
f	Frequency of revolution; fraction of incident particles that are scattered
$f(u)$	Occupation index of state of energy u
g	A priori probability
I	Moment of inertia
k	Propagation constant
K	Rotational quantum number
$K, L, M, N, \ldots$	Atomic shells of total quantum number $n = 1, 2, 3, 4, \ldots$
l	Orbital quantum number; mean free path
L	Lepton number
$\mathbf{L}$	Angular momentum
L_r	Angular momentum of rotating molecule
L_s	Spin angular momentum
m	Mass
m'	Reduced mass
m_0	Rest mass
m_l	Magnetic quantum number
m_s	Spin magnetic quantum number
n	Total quantum number; atomic density; index of refraction; order of scattered beam in X-ray diffraction
N	Neutron number; total number of particles in an assembly of particles
p	Electric dipole moment
$\mathbf{p}$	Momentum
P	Probability
Q	Energy released in nuclear reaction

Note: Symbols for elementary particles are given in Table 15.2, page 378.

R	Activity
s	Spin quantum number
$s, p, d, f, g, \ldots$	Angular momentum states of $l = 0, 1, 2, 3, 4, \ldots$
S	Strangeness number
T	Kinetic energy; absolute temperature
$T_{1/2}$	Half life
u	Group velocity; particle energy
u_F	Fermi energy
U	Total energy of assembly of particles
v	Vibrational quantum number
v_d	Drift velocity
V	Potential energy
V_m	Magnetic potential energy
w	Atomic weight; wave (phase) velocity
W	Total probability of a distribution
Z	Atomic number
α	Polarizability
η	Free electron density
γ	Gamma ray
λ	Wavelength; decay constant
$\boldsymbol{\mu}$	Magnetic moment
μ_s	Spin magnetic moment
ν	Frequency
ω	Angular frequency
$\psi\ (x, y, z)$	Time-independent wave function
ψ^*	Complex conjugate of ψ
ψ_n	Wave function corresponding to quantum number n
$\Psi\ (x, y, z, t)$	Time-dependent wave function
Ψ^*	Complex conjugate of Ψ
ρ	Mass density
σ	Cross section
τ	Volume of cell in phase space
$\boldsymbol{\tau}$	Torque

INDEX